AI for Marine, Ocean and Climate Change Monitoring

AI for Marine, Ocean and Climate Change Monitoring

Editors

Veronica Nieves
Ana B. Ruescas
Raphaëlle Sauzède

Basel • Beijing • Wuhan • Barcelona • Belgrade • Novi Sad • Cluj • Manchester

Editors

Veronica Nieves
Image Processing Laboratory
University of Valencia
Valencia
Spain

Ana B. Ruescas
Department of Geography
University of Valencia
Valencia
Spain

Raphaëlle Sauzède
Laboratoire d'Océanographie
de Villefranche
Institut de la Mer de Villefranche
Villefranche-sur-Mer
France

Editorial Office
MDPI
St. Alban-Anlage 66
4052 Basel, Switzerland

This is a reprint of articles from the Special Issue published online in the open access journal *Remote Sensing* (ISSN 2072-4292) (available at: www.mdpi.com/journal/remotesensing/special_issues/AI_Marine_Climate).

For citation purposes, cite each article independently as indicated on the article page online and as indicated below:

Lastname, A.A.; Lastname, B.B. Article Title. *Journal Name* **Year**, *Volume Number*, Page Range.

ISBN 978-3-0365-9998-4 (Hbk)
ISBN 978-3-0365-9997-7 (PDF)
doi.org/10.3390/books978-3-0365-9997-7

Contents

About the Editors

Veronica Nieves

Veronica Nieves is a distinguished researcher currently serving as the head of the AI4OCEANS Research Group at the University of Valencia. Prior to this role, she held various positions at the University of California and NASA's Jet Propulsion Laboratory from 2009 to 2019. She obtained her PhD in Applied Physics and Scientific Simulation from the Polytechnic University of Catalonia, Spain. With a broad and interdisciplinary background, her expertise spans artificial intelligence (AI), climate science, oceanography, and Earth system modeling. In recent years, her primary research focus has centered on harnessing advanced AI tools to analyze complex remote sensing and in situ data, aiming to gain a deeper understanding of the ocean's role in the global climate system at both local and global scales. Since 2021, she has been a member of the Scientific Advisory Committee for the EU Copernicus Marine Service, and since 2022, she has been a member of the European Laboratory for Learning and Intelligent Systems Society. Furthermore, she has served as a member of the NASA Sea Level Change Team and has acted as an expert for the United Nations' Regular Process for Global Reporting and Assessment of the State of the Marine Environment. Additionally, she has worked as an expert assessor for the US National Academy of Sciences and the US Department of Energy.

Ana B. Ruescas

Ana Ruescas holds a PhD in Physical Geography with a specialization in Remote Sensing from the University of Valencia (UV). She has extensive experience in processing and analyzing Earth observation data within both thermal and optical ranges. With a background in History and Geography, she began her career in Remote Sensing and Geographic Information Systems following her graduation. Her expertise led her to the European Space Agency in Italy, ESRIN, where she established a robust network and initiated an international career. Since 2011, she has been employed at Brockmann Consult GmbH, actively participating in various marine projects related to the European Union and the European Space Agency (ESA). In 2018, she joined the Department of Geography at UV as an associate professor, teaching subjects related to physical geography, biogeography, and cartography. Her research collaboration with the Image Processing Laboratory at UV, particularly within the Image and Signal Processing groups, further enhances her scientific contributions in the field of machine learning applications to remote sensing topics. Additionally, she has played an active role in organizing and delivering courses for ESA and EUMETSAT, with a focus on training users in the SeNtinel Application Platform (SNAP). Her current position includes serving as a lecturer on the ESA-Academy team and the EUMETSAT Marine Training Service.

Raphaëlle Sauzède

Raphaëlle Sauzède earned her PhD in Biogeochemical Oceanography from Sorbonne University in 2015. Following two years in French Polynesia as a postdoctoral fellow, she joined the CNRS as a research engineer at the Institut de la Mer de Villefranche. Her primary research focus revolves around developing synergic approaches based on machine learning that integrate satellite data, autonomous platforms, and historical in situ databases to enhance the characterization of global phytoplankton biomass, its community composition, and associated carbon fluxes on a global scale. The resulting products from these innovative methods are operationally released by the European Copernicus Marine Service, where she actively participates in the MULTIOBS thematic assembly center.

Additionally, she is a member of the Argo Data Management Team for the BioGeoChemical-Argo international program, where she applies machine learning to improve data quality, with a specific emphasis on delayed mode data processing.

Editorial

AI for Marine, Ocean and Climate Change Monitoring

Veronica Nieves [1,*], Ana Ruescas [2] and Raphaëlle Sauzède [3]

1 Image Processing Laboratory, University of Valencia, 46980 Valencia, Spain
2 Department of Geography, University of Valencia, 46010 Valencia, Spain; ana.b.ruescas@uv.es
3 Laboratoire d'Océanographie de Villefranche, Institut de la Mer de Villefranche, 06230 Villefranche-Sur-Mer, France; raphaelle.sauzede@imev-mer.fr
* Correspondence: veronica.nieves@uv.es

Citation: Nieves, V.; Ruescas, A.; Sauzède, R. AI for Marine, Ocean and Climate Change Monitoring. *Remote Sens.* **2024**, *16*, 15. https://doi.org/10.3390/rs16010015

Received: 17 October 2023
Accepted: 18 December 2023
Published: 20 December 2023

1. Introduction

In the ever-evolving landscape of marine, oceanic, and climate change monitoring, the intersection of cutting-edge artificial intelligence (AI), machine learning (ML), and data analytics has emerged as a pivotal catalyst for transformative advancements. Within this article compilation, we embark on a journey through a diverse array of innovative AI-driven applications, each meticulously crafted to address the critical challenges inherent in understanding and managing our world's aquatic realms and climate dynamics. This exploration delves into the current status quo of marine and climate monitoring, the evolutionary processes driving us forward, and the mechanisms that underpin these advancements.

The selection of articles underscores the pivotal role of AI-driven solutions in tackling environmental challenges. These encompass a spectrum of tasks, from predicting sea-surface temperature (SST) to detecting Sargassum aggregations and addressing water eutrophication. AI's impact extends across these applications, with a significant role in unraveling the intricacies of cloud behavior—a vital component in addressing climate uncertainties. Moreover, AI offers predictive capabilities with regard to chlorophyll-a distribution and enhances our understanding of oceanic light models and thermal variations beneath the ocean's surface. These advancements are of utmost importance for understanding the biogeochemical cycle, the evolving dynamics of our oceans, and their implications for climate change at both regional and global scales. Additionally, AI and data analytics are being leveraged to enhance the precision of near-surface humidity data, effectively addressing critical climate research needs. Lastly, ML algorithms are being proposed to correct errors in satellite-derived sea-surface salinity data, promising increased accuracy and deeper insights into ocean salinity patterns. These technologies enable us to gain a more comprehensive understanding and effectively manage our natural resources.

These featured applications collectively serve as a testament to the profound significance of AI, ML, and data analytics in advancing our comprehension and stewardship of marine, oceanic, and climate dynamics. They emphasize the transformative potential of AI in reshaping our approach to environmental monitoring and bolster strategies for conservation. This represents a significant stride toward implementing more efficient solutions that can account for the intricate complexities inherent in data products and the climate system, addressing both present and forthcoming challenges associated with climate change.

2. Articles

A total of eleven papers are featured in this Special Issue. Refer to Table 1 for a concise overview of the article titles, authors, and keywords.

These research papers delve into various subjects and investigate efforts that harness the potential of artificial intelligence to enhance our understanding of marine and oceanic environments while tackling the complex challenges posed by climate change. These papers can be categorized into several core themes as previously outlined:

AI-Enhanced Sea-Surface Temperature Prediction: Researchers have been working on innovative AI/ML techniques to improve SST predictions, which have significant implications for various fields, including climate research, ecological preservation, and economic progress. These advancements include the use of graph memory neural networks (GMNNs) to encode irregular SST data effectively [1] and long-term and short-term memory neural networks (LSTMs) for SST prediction [2].

Table 1. Overview of the papers featured in the Special Issue entitled "AI for Marine, Ocean, and Climate Change Monitoring" in *Remote Sensing*.

Title	Authors	Keywords
A Graph Memory Neural Network for Sea Surface Temperature Prediction	Shuchen Liang Anming Zhao Mengjiao Qin Linshu Hu Sensen Wu Zhenhong Du Renyi Liu	sea surface temperature spatiotemporal prediction deep learning graph neural network
Prediction of Sea Surface Temperature in the East China Sea Based on LSTM Neural Network	Xiaoyan Jia Qiyan Ji Lei Han Yu Liu Guoqing Han Xiayan Lin	long short-term memory (LSTM) sea surface temperature (SST) East China Sea
Detection of Sargassum from Sentinel Satellite Sensors Using Deep Learning Approach	Marine Laval Abdelbadie Belmouhcine Luc Courtrai Jacques Descloitres Adán Salazar-Garibay Léa Schamberger Audrey Minghelli Thierry Thibaut René Dorville Camille Mazoyer Pascal Zongo Cristèle Chevalier	ocean color Sargassum MODIS MSI OLCI Sentinel-2 Sentinel-3 convolutional neural network deep learning
End-to-End Neural Interpolation of Satellite-Derived Sea Surface Suspended Sediment Concentrations	Jean-Marie Vient Ronan Fablet Frédéric Jourdin Christophe Delacourt	Interpolation data-driven model neural networks variational data assimilation missing data suspended particulate matter observing system experiment Bay of Biscay
AICCA: AI-Driven Cloud Classification Atlas	Takuya Kurihana Elisabeth Moyer Ian T. Foster	cloud classification MODIS artificial intelligence deep learning machine learning
Applying Deep Learning in the Prediction of Chlorophyll-a in the East China Sea	Haobin Cen Jiahan Jiang Guoqing Han Xiayan Lin Yu Liu Xiaoyan Jia Qiyan Ji Bo Li	LSTM chlorophyll-a East China Sea

Table 1. *Cont.*

Title	Authors	Keywords
Vertically Resolved Global Ocean Light Models Using Machine Learning	Pannimpullath Remanan Renosh Jie Zhang Raphaëlle Sauzède Hervé Claustre	BGC-Argo ED380 ED412 ED490 global ocean light models neural network PAR
Spatiotemporal Prediction of Monthly Sea Subsurface Temperature Fields Using a 3D U-Net-Based Model	Nengli Sun Zeming Zhou Qian Li Xuan Zhou	sea temperature prediction reconstructed sea subsurface temperature data 3D U-Net
Deep Learning to Near-Surface Humidity Retrieval from Multi-Sensor Remote Sensing Data over the China Seas	Rongwang Zhang Weihao Guo Xin Wang	near-surface humidity remote sensing deep learning China Seas
An Algorithm to Bias-Correct and Transform Arctic SMAP-Derived Skin Salinities into Bulk Surface Salinities	David Trossman Eric Bayler	Salinity SMAP skin-effect bias air-sea Arctic ocean machine-learning
Super-Resolving Ocean Dynamics from Space with Computer Vision Algorithms	Bruno Buongiorno Nardelli Davide Cavaliere Elodie Charles Daniele Ciani	earth observations ocean dynamics satellite altimetry sea surface temperature artificial intelligence machine learning deep learning neural networks

Satellite-Based AI Monitoring for Environmental Challenges: Satellite-based monitoring is crucial for addressing environmental challenges such as Sargassum aggregations and suspended sediment dynamics. Novel deep learning models have been developed to detect Sargassum aggregations with higher accuracy compared to traditional index-based techniques [3]. Additionally, end-to-end deep learning schemes like 4DVarNet have been employed to improve the interpolation of sea-surface sediment concentration fields from satellite data [4].

Advancements in Cloud Classification and Climate Uncertainty Reduction Using AI: Understanding cloud behavior and reducing uncertainties in climate projections is vital. Researchers have introduced novel AI-driven techniques for cloud classification based on convolutional autoencoders [5]. These techniques aim to reduce the dimensionality of satellite cloud observations and provide valuable insights into cloud patterns, helping to address climate uncertainties.

AI-Driven Ocean Chlorophyll-a Concentration Modeling for Eutrophication Mitigation: Predicting ocean chlorophyll-a concentrations is critical for mitigating issues like water eutrophication. AI methods, particularly neural networks, have been utilized to predict chlorophyll-a distribution in marine environments, with a focus on the East China Sea [6].

ML Advances Oceanic Light Models for Comprehensive Global Biogeochemical Insights: Authors developed SOCA-light, a machine learning model, predicting oceanic light profiles globally using BGC-Argo data. The study highlights the model's accuracy, addresses data gaps, and suggests versatile applications for improving biogeochemical databases [7].

Harnessing AI for Subsurface Ocean Temperature and Oceanic Impact: Researchers have explored the prediction of subsurface ocean temperature (SSbT), an essential indicator of the ocean's thermal state. A 3D U-Net model has been employed to predict SSbT, enhancing our understanding of ocean temperature variations [8].

Data Analytics for Near-Surface Humidity Monitoring and Climate Implications: Near-surface humidity monitoring is crucial for climate research. AI-driven approaches like Ensemble Mean of Target deep neural networks (EMTnets) have been introduced to improve the accuracy of near-surface humidity data [9]. These methods have implications for understanding the impact of global warming on humidity levels.

Machine Learning for Sea-Surface Salinity Correction in Subpolar and Arctic Oceans: The correction of errors in satellite-derived sea-surface salinity (SSS) data, particularly in subpolar and Arctic Oceans, has been addressed using ML algorithms [10]. These corrections aim to improve the accuracy of salinity measurements and enhance our understanding of ocean salinity patterns.

Surface Ocean Dynamics and Climate Regulation with Advanced Data Analysis: Researchers have introduced innovative neural network architectures to improve the reconstruction of absolute dynamic topography from satellite altimeter data [11]. These advancements offer insights into surface ocean dynamics and their role in climate regulation.

3. Conclusions

In this collective reprint, multiple studies and models are presented, each addressing different aspects of marine, ocean, and climate data analysis and prediction. These investigations utilize a range of methodologies, including deep learning, neural networks, and data assimilation, with the aim of enhancing our understanding of various phenomena such as sea-surface temperature, Sargassum detection, cloud classification, chlorophyll-a concentration, subsurface ocean temperature, skin salinity, and ocean dynamic topography. Overall, these studies demonstrate the potential of AI/ML and deep learning techniques to enhance the accuracy and efficiency of data analysis, while also acknowledging the imperative for continued research and advancements in areas such as model input data, interpretation, and the refinement of more sophisticated network architectures. These papers contribute significantly to advancing our comprehension and the effective management of crucial environmental factors that impact our planet.

Funding: This research received no external funding.

Institutional Review Board Statement: Not applicable.

Informed Consent Statement: Not applicable.

Data Availability Statement: Not applicable.

Acknowledgments: We express our sincere thanks for all contributions to this Special Issue, the time invested by each author, as well as the anonymous reviewers who contributed to the development and improvement of the articles. We greatly appreciate the efficient and accountable review processes and management of this Special Issue.

Conflicts of Interest: The authors declare no conflict of interest.

References

1. Liang, S.; Zhao, A.; Qin, M.; Hu, L.; Wu, S.; Du, Z.; Liu, R. A Graph Memory Neural Network for Sea Surface Temperature Prediction. *Remote Sens.* **2023**, *15*, 3539. [CrossRef]
2. Jia, X.; Ji, Q.; Han, L.; Liu, Y.; Han, G.; Lin, X. Prediction of Sea Surface Temperature in the East China Sea Based on LSTM Neural Network. *Remote Sens.* **2022**, *14*, 3300. [CrossRef]

3. Laval, M.; Belmouhcine, A.; Courtrai, L.; Descloitres, J.; Salazar-Garibay, A.; Schamberger, L.; Minghelli, A.; Thibaut, T.; Dorville, R.; Mazoyer, C.; et al. Detection of *Sargassum* from Sentinel Satellite Sensors Using Deep Learning Approach. *Remote Sens.* **2023**, *15*, 1104. [CrossRef]

4. Vient, J.-M.; Fablet, R.; Jourdin, F.; Delacourt, C. End-to-End Neural Interpolation of Satellite-Derived Sea Surface Suspended Sediment Concentrations. *Remote Sens.* **2022**, *14*, 4024. [CrossRef]

5. Kurihana, T.; Moyer, E.J.; Foster, I.T. AICCA: AI-Driven Cloud Classification Atlas. *Remote Sens.* **2022**, *14*, 5690. [CrossRef]

6. Cen, H.; Jiang, J.; Han, G.; Lin, X.; Liu, Y.; Jia, X.; Ji, Q.; Li, B. Applying Deep Learning in the Prediction of Chlorophyll-a in the East China Sea. *Remote Sens.* **2022**, *14*, 5461. [CrossRef]

7. Renosh, P.R.; Zhang, J.; Sauzède, R.; Claustre, H. Vertically Resolved Global Ocean Light Models Using Machine Learning. *Remote Sens.* **2023**, *15*, 5663. [CrossRef]

8. Sun, N.; Zhou, Z.; Li, Q.; Zhou, X. Spatiotemporal Prediction of Monthly Sea Subsurface Temperature Fields Using a 3D U-Net-Based Model. *Remote Sens.* **2022**, *14*, 4890. [CrossRef]

9. Zhang, R.; Guo, W.; Wang, X. Deep Learning to Near-Surface Humidity Retrieval from Multi-Sensor Remote Sensing Data over the China Seas. *Remote Sens.* **2022**, *14*, 4353. [CrossRef]

10. Trossman, D.; Bayler, E. An Algorithm to Bias-Correct and Transform Arctic SMAP-Derived Skin Salinities into Bulk Surface Salinities. *Remote Sens.* **2022**, *14*, 1418. [CrossRef]

11. Nardelli, B.B.; Cavaliere, D.; Charles, E.; Ciani, D. Super-Resolving Ocean Dynamics from Space with Computer Vision Algorithms. *Remote Sens.* **2022**, *14*, 1159. [CrossRef]

Article

A Graph Memory Neural Network for Sea Surface Temperature Prediction

Shuchen Liang [1,2], Anming Zhao [1,2], Mengjiao Qin [1,2], Linshu Hu [1,2,*], Sensen Wu [1,2], Zhenhong Du [1,2] and Renyi Liu [1,2]

1   School of Earth Sciences, Zhejiang University, 866 Yuhangtang Rd, Hangzhou 310058, China
2   Zhejiang Provincial Key Laboratory of Geographic Information Scaience, 866 Yuhangtang Rd, Hangzhou 310058, China
*   Correspondence: hulinshu1010@zju.edu.cn

Abstract: Sea surface temperature (SST) is a key factor in the marine environment, and its accurate forecasting is important for climatic research, ecological preservation, and economic progression. Existing methods mostly rely on convolutional networks, which encounter difficulties in encoding irregular data. In this paper, allowing for comprehensive encoding of irregular data containing land and islands, we construct a graph structure to represent SST data and propose a graph memory neural network (GMNN). The GMNN includes a graph encoder built upon the iterative graph neural network (GNN) idea to extract spatial relationships within SST data. It not only considers node but also edge information, thereby adequately characterizing spatial correlations. Then, a long short-term memory (LSTM) network is used to capture temporal dynamics in the SST variation process. We choose the data from the Northwest Pacific Ocean to validate GMNN's effectiveness for SST prediction in different partitions, time scales, and prediction steps. The results show that our model has better performance for both complete and incomplete sea areas compared to other models.

Keywords: sea surface temperature; spatiotemporal prediction; deep learning; graph neural network

Citation: Liang, S.; Zhao, A.; Qin, M.; Hu, L.; Wu, S.; Du, Z.; Liu, R. A Graph Memory Neural Network for Sea Surface Temperature Prediction. *Remote Sens.* 2023, 15, 3539. https:// doi.org/10.3390/rs15143539

Academic Editors: Ana B. Ruescas, Veronica Nieves and Raphaëlle Sauzède

Received: 12 May 2023
Revised: 2 July 2023
Accepted: 11 July 2023
Published: 14 July 2023

1. Introduction

Sea surface temperature (SST) is a crucial variable in marine environments [1]. Changes in SST can greatly impact the climate. Persistent anomalies in SST, characterized by unusually warm or cold conditions, may give rise to phenomena such as El Niño and La Niña [2,3]. Additionally, SST serves to guide marine activities by analyzing its influence on fish migration, which in turn informs fishery distribution and policy formulation [4,5]. It also plays an important role in forecasting marine disasters such as storm surges and red tides [6,7]. Thus, it is evident that accurate prediction of SST has great significance for the marine economy, ecology, and disaster forecasting.

Existing SST prediction methods can be divided into two major categories: numerical methods and data-driven methods. Numerical methods are based on a series of physicochemical parameters, constructing complex equations according to the principles of dynamics and thermodynamics [8–10]. However, they demand substantial computational resources and accurate parameter selection for precise results. Data-driven methods, on the other hand, learn patterns directly from the data [11] and have evolved from traditional statistical approaches to machine learning and deep learning techniques. Markov, canonical correlation analysis (CCA), and other statistical approaches are widely used to predict SST [12–14], but these models may lack accuracy when dealing with complex nonlinear problems due to their weak nonlinear fitting ability [15]. Therefore, machine learning approaches capable of addressing nonlinear problems are garnering attention in SST prediction research. For example, researchers use support vector machine (SVM) and artificial neural networks (ANN) and achieve promising results [16–18].

Machine learning methods require manual feature engineering which can be time-consuming, with their accuracy dependent on the quality of features. Furthermore, deep learning methods automatically extract useful features from big data and can achieve higher accuracy than traditional machine learning methods. As a result, deep learning techniques are becoming increasingly popular for SST prediction. Recurrent neural networks (RNN), including long short-term memory (LSTM) and gated recurrent unit (GRU) variants, excel in processing sequences, making them suitable for time series prediction tasks. Zhang et al. [19] pioneered the use of deep learning in SST prediction by developing an FC-LSTM model, which combined an LSTM layer with a fully connected layer. This approach outperformed support vector regression (SVR) and multilayer perceptron (MLP) in terms of prediction accuracy.

In fact, SST is a variable with spatiotemporal properties, showing dynamic and non-linear characteristics. However, previous works overlook the spatial features of SST, which limits the prediction accuracy of SST [20]. To fully consider spatial information, researchers generally adopt two approaches. The first one is to use spatial data, such as latitude, longitude, and regional features, as input for the model [21,22]. The second approach is to employ convolutional neural networks (CNN) to extract spatial features at different scales, and integrates them with time series prediction models to form a comprehensive spatiotemporal forecasting method [23–25].

Among methods based on convolutional idea, ConvLSTM and its variant, ConvGRU, proposed by Shi et al. [26] in 2015 for precipitation forecasting are widely applied in SST prediction tasks [21,27–29], owing to their effectiveness in capturing spatiotemporal correlations. These methods treat SST data as regular images, but in actual research, areas containing land or islands may lack valid data. Standard matrix convolution kernels cannot directly extract information from these locations, and filling in missing values may impact the prediction accuracy at the land-sea boundaries [30].

In recent years, graph neural networks (GNN) have succeeded in areas such as traffic flow prediction, weather forecasting, and disease risk assessment [31]. Graph structures are well-suited for irregular data, and GNNs' message-passing mechanisms [32] capture adjacency relationships better than CNN, effectively extracting data features. Therefore, in SST prediction, researchers start to explore how to learn SST's spatial relationships based on graph structures [30,33–35]. Among them, most methods use graph convolutional networks (GCN) to update and aggregate the representations of nodes along with their neighboring nodes.

In this study, we propose a graph memory neural network (GMNN) for SST prediction based on GNN idea. First, we develop an SST graph representation using distance threshold and Pearson correlation coefficient to fully express spatial information in irregular regions. An innovation of our model lies in adequately expressing spatial information for these incomplete areas using graph representations. Next, we design a graph encoder using iterative GNN to encode spatial relationships that take into account not only node but also edge features.

Finally, a GMNN model consisting of a graph encoder, a temporal encoder, and a decoder is constructed, offering a novel perspective for SST prediction. We validate the effectiveness of our model through diverse experiments in the Northwest Pacific region, considering different partitions, time scales, and prediction steps.

The remainder of this paper is organized as follows. Section 2 shows the data used in the study. Section 3 describes the details of the proposed method. Section 4 presents the experimental results. Section 4 provides the discussion of the results. Finally, Section 5 offers the conclusion of the paper.

2. Materials

2.1. Datasets

As shown in Figure 1, the study area is the Northwest Pacific, from 0° to 60°N and 100° to 180°E. The Northwest Pacific Ocean exhibits an intricate array of climate features, including tropical, subtropical, and temperate climates. The marine environment in this area is influenced by various natural factors, such as monsoons, ocean currents, and typhoons. Due to its diverse climate conditions and complex oceanic processes, this region is representative in SST prediction research.

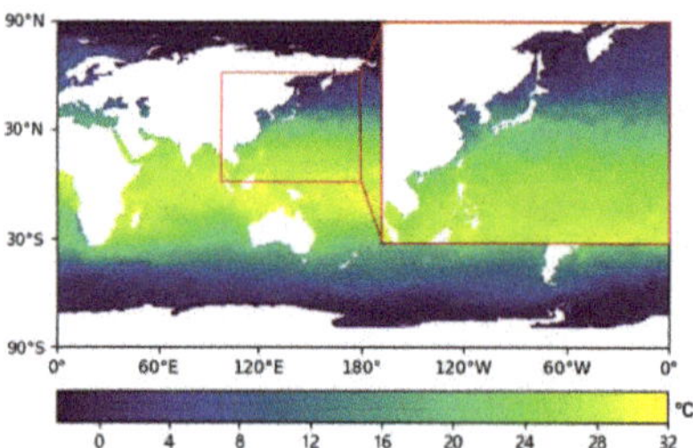

Figure 1. Study area and heat map of SST on 1 January 1993.

The SST data used in this study is from the optimum interpolation sea surface temperature (OISST) v2.1 product, produced by the United States National Oceanic and Atmospheric Administration (NOAA), with a spatial resolution of 1/4° latitude by 1/4° longitude. The time scale of the predictions is daily, weekly, and monthly. The OISST for the study area covers temporal range from 1 January 1993 to 31 December 2020.

More information can be found at the following link: https://psl.noaa.gov/data/gridded/data.noaa.oisst.v2.highres.html, accessed on 11 May 2023.

2.2. Pre-Processing

The data has a spatial resolution of 0.25°, with a corresponding grid size of 320 × 240 (8° × 6°) for the study area. Considering model parameter size, hardware and software environments, as well as the limited accuracy at large scales, we divide the study area into 8 × 6 subregions, each with a 40 × 40 (10° × 10°) grid. Subregions without ocean are excluded, leaving 41 subregions as experimental data, numbered sequentially (Figure 2).

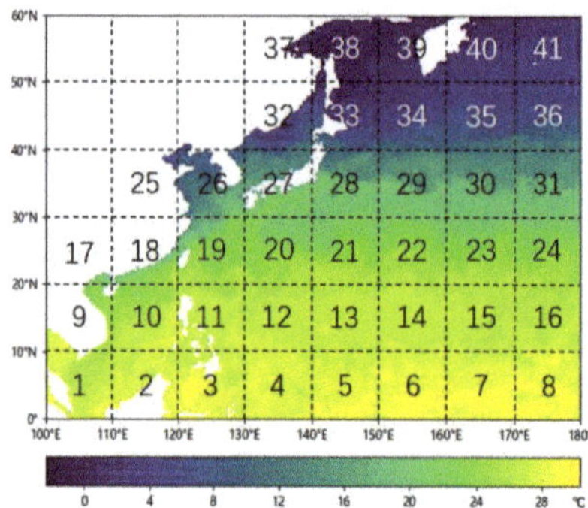

Figure 2. Division of the study area.

Among the 41 subregions, 19 of them (1, 2, 3, 9, 10, 11, 17, 18, 19, 25, 26, 27, 28, 32, 33, 37, 38, 39, and 40) contain land or islands, forming incomplete sea areas. Therefore, the constructed subregion samples are representative.

Then, we divide the datasets into daily, weekly, and monthly mean. We allocate 60% of the data for training, 20% for testing, and 20% for validation to prevent overfitting. The specific time ranges for each set are presented in Table 1.

Table 1. Datasets.

Temporal Resolution	Dataset	Time Range
Daily Mean	Training Set	1 January 1993~31 December 2010
	Validation Set	1 January 2011~31 December 2015
	Testing Set	1 January 2016~31 December 2020
Weekly Mean	Training Set	3 January 1993~26 December 2010
	Validation Set	2 January 2011~27 December 2015
	Testing Set	3 January 2016~27 December 2020
Monthly Mean	Training Set	January 1993~December 2010
	Validation Set	January 2011~December 2015
	Testing Set	January 2016~December 2020

3. Methods

The complete framework of the graph memory neural network (GMNN) is presented in Figure 3. Initially, historical SST data are preprocessed and transformed into a series of time-sorted graphs with fixed time intervals. These graphs encompass temporal, spatial, and attribute features as the model input. Next, a neural network is constructed for the graph sequence, featuring an encoder with both graph and temporal encoder modules to learn spatial and temporal patterns. The graph encoder is composed of multiple iterative GNN layers, each aggregating and updating the graph's nodes and edges to extract spatial features. The temporal encoder employs LSTM to capture temporal dynamics. Finally, by integrating the multi-output strategy and a fully connected layer decoder, the extracted spatiotemporal features are transformed into future SST prediction results.

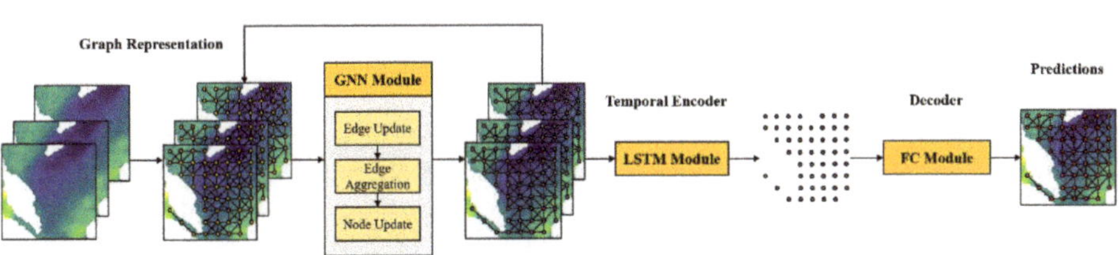

Figure 3. Framework of GMNN.

3.1. Graph Representation

In research with defined coordinate systems, locations are represented by longitude and latitude pairs. For SST data, each point at the sea surface defined by a coordinate pair generates an SST record at each time step. Records from different locations form a spatially correlated snapshot, and a series of snapshots over time create a temporally connected sequence. An SST image sequence of length T can be denoted as $S = (S_1, S_2, \ldots, S_T)$.

In the study area with land and islands, some locations lack SST observations, leading to empty pixels. Consequently, each image in the time series contains N valid pixels, where $N \leq row * col$, row and col represent the number of rows and columns in the image, respectively.

To express the connectivity between pixels, we construct edges for each valid pixel based on distance threshold and Pearson correlation coefficient.

As shown in Equation (1), e_{ij} represents the connectivity between points i and j based on distance threshold, where 1 means connected and 0 means unconnected. d_{ij} represents the Euclidean distance between points i and j, with d_{min} being the set distance threshold.

When d_{ij} is greater than d_{min}, the spatial association between points i and j is considered weak, and no edge is formed between them.

$$e_{ij} = \begin{cases} 1, & if\ d_{ij} \leq d_{min} \\ 0, & otherwise \end{cases} \tag{1}$$

Figure 4 demonstrates the effect of edge construction based on distance threshold. Thicker solid lines represent edges with a distance threshold d_{min} of 1, while thinner solid lines correspond to edges with a d_{min} of $\sqrt{2}$.

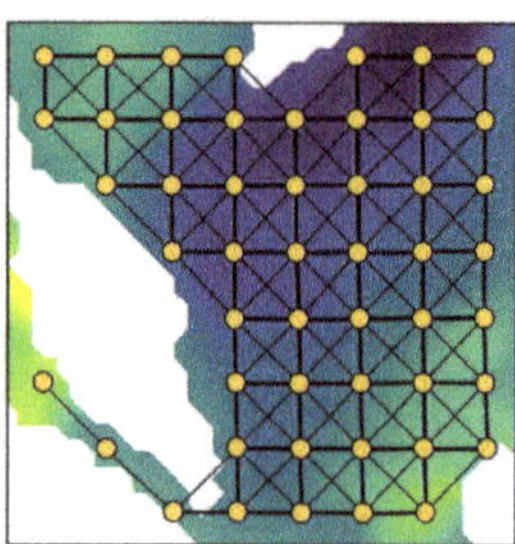

Figure 4. Edge construction results based on distance threshold.

The Pearson correlation coefficient (PCC) measures the linear relationship between two variables. Its value lies between -1 and 1, with larger absolute values indicating stronger correlations. The formula for the PCC, r, is shown in Equation (2).

$$r_{ij} = \frac{\sum_{t=1}^{T} \left(I_t - \bar{I}\right)\left(J_t - \bar{J}\right)}{\sqrt{\sum_{t=1}^{T}\left(I_t - \bar{I}\right)^2}\sqrt{\sum_{t=1}^{T}\left(J_t - \bar{J}\right)^2}} \tag{2}$$

For SST prediction, I and J represent the SST value sets for points i and j on the sea surface, each containing T samples corresponding to the time series length. I_t and J_t denote the SST values at time t, and $\bar{I}$ and $\bar{J}$ are the average SST values of the two sets.

Equation (3) shows the edge construction based on the Pearson correlation coefficient, where e_{ij} represents the connectivity between points i and j based on Pearson correlation coefficient threshold, where 1 means connected and 0 means unconnected. r_{min} is the threshold.

$$e_{ij} = \begin{cases} 1, & if\ |r_{ij}| > r_{min} \\ 0, & otherwise \end{cases} \tag{3}$$

The distance threshold and Pearson correlation coefficient evaluate the spatial relationship between any two points on the graph from the perspectives of position relation and attribute correlation. By combining these two factors, we create an edge construction method, as shown in Equation (4).

$$e_{ij} = \begin{cases} 1, & if\ d_{ij} \leq d_{min}\ and\ |r_{ij}| > r_{min} \\ 0, & otherwise \end{cases} \tag{4}$$

Figure 5 illustrates the edge construction process for a node in an SST image. The left image shows the edge connections when $d_{min} = \sqrt{5}$ grid (grid equals $1/4°$). Next, we calculate the Pearson correlation coefficient between the connected nodes, as depicted in the middle image. Solid lines represent calculations exceeding r_{min}, while dashed lines represent calculations less than or equal to r_{min}. By removing the dashed lines, we achieve the final result in the right image.

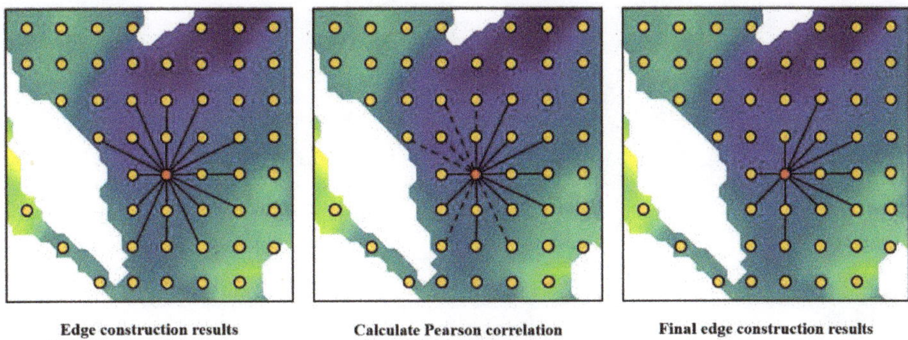

Figure 5. Edge construction results based on distance threshold and Pearson correlation coefficient.

The values of d_{min} and r_{min} in our study are 1.5 grid and 0.8 grid, respectively. By applying the edge construction method to each node, we obtain the node and edge representation of SST data.

The SST image sequence $S = (S_1, S_2, \ldots, S_T)$ is converted into a graph sequence $G = (G_1, G_2, \ldots, G_T)$. Each graph G in the sequence is represented as a collection of nodes V and edges E connecting them, denoted as $G = (V, E)$, where $v_i \in V$ represents node i, and $e_{ij} = (v_i, v_j) \in E$ represents the edge from i to j.

Compared to the pixel image representation, the graph representation offers greater flexibility, as it directly omits points corresponding to missing values.

3.2. Graph Encoder

GNN is a neural network that learns target objects by propagating neighbor information based on graph structures [36]. Compared to CNN, GNN excels at handling irregular data and is better suited for tasks with strong interdependencies [37,38], making them applicable for encoding SST variation process.

To incorporate features of nodes, edges, and their relationships, we adopt the multistage aggregation-update framework by Sanchez-Gonzalez et al. [39] and design a GNN module consisting of edge update, edge aggregation, and node update, as shown in Figure 6.

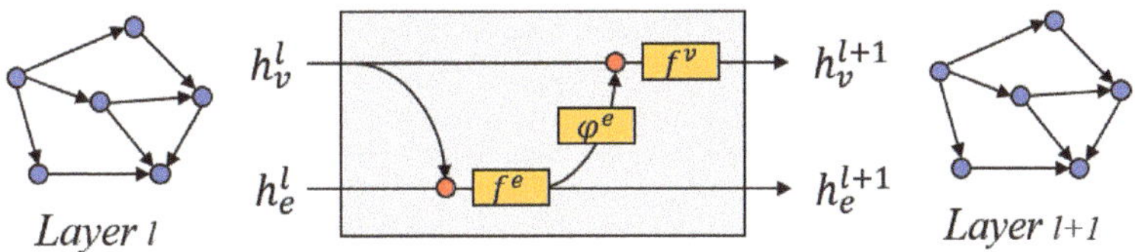

Figure 6. GNN module. h_e^l represents the hidden state of node e at layer l.

Here, φ is the aggregation function, designed to transfer edge states to nodes, thereby extracting more neighborhood information. f is the update function, responsible for further updating the aggregated representations.

Then, we embed the GNN module into the model to form a graph encoder. Figure 7 displays its structure. The features shown are for a node and its neighborhood in the graph G_t at time t.

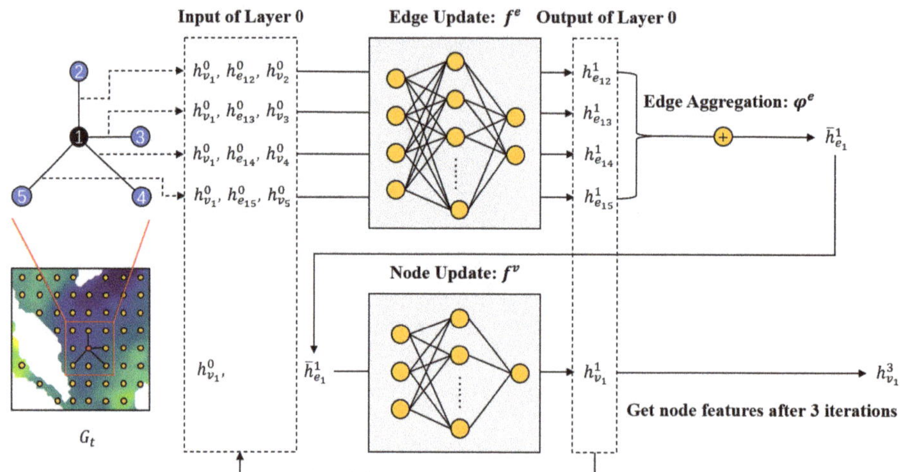

Figure 7. Graph encoder. The feature indicated in the graph is an example of a certain node and its neighborhood in the graph G_t at time t. The static image encoder encodes all nodes and edges in the graph in the same way.

For graph G_t, the node feature matrix is $X \in R^{n \times 3}$, with $x_i \in R^3$ as the feature vector for node v_i. The edge feature matrix is $Y \in R^{m \times 2}$, and $y_{ij} \in R^2$ is the feature vector for edge e_{ij}. Here, n is the number of nodes in G_t, and m denotes the number of edges. Node features have three dimensions, SST, longitude, and latitude. Edge features have two dimensions: direction and length, length represents the shortest path between two edges, and direction is a measure of its angle to the North. The hidden state of v_i at layer l is $h_{v_i}^l \in R^1$, and the hidden state of e_{ij} at layer l is $h_{e_{ij}}^l \in R^2$. Thus, the initial value is $h_{v_i}^0 = x_i$ and $h_{e_{ij}}^0 = y_{ij}$.

- Edge update: As shown in Equation (5), we gather the current edge state and the states of its adjacent nodes, and pass them through the edge update function f^e to obtain the updated result. This output will be used in the edge aggregation and the next iteration. The f^e is a multilayer perceptron and a ReLU activation function to capture nonlinear features.

$$h_{e_{ij}}^{l+1} = f^e\left(h_{e_{ij}}^l, h_{v_i}^l, h_{v_j}^l\right) \tag{5}$$

- Edge aggregation: Next, as shown in Equation (6), we use the function φ^e to aggregate the updated edge states of all connected edges for each node. Common aggregation methods include sum, mean, and max. Considering that for a point on the sea surface, heat changes manifest as a convergence or dissipation process, we choose the sum aggregation method.

$$\bar{h}_{e_i}^{l+1} = \varphi_{v_j \in N(v_i)}^{e \to v}\left(h_{e_{ij}}^{l+1}\right) \tag{6}$$

- Node update: Finally, we gather the previous aggregation outputs and their current states and put them into the update function f^v. Similar to f^e, f^v is also a combination of a multilayer perceptron and a ReLU activation function.

$$h_{v_i}^{l+1} = f^v\left(\bar{h}_{e_i}^{l+1}, h_{v_i}^l\right) \tag{7}$$

The three stages described above constitute a single iteration. By stacking multiple GNN layers and performing iterative updates, information can propagate within the graph, enabling the model to learn more abstract and complex features. In this study, we set the iteration times to 3.

3.3. Temporal Encoder

Figure 8 shows the structure and encoding process of the temporal encoder. A sequence of graphs with extracted spatial features is obtained after the graph encoder, which contains updated node and edge states. Then, we use the node state sequence as the input for the LSTM layer, and the encoded hidden state h_t is acquired after temporal feature extraction.

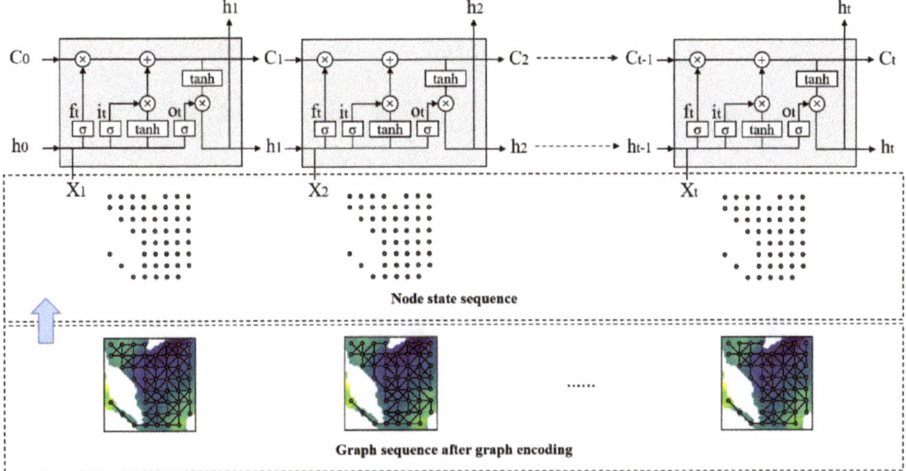

Figure 8. Temporal encoder. X_t, h_t and C_t represent the input, output and memory cell state at the current timestep t, respectively. ft, it and Ot represent forget gate, input gate and output gate, respectively.

The LSTM layer contains multiple LSTM units, which have the ability to selectively remember important information while filtering out noise [40]. This ability is attributed to the gating mechanism, which includes forget gate ft, input gate it, and output gate Ot, helps control gradients and addresses the vanishing and exploding gradient problems in RNNs.

3.4. Decoder and Loss Function

After the graph and temporal encoders, we obtain the node state h_t. In this study, we aim to predict multi-step future SST values based on historical observations. Accordingly, we apply a direct multi-output prediction strategy to convert h_t into a prediction sequence with a length equal to the prediction steps. The prediction steps are consistent with the time scale of the input data, for example, the time scale of the input data is daily, the prediction for each step is one day.

Then, we use the mean squared error (MSE) as the loss function in this study, as shown in Equation (8). T denotes the total prediction steps, y_t represents the actual value at time t, and $\hat{y}_t$ is the predicted value.

$$L_{sup} = \frac{\sum_{t=1}^{T}(y_t - \hat{y}_t)^2}{T} \tag{8}$$

4. Experiments

4.1. Metrics

SST prediction is inherently a regression task. To accurately assess the performance of each model, we consider two perspectives: the deviation between predictions and observations, and data fitting. We choose three evaluation metrics: Root Mean Squared Error (RMSE), Mean Absolute Error (MAE), and R-squared. For a sequence of length T, with y_t as the observations at time t and $\hat{y}_t$ as the predictions, and $\bar{y}$ as the average of observations, the formulas for each metric are provided below.

$$RMSE = \sqrt{\frac{\sum_{t=1}^{T}(y_t - \hat{y}_t)^2}{T}} \tag{9}$$

$$MAE = \frac{\sum_{t=1}^{T}|y_t - \hat{y}_t|}{T} \tag{10}$$

$$R^2 = \frac{\sum_{t=1}^{T}(\hat{y}_t - \bar{y})^2}{\sum_{t=1}^{T}(y_t - \bar{y})^2} \tag{11}$$

4.2. Compared Models

To evaluate the performance of GMNN, we selected three types of comparison models:

- FC-LSTM and FC-GRU: They are time series prediction models, which integrate LSTM or GRU layers with fully connected layers for feature extraction and improved representation capability.
- ConvLSTM: This is a spatiotemporal model utilizing CNN idea with LSTM, which incorporates convolution operations into input data and hidden states, allowing for the capture of spatial information and complex spatiotemporal features.
- GCN-LSTM: This is a spatiotemporal model employing GNN idea, which combines graph convolutional networks (GCN) with LSTM for graph sequence prediction, effectively extracting features from nodes and their multi-order neighbors and integrating them into the LSTM layer for temporal information processing.

4.3. Results of Different Subregions

To verify the generalization ability of GMNN in different regions, we select several subregions in the daily mean dataset and predict the SST for the next 1, 3, and 7 days.

GMNN is applicable to both complete and incomplete sea areas (with land or islands). In contrast, ConvLSTM based on CNN idea, is suitable only for complete sea areas, the missing values in incomplete sea areas must be filled using interpolation, which introduces noise and can affect model accuracy. Therefore, we select data from three incomplete sea area subregions (No. 1, 2, and 3) and three complete sea area subregions (No. 4, 5, and 6) at the same latitude for comparison (Figure 2). The models for incomplete sea area subregions include FC-LSTM, FC-GRU, and GCN-LSTM. For complete sea area subregions, FC-LSTM, FC-GRU, ConvLSTM, and GCN-LSTM are used.

4.3.1. Results of Incomplete Sea Areas

We analyze the effectiveness of our model in incomplete sea areas, taking subregion 1 as an example. The constructed graph in this region contains 1245 nodes. Table 2 shows the experiment results.

Table 2. Daily prediction results on the incomplete sea area dataset (subregion 1).

Method	Metric	Daily		
		1	3	7
FC-LSTM	RMSE	0.084	0.184	0.311
	MAE	0.020	0.071	0.160
	R-squared	0.993	0.952	0.911
FC-GRU	RMSE	0.084	0.186	0.312
	MAE	0.209	0.074	0.163
	R-squared	0.994	0.933	0.909
GCN-LSTM	RMSE	0.081	0.178	0.292
	MAE	0.019	0.070	0.153
	R-squared	0.996	0.965	0.924
GMNN	RMSE	0.080	0.177	0.288
	MAE	0.019	0.070	0.152
	R-squared	0.999	0.968	0.924

The two worst-performing models are FC-LSTM and FC-GRU, with the maximum RMSE, MAE and the minimum R- squared (Table 2), indicating that ignoring spatial correlation can significantly affect prediction accuracy. There is little difference between these two models and FC-GRU's is slightly worse than FC-LSTM's when predicting the future 3 and 7 days. This suggests that in this study, using LSTM for time feature extraction is more suitable. Both graph-based models exhibit good performance, with GMNN performing the best in all metrics. For instance, in terms of RMSE for seven-day prediction, GMNN's 0.288 is 7.7% lower than FC-LSTM and 1.6% lower than GCN-LSTM. This indicates that the iterative GNN idea can effectively capture the spatial information of SST data.

To visually compare the results, we take the node with longitude 109.875°E and latitude 0.125°N in subregion 1 as an example. The predictions and observations of each model were compared using a line chart for a 7-step prediction, as shown in Figure 9.

It can be seen that the main difference in the prediction results of each model lies in the degree of fitting to the peak values. Therefore, we select four peak areas, a, b, c, and d for detailed analysis. Among them, a and c are steep peak areas, while b is a gentle peak area, and d is a low peak area.

FC-GRU predicts well in b and d, but has the worst performance among all models in the steep peak areas a and c. FC-LSTM performs slightly better than FC-GRU in the steep peak areas, but its fitting degree in the low peak area is low. GCN-LSTM's predictions can already fit the observations well, but there is still room for improvement in the steep peak areas. GMNN has the best overall prediction accuracy, showing a high degree of fitting in these peak areas with different characteristics. Especially in the steep peak areas, the performance is significantly better than the other compared models.

4.3.2. Results of Complete Sea Areas

Similarly, we analyze the effectiveness of GMNN in complete sea areas using the example of subregion 4 which contains 1600 nodes. The results are shown in Table 3.

FC-LSTM and FC-GR performed the worst. ConvLSTM which uses CNN idea, exhibits good performance in complete sea areas where data can be expressed in pixel form, and its prediction accuracy is slightly better than that of the GCN-LSTM model, which uses graph idea. Among prediction models, GMNN is better than other models in the metrics. GMNN's RMSE value for seven-day prediction decreased by 5.5% compared to FC-LSTM, 0.9% compared to ConvLSTM, and 2.0% compared to GCN-LSTM.

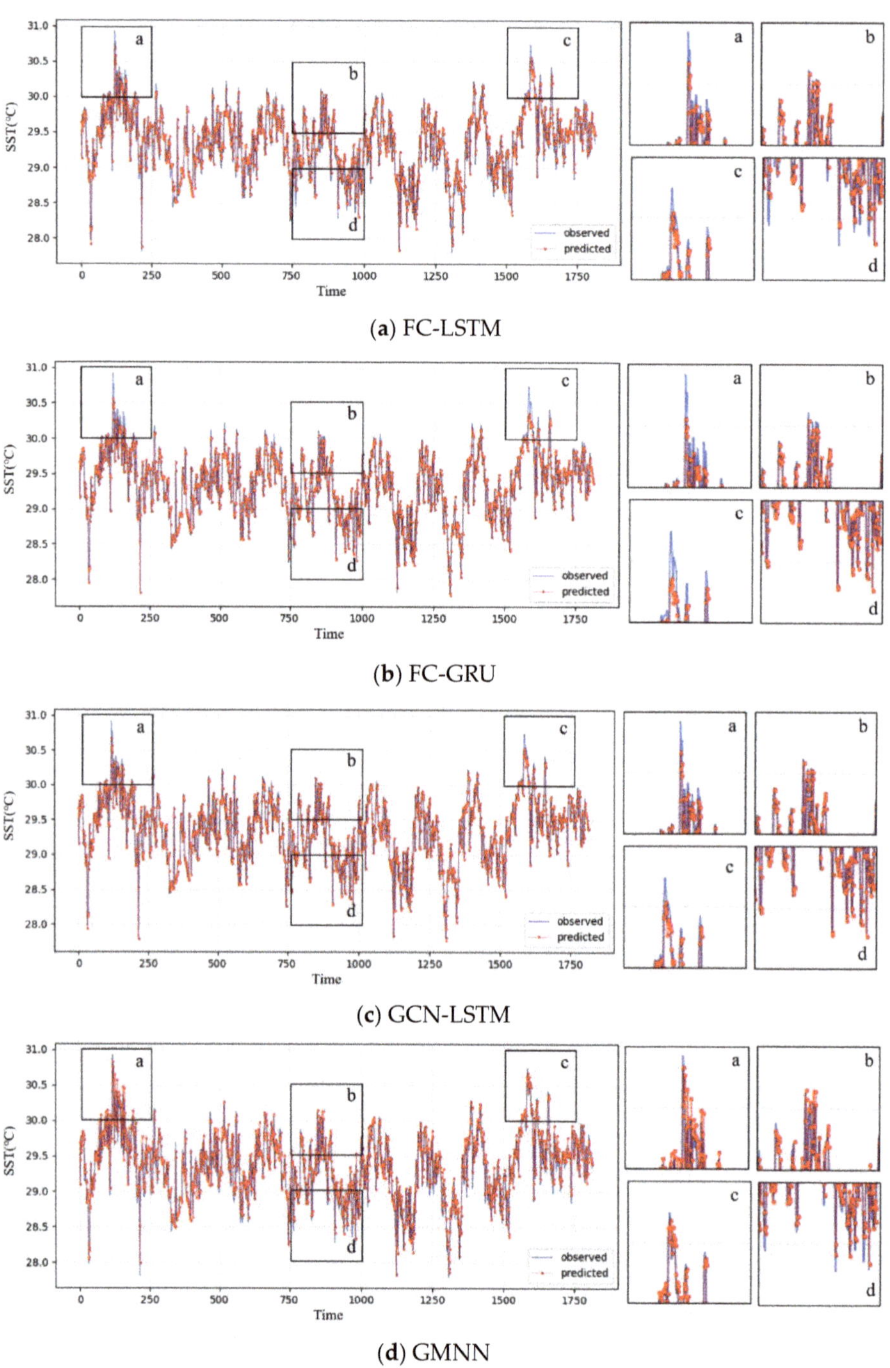

(**a**) FC-LSTM

(**b**) FC-GRU

(**c**) GCN-LSTM

(**d**) GMNN

Figure 9. Line charts of observations and predictions in seven-day of different models on the incomplete sea area dataset (subregion 1): (**a**) FC-LSTM; (**b**) FC-GRU; (**c**) GCN-LSTM; (**d**) GMNN.

Table 3. Daily prediction results on the complete sea area dataset (subregion 4).

Method	Metric	Daily		
		1	**3**	**7**
FC-LSTM	RMSE	0.078	0.164	0.252
	MAE	0.019	0.069	0.134
	R-squared	0.979	0.948	0.807
FC-GRU	RMSE	0.076	0.169	0.252
	MAE	0.019	0.070	0.134
	R-squared	0.979	0.949	0.798
ConvLSTM	RMSE	0.079	0.154	0.241
	MAE	0.018	0.062	0.127
	R-squared	0.982	0.940	0.834
GCN-LSTM	RMSE	0.075	0.156	0.243
	MAE	0.018	0.062	0.129
	R-squared	0.982	0.939	0.834
GMNN	RMSE	0.073	0.154	0.238
	MAE	0.018	0.062	0.127
	R-squared	0.983	0.956	0.855

As shown in Figure 10, a comparison chart of the seven-day predictions and observations is created for node located at 130.125°E and 0.125°N in subregion 4. We analyze two high peaks (a, d) and two low peaks (b, c) in detail. The performance of FC-LSTM and FC-GRU is quite similar, with poor predictions for the highest and lowest points in all four areas. ConvLSTM and GCN-LSTM show significant improvement in the prediction of areas a, b, and d, with ConvLSTM showing better fitting. GMNN performs well in all four areas with excellent prediction ability.

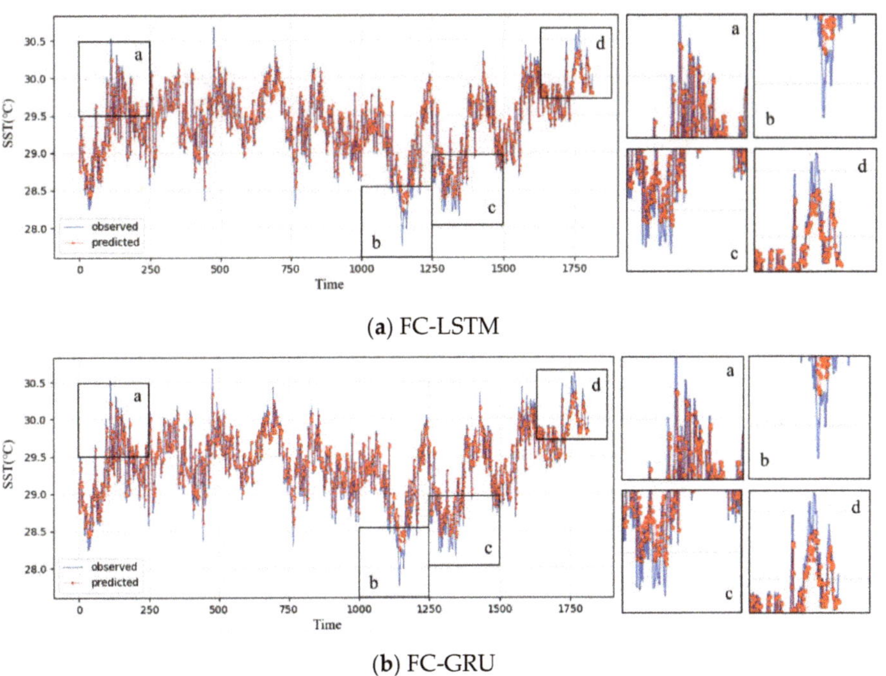

(**a**) FC-LSTM

(**b**) FC-GRU

Figure 10. *Cont.*

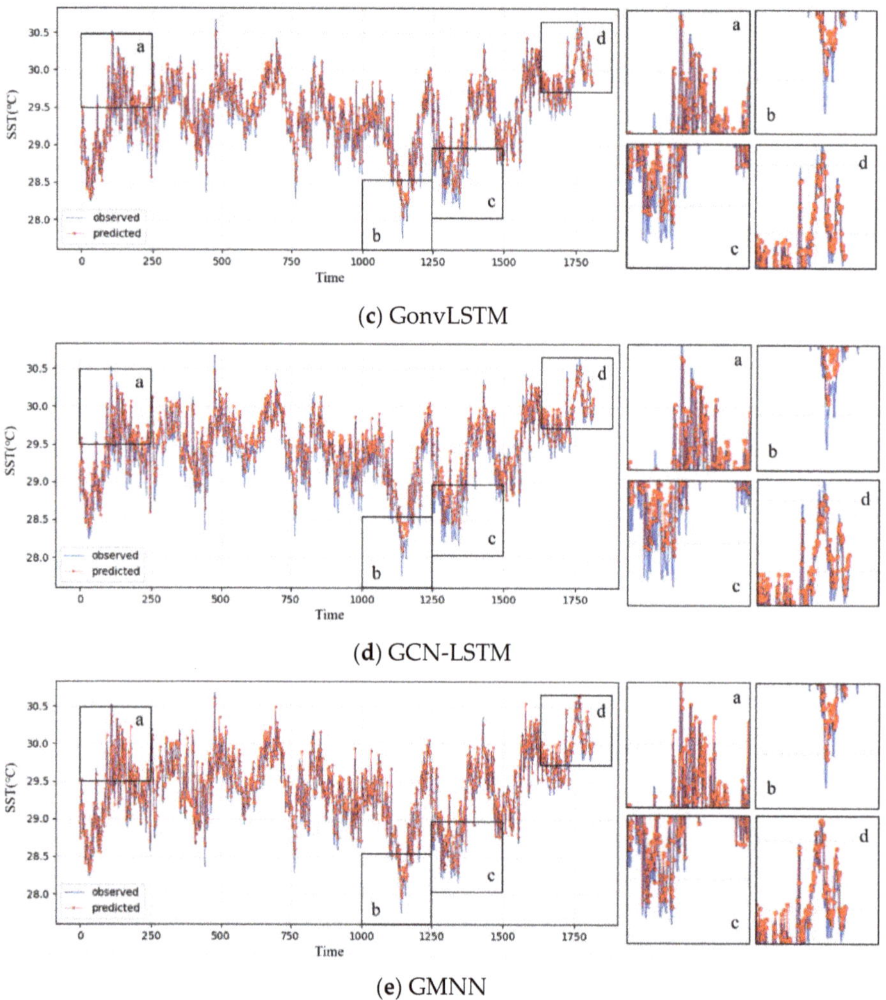

Figure 10. Line charts of observations and predictions in seven-day of different models on the complete sea area dataset (subregion 4): (**a**) FC-LSTM; (**b**) FC-GRU; (**c**) GonvLSTM; (**d**) GCN-LSTM; (**e**) GMNN.

The results prove that GMNN has excellent prediction ability in both complete and incomplete sea areas.

4.4. Results of Different Time Scales

To verify the accuracy and stability of GMNN for different time scales and prediction steps, we conduct comparison experiments for future 1 step, 3 steps, and 7 steps on three types of datasets: daily, weekly, and monthly mean, using the example of subregion 5. The results are presented in Figure 11. The y-axis of each metric is standardized across different time scales for comparison.

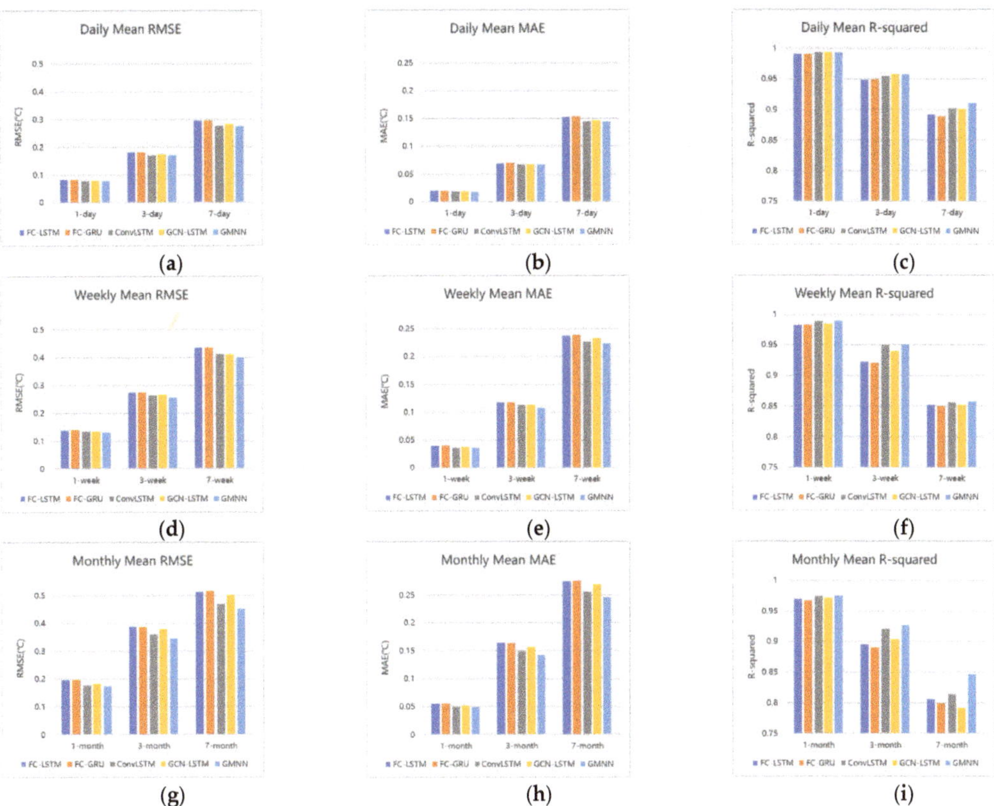

Figure 11. Comparison of prediction results at different time scales and prediction steps: (**a**) Daily Mean RMSE; (**b**) Daily Mean MAE; (**c**) Daily Mean R-squared; (**d**) Weekly Mean RMSE; (**e**) Weekly Mean MAE; (**f**) Weekly Mean R-squared; (**g**) Monthly Mean RMSE; (**h**) Monthly Mean MAE; (**i**) Monthly Mean R-squared.

From the perspective of fixed time scales, as the prediction step increases, the performance of each model declines, with RMSE and MAE increasing and R-squared decreasing. Taking daily predictions as an example, the R-squared, RMSE, and MAE for predicting one day ahead are 0.994, 0.078, and 0.018, respectively. When predicting three days ahead, R-squared decreased by 0.037, while RMSE and MAE both increased by more than double. When predicting seven days ahead, R-squared continued to decrease by 0.047, with RMSE and MAE increasing by 0.62 and 1.10 times, respectively. This suggests that multi-step prediction incurs greater errors than single-step prediction. With an increasing number of prediction steps, more relationships need to be learned, and models become more challenged in capturing the changing trends and periodicity of time series, which results in increased errors.

From the perspective of fixed prediction steps, as the time scale increases, the performance of each model also declines. Taking RMSE as an example, on a daily scale, the RMSE is 0.078, 0.170, and 0.275, when predicting the future 1, 3, and 7 step. On a weekly scale, the RMSE increases by 0.67 times, 0.50 times, and 0.46 times when predicting 1, 3, and 7 steps in the future. On a monthly scale, compared with the weekly scale, the RMSE increases by 0.33 times, 0.34 times, and 0.13 times when predicting 1, 3, and 7 steps in the future. The reasons for this phenomenon are mainly twofold. First, the time series of daily, weekly, and monthly mean datasets used in this study contain 10,227, 1461, and 336 time steps, respectively, which means that the data available for training are sparser at larger time scales, and affect the prediction accuracy. Second, from daily to weekly to monthly, the smoothness of the SST changes gradually decreases, and the changing trend and periodicity of the time series become less obvious, making it difficult to capture the nonlinear features, resulting in a decrease in prediction accuracy.

As shown in Figure 11, GMNN has better prediction accuracy than the comparison models at different time scales and prediction steps. In order to more clearly show the performance improvement, we use FC-LSTM as the baseline and calculate the percentage of RMSE reduction of GMNN relative to the baseline under different time scales and prediction steps (Figure 12), which serves as the performance improvement ratio.

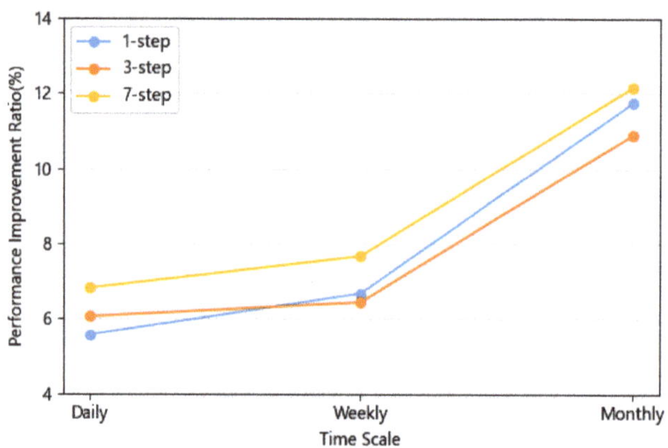

Figure 12. Comparison of GMNN performance improvement ratios at different time scales and prediction steps.

At the same time scale, changes in the prediction step do not result in significant changes in the improvement ratio. However, when the time scale changes to the monthly scale at the same prediction step, the improvement ratio increases significantly. This suggests that GMNN can capture hidden spatiotemporal features on large scales.

5. Discussion

5.1. Model Comparison

Through experiments in different partitions, time scales, and prediction steps, we find that our GMNN is better than other comparison models, which can be categorized into time series models (FC-LSTM, FC-GRU), convolution-based model (ConvLSTM), and graph-based model (GCN-LSTM). The results provide insights into the applicability and effectiveness of different ideas for SST prediction tasks.

The inferior performance of time series models suggests the impact of neglecting spatial information on prediction accuracy. Convolution-based models and graph-based models differ in their learning styles and applicable structures. In terms of learning styles, CNN extracts feature by sliding convolution kernels, thus exhibiting strong capabilities in extracting multi-scale local spatial features [41]. GNN focus more on adjacency relationships, with their message-passing mechanism providing better abilities for tasks with strong object interrelations. As SST is influenced by ocean currents, winds, and heat exchange processes in nearby regions, graph-based models can well represent temperature variation processes. Regarding applicable structures, CNN is based on traditional grid structures and is suitable for regular datasets. Therefore, convolution-based model (ConvLSTM) demonstrates excellent forecasting performance in experiments with complete sea area datasets. Graph-based models, on the other hand, are not restricted by data regularity, offering greater flexibility.

Both being graph-based methods, the iterative GNN used in this study and the GCN adopted by the comparison model GCN-LSTM differ in the information they emphasize. GCN primarily focuses on node information, with its convolution operation aggregating features of nodes and their adjacent counterparts [42,43]. Although GCN considers node information and adjacency relationships, edge attributes are typically not directly incorporated into calculations. In contrast, the iterative GNN takes both node and edge information into account through its designed aggregation and update functions. As a result, in comparative experiments, GMNN consistently achieves better prediction outcomes than GCN-LSTM.

5.2. Error Distribution

To clearly show the prediction performance and error distribution of our model, we use the future seven-day prediction results of GMNN and select 12 subregions with observations, predictions, and errors on 26 February 2016 for analysis. Among them, two regions are selected within every 10° latitude range, corresponding to incomplete and complete sea areas, respectively. The experiment results of the 12 regions are shown in Figure S1.

When comparing the error between two regions at the same latitude, there is no significant difference in the prediction accuracy of the model between incomplete and complete sea areas, indicating a good prediction performance in both types of regions.

Comparing the errors of regions at different latitudes, the regions with latitudes between 30°N and 50°N have the largest errors, followed by the regions between 20°N and 30°N and between 50°N and 60°N, while the regions between 0°–20°N have relatively smaller errors. The complexity of the meteorological and oceanic environment is the main reason for the differences in the prediction performance among these regions. Regions between 30° and 50°N belong to the North Temperate Zone and are influenced by subtropical high-pressure zones, westerlies, monsoons, and continental climates. They are also affected by multiple ocean currents such as the Kuroshio Current, the Oyashio Current, and the North Pacific Warm Current, resulting in complex spatiotemporal characteristics and making predictions difficult. In contrast, regions between 0° and 20°N are mainly affected by tropical and subtropical climates, with relatively simple spatiotemporal characteristics and thus easier to predict. Regions between 20° and 30°N and between 50° 60°N have moderate environmental complexity and prediction difficulty.

6. Conclusions

In this paper, we propose a GMNN to predict future SST. The model uses a graph representation method based on distance threshold and Pearson correlation coefficient to transform SST data into a graph structure, thus overcoming the limitations of convolution-based methods in encoding irregular data that includes land or islands. We also design a graph encoder based on iterative GNN, incorporating edge information to fully express the heat transfer process at the sea surface. To validate the effectiveness of GMNN, we choose time series prediction models (FC-LSTM, FC-GRU), convolution-based model (ConvLSTM), and graph-based model without considering edge information (GCN-LSTM) as comparison. We conduct experiments of these models in incomplete and complete sea area partitions, daily, weekly and monthly time scales, as well as 1-step, 3-step, and 7-step prediction steps, and our model exhibits superior prediction ability compared to the others, reflecting its accuracy and stability.

In addition, we find that with increasing time scales and prediction steps, the prediction accuracy decreases. GMNN shows a higher performance improvement at the monthly time scale than at the daily and weekly time scales. Error analysis reveals that GMNN has larger prediction errors for areas with greater temperature variations. The errors also have a certain correlation with latitude, with higher errors for the region of 30–50°N due to the complex ocean and meteorological environment, and lower errors for the region of 0–20°N with relatively stable temperature changes.

However, there are still some limitations in our work. Although we use SST as the input for prediction, in the future, other factors will be considered and collected for systematic analysis so as to explore the impact of these factors on SST prediction. Moreover, the study area in this case was the Northwest Pacific. To generalize the ability of our model, we will select different ocean basins with various dynamic features and make improvements in subsequent studies to explore large-scale SST prediction.

Supplementary Materials: The following supporting information can be downloaded at: https://www.mdpi.com/article/10.3390/rs15143539/s1, Figure S1: GMNN prediction results for SST on 26 February 2016, in 12 subregions: (a) Subregion 3; (b) Subregion 4; (c) Subregion 11; (d) Subregion 15; (e) Subregion 18; (f) Subregion 22; (g) Subregion 27; (h) Subregion 30; (i) Subregion 32; (j) Subregion 36; (k) Subregion 40; (l) Subregion 41.

Author Contributions: Conceptualization, L.H.; Data curation, L.H.; Formal analysis, M.Q.; Funding acquisition, Z.D. and R.L.; Methodology, M.Q.; Project administration, Z.D. and R.L.; Resources, Z.D. and R.L.; Software, S.L. and A.Z.; Supervision, L.H. and S.W.; Validation, S.W.; Visualization, S.L. and A.Z.; Writing—original draft, S.L.; Writing—review & editing, A.Z., M.Q. and S.W. All authors have read and agreed to the published version of the manuscript.

Funding: This research was supported by the National Natural Science Foundation of China (42225605) and China Postdoctoral Science Foundation (2022M720121).

Data Availability Statement: Publicly available datasets were analyzed in this study. The SST data can be downloaded from https://psl.noaa.gov/data/gridded/data.noaa.oisst.v2.highres.html, accessed on 11 May 2023.

Conflicts of Interest: The authors declare no conflict of interest.

References

1. Wang, G.; Cheng, L.; Abraham, J.; Li, C. Consensuses and discrepancies of basin-scale ocean heat content changes in different ocean analyses. *Clim. Dyn.* **2018**, *50*, 2471–2487. [CrossRef]
2. Ham, Y.; Kug, J.; Park, J.; Jin, F. Sea surface temperature in the north tropical Atlantic as a trigger for El Niño/Southern Oscillation events. *Nat. Geosci.* **2013**, *6*, 112–116. [CrossRef]
3. Chen, Z.; Wen, Z.; Wu, R.; Lin, X.; Wang, J. Relative importance of tropical SST anomalies in maintaining the Western North Pacific anomalous anticyclone during El Niño to La Niña transition years. *Clim. Dyn.* **2016**, *46*, 1027–1041. [CrossRef]
4. Andrade, H.A.; Garcia, C.A.E. Skipjack tuna fishery in relation to sea surface temperature off the southern Brazilian coast. *Fish Oceanogr.* **1999**, *8*, 245–254. [CrossRef]
5. Wang, W.; Zhou, C.; Shao, Q.; Mulla, D.J. Remote sensing of sea surface temperature and chlorophyll-a: Implications for squid fisheries in the north-west Pacific Ocean. *Int. J. Remote Sens.* **2010**, *31*, 4515–4530. [CrossRef]
6. Khan, T.M.A.; Singh, O.P.; Rahman, M.S. Recent sea level and sea surface temperature trends along the Bangladesh coast in relation to the frequency of intense cyclones. *Mar. Geod.* **2000**, *23*, 103–116. [CrossRef]
7. Emanuel, K.; Sobel, A. Response of tropical sea surface temperature, precipitation, and tropical cyclone-related variables to changes in global and local forcing. *J. Adv. Model. Earth Syst.* **2013**, *5*, 447–458. [CrossRef]
8. Chassignet, E.P.; Hurlbert, H.E.; Smedstad, O.M.; Halliwell, G.R.; Hogan, P.J.; Wallcraft, A.J.; Bleck, R. Ocean prediction with the hybrid coordinate ocean model (HYCOM). In *Ocean Weather Forecasting: An Integrated View of Oceanography*; Springer: Berlin/Heidelberg, Germany, 2006; pp. 413–426. [CrossRef]
9. Chassignet, E.P.; Hurlburt, H.E.; Metzger, E.J.; Smedstad, O.M.; Cummings, J.A.; Halliwell, G.R. US GODAE: Global ocean prediction with the HYbrid Coordinate Ocean Model (HYCOM). *Oceanography* **2009**, *22*, 64–75. [CrossRef]
10. Qian, C.; Huang, B.; Yang, X.; Chen, G. Data science for oceanography: From small data to big data. *Big Earth Data* **2022**, *6*, 236–250. [CrossRef]
11. Kartal, S. Assessment of the spatiotemporal prediction capabilities of machine learning algorithms on Sea Surface Temperature data: A comprehensive study. *Eng. Appl. Artif. Intell.* **2023**, *118*, 105675. [CrossRef]
12. Xue, Y.; Leetmaa, A. Forecasts of Tropical Pacific SST and Sea Level Using a Markov Model. *Geophys. Res. Lett.* **2000**, *27*, 2701–2704. [CrossRef]
13. Collins, D.C.; Reason, C.J.C. Predictability of Indian Ocean Sea Surface Temperature Using Canonical Correlation Analysis. *Clim. Dyn.* **2004**, *22*, 481–497. [CrossRef]

14. Neetu, S.R.; Basu, S.; Sarkar, A.; Pal, P.K. Data-Adaptive Prediction of Sea-Surface Temperature in the Arabian Sea. *Ieee Geosci. Remote Sens. Lett.* **2010**, *8*, 9–13. [CrossRef]

15. Shao, Q.; Li, W.; Han, G.; Hou, G.; Liu, S.; Gong, Y.; Qu, P. A Deep Learning Model for Forecasting Sea Surface Height Anomalies and Temperatures in the South China Sea. *J. Geophys. Res. Ocean.* **2021**, *126*, e2021JC017515. [CrossRef]

16. Garcia-Gorriz, E.; Garcia-Sanchez, J. Prediction of Sea Surface Temperatures in the Western Mediterranean Sea by Neural Networks Using Satellite Observations. *Geophys. Res. Lett.* **2007**, *34*. [CrossRef]

17. Lee, Y.; Ho, C.; Su, F.; Kuo, N.; Cheng, Y. The Use of Neural Networks in Identifying Error Sources in Satellite-Derived Tropical SST Estimates. *Sensors* **2011**, *11*, 7530–7544. [CrossRef]

18. Lins, I.D.; Araujo, M.; Moura, M.D.C.; Silva, M.A.; Droguett, E.L. Prediction of Sea Surface Temperature in the Tropical Atlantic by Support Vector Machines. *Comput. Stat. Data Anal.* **2013**, *61*, 187–198. [CrossRef]

19. Zhang, Q.; Wang, H.; Dong, J.; Zhong, G.; Sun, X. Prediction of Sea Surface Temperature Using Long Short-Term Memory. *IEEE Geosci. Remote Sens. Lett.* **2017**, *14*, 1745–1749. [CrossRef]

20. Sun, T.; Feng, Y.; Li, C.; Zhang, X. High Precision Sea Surface Temperature Prediction of Long Period and Large Area in the Indian Ocean Based on the Temporal Convolutional Network and Internet of Things. *Sensors* **2022**, *22*, 1636. [CrossRef]

21. Su, H.; Zhang, T.; Lin, M.; Lu, W.; Yan, X. Predicting subsurface thermohaline structure from remote sensing data based on long short-term memory neural networks. *Remote Sens. Environ.* **2021**, *260*, 112465. [CrossRef]

22. Guo, H.; Xie, C. Multi-Feature Attention Based LSTM Network for Sea Surface Temperature Prediction. In Proceedings of the 3rd International Conference on Computer Information and Big Data Applications, Wuhan, China, 25–27 March 2022.

23. Yang, Y.; Dong, J.; Sun, X.; Lima, E.; Mu, Q.; Wang, X. A CFCC-LSTM Model for Sea Surface Temperature Prediction. *IEEE Geosci. Remote Sens. Lett.* **2018**, *15*, 207–211. [CrossRef]

24. Yu, X.; Shi, S.; Xu, L.; Liu, Y.; Miao, Q.; Sun, M. A Novel Method for Sea Surface Temperature Prediction Based on Deep Learning. *Math. Probl. Eng.* **2020**, *2020*, 6387173. [CrossRef]

25. Qiao, B.; Wu, Z.; Tang, Z.; Wu, G. Sea Surface Temperature Prediction Approach Based on 3D CNN and LSTM with Attention Mechanism. In Proceedings of the 23rd International Conference on Advanced Communication Technology (ICACT), PyeongChang, Republic of Korea, 7–10 February 2021. [CrossRef]

26. Shi, X.; Chen, Z.; Wang, H.; Yeung, D.; Wong, W. Convolutional LSTM Network: A Machine Learning Approach for Precipitation Nowcasting. *Adv. Neural Inf. Process. Syst.* **2015**, *28*, 1–9. [CrossRef]

27. Xiao, C.; Chen, N.; Hu, C.; Wang, K.; Xu, Z.; Cai, Y.; Xu, L.; Chen, Z.; Gong, J. A spatiotemporal deep learning model for sea surface temperature field prediction using time-series satellite data. *Environ. Modell. Softw.* **2019**, *120*, 104502. [CrossRef]

28. Jung, S.; Kim, Y.J.; Park, S.; Im, J. Prediction of Sea Surface Temperature and Detection of Ocean Heat Wave in the South Sea of Korea Using Time-series Deep-learning Approaches. *Korean J. Remote Sens.* **2020**, *36*, 1077–1093. [CrossRef]

29. Zhang, K.; Geng, X.; Yan, X.H. Prediction of 3-D Ocean Temperature by Multilayer Convolutional LSTM. *IEEE Geosci. Remote Sens. Lett.* **2020**, *17*, 1303–1307. [CrossRef]

30. Zhang, X.; Li, Y.; Frery, A.C.; Ren, P. Sea Surface Temperature Prediction With Memory Graph Convolutional Networks. *IEEE Geosci. Remote Sens. Lett.* **2022**, *19*, 1–5. [CrossRef]

31. Zhou, J.; Cui, G.; Hu, S.; Zhang, Z.; Yang, C.; Liu, Z.; Wang, L.; Li, C.; Sun, M. Graph neural networks: A review of methods and applications. *AI Open* **2020**, *1*, 57–81. [CrossRef]

32. Gilmer, J.; Schoenholz, S.S.; Riley, P.F.; Vinyals, O.; Dahl, G.E. Neural Message Passing for Quantum Chemistry. *arXiv* **2017**, arXiv:1704.01212. [CrossRef]

33. Sun, Y.; Yao, X.; Bi, X.; Huang, X.; Zhao, X.; Qiao, B. Time-Series Graph Network for Sea Surface Temperature Prediction. *Big Data Res.* **2021**, *25*, 100237. [CrossRef]

34. Wang, T.; Li, Z.; Geng, X.; Jin, B.; Xu, L. Time Series Prediction of Sea Surface Temperature Based on an Adaptive Graph Learning Neural Model. *Future Internet* **2022**, *14*, 171. [CrossRef]

35. Pan, J.; Li, Z.; Shi, S.; Xu, L.; Yu, J.; Wu, X. Adaptive graph neural network based South China Sea seawater temperature prediction and multivariate uncertainty correlation analysis. *Stoch. Environ. Res. Risk Assess.* **2022**, *37*, 1877–1896. [CrossRef]

36. Xia, F.; Sun, K.; Yu, S.; Aziz, A.; Wan, L.; Pan, S.; Liu, H. Graph Learning: A Survey. *IEEE Trans. Artif. Intell.* **2021**, *2*, 109–127. [CrossRef]

37. LeCun, Y.; Bengio, Y. Convolutional Networks for Images, Speech, and Time-Series. In *The Handbook of Brain Theory and Neural Networks*; MIT Press: Cambridge, MA, USA, 1995; pp. 255–258.

38. Ye, Z.; Kumar, Y.J.; Sing, G.O.; Song, F.; Wang, J. A Comprehensive Survey on Graph Neural Networks for Knowledge Graphs. *IEEE Access* **2019**, *10*, 75729–75741. [CrossRef]

39. Scarselli, F.; Gori, M.; Tsoi, A.; Hagenbuchner, M.; Monfardini, G. The Graph Neural Network Model. *IEEE Trans. Neural Netw.* **2009**, *20*, 61–80. [CrossRef] [PubMed]

40. Hochreiter, S.; Schmidhuber, J. Long Short-Term Memory. *Neural Comput.* **1997**, *9*, 1735–1780. [CrossRef]

41. Li, Z.; Liu, F.; Yang, W.; Peng, S.; Zhou, J. A Survey of Convolutional Neural Networks: Analysis, Applications, and Prospects. *IEEE Trans. Neural Netw. Learn. Syst.* **2022**, *33*, 6999–7019. [CrossRef]

42. Bruna, J.; Zaremba, W.; Szlam, A.; LeCun, Y. Spectral Networks and Deep Locally Connected Networks on Graphs. *arXiv* **2014**, arXiv:1312.6203. [CrossRef]
43. Micheli, A. Neural Network for Graphs: A Contextual Constructive Approach. *IEEE Trans. Neural Netw.* **2009**, *20*, 498–511. [CrossRef]

remote sensing

Article

Prediction of Sea Surface Temperature in the East China Sea Based on LSTM Neural Network

Xiaoyan Jia [1], Qiyan Ji [1,*], Lei Han [1], Yu Liu [1,2], Guoqing Han [1] and Xiayan Lin [1]

1   Marine Science and Technology College, Zhejiang Ocean University, Zhoushan 316022, China; s20070700001@zjou.edu.cn (X.J.); s20070700026@zjou.edu.cn (L.H.); liuyuhk@zjou.edu.cn (Y.L.); hanguoqing@zjou.edu.cn (G.H.); linxiayan@zjou.edu.cn (X.L.)

2   Southern Marine Science and Engineering Guangdong Laboratory (Zhuhai), Zhuhai 519000, China

*   Correspondence: jiqiyan@zjou.edu.cn

Abstract: Sea surface temperature (SST) is an important physical factor in the interaction between the ocean and the atmosphere. Accurate monitoring and prediction of the temporal and spatial distribution of SST are of great significance in dealing with climate change, disaster prevention, disaster reduction, and marine ecological protection. This study establishes a prediction model of sea surface temperature for the next five days in the East China Sea using long-term and short-term memory neural networks (LSTM). It investigates the influence of different parameters on prediction accuracy. The sensitivity experiment results show that, based on the same training data, the length of the input data of the LSTM model can improve the model's prediction performance to a certain extent. However, no obvious positive correlation is observed between the increase in the input data length and the improvement of the model's prediction accuracy. On the contrary, the LSTM model's performance decreases with the prediction length increase. Furthermore, the single-point prediction results of the LSTM model for the estuary of the Yangtze River, Kuroshio, and the Pacific Ocean are accurate. In particular, the prediction results of the point in the Pacific Ocean are the most accurate at the selected four points, with an *RMSE* of 0.0698 °C and an R^2 of 99.95%. At the same time, the model in the Pacific region is migrated to the East China Sea. The model was found to have good mobility and can well represent the long-term and seasonal trends of SST in the East China Sea.

Keywords: long short-term memory (LSTM); sea surface temperature (SST); East China Sea

Citation: Jia, X.; Ji, Q.; Han, L.; Liu, Y.; Han, G.; Lin, X. Prediction of Sea Surface Temperature in the East China Sea Based on LSTM Neural Network. *Remote Sens.* **2022**, *14*, 3300. https://doi.org/10.3390/rs14143300

Academic Editors: Ana B. Ruescas, Veronica Nieves and Raphaëlle Sauzède

Received: 21 May 2022
Accepted: 5 July 2022
Published: 8 July 2022

Publisher's Note: MDPI stays neutral with regard to jurisdictional claims in published maps and institutional affiliations.

1. Introduction

Sea surface temperature (SST) plays a vital role in the energy balance of the earth's surface and the exchange of energy, momentum, and moisture between the ocean and atmosphere [1,2]. It could affect the precipitation distribution, leading to extreme weather events, such as droughts and floods [3,4]. The variation of SST would also affect biological processes, such as the distribution and reproduction of marine organisms; it can also impact marine ecosystems [5–10]. The accurate prediction of SST is of great significance in marine disaster prevention and mitigation, ecological protection, and response to global climate change.

The East China Sea (ECS) is a marginal sea of the Northwest Pacific [11]. It is located east of the China mainland, south of the Yellow Sea, and north of the South China Sea, with an area of about 770,000 km^2. The SST of ECS is affected by the East Asian monsoon system, with an annual average water temperature between 20 °C and 24 °C and an annual temperature difference between 7 °C and 9 °C. In addition to being affected by the monsoon climate, the SST of the ECS is also affected by the tidal system and the complex circulation in the ECS, such as the Kuroshio Current, the Taiwan Warm Current, the Zhejiang-Fujian Coastal Current, and the Tsushima Warm Current [12]. The change of SST is extremely complex. Moreover, the ECS is also one of the most important areas for marine heat wave disasters [13]. The prediction of SST is of great significance to the local hydrology and ecological environment. It also provides an important basis for predicting and warning of marine heat wave disasters in the ECS under climate warming.

Currently, the SST prediction methods are mainly divided into two categories. One is to use ocean numerical models. For example, Gao et al. [13] used the Finite-Volume Coastal Ocean Model (FVCOM) to study marine heatwaves in the East China Sea and the South Yellow Sea. Tiwari et al. [14] used the Regional Ocean Modeling System (ROMS) to study the sea surface temperature of the Indian Ocean. Gao et al. [15] used the HYbrid Coordinate Ocean Model (HYCOM) to simulate the sea surface temperature of the tropical and North Pacific basins. These oceanic numerical models had been established through kinetic and thermal equations and obtained the numerical solution with initial conditions and boundary conditions [16]. In terms of improving the accuracy of numerical models, the higher the accuracy, the more complex the numerical model and the higher the computational cost, which leads to the need for a large number of computing resources and relevant professional personnel to carry out the operational SST prediction work [17]. The second is adopting a data-driven approach, including traditional statistical methods and the latest machine learning methods. Traditional statistical methods, such as the Markov model [18], regression model [19], and empirical canonical correlation analysis, etc., [20], can reflect the changing law of data to a certain extent based on specific observation data and have the characteristics of a small calculation amount, but they are difficult to improve the prediction accuracy [21]. In recent years, machine learning methods have gradually become popular with the increased SST data and the rapid advance in computer technology. The current popular machine learning methods include decision trees [22], random forests [23], artificial neural networks [24], and support vector machines, etc., [25]. The machine learning method is done to discover the law of data changes from a large amount of observation data. Compared with the traditional statistical method, the prediction accuracy is significantly improved. It also has the advantages of low computational cost and easy parameterization to other geographic locations. Furthermore, the demand for this method for marine professional knowledge is not as high as that of marine numerical prediction.

Among the popular machine learning methods, neural network models are widely used because of their flexibility and powerful modeling ability [26,27]. Tang et al. [28–30] applied the neural network method to the prediction of SST for the first time. They used a feed-forward neural network to predict the average sea surface temperature anomalies in the Niño region, showing that the neural network is excellent in capturing nonlinear relationships. Then, Wu et al. [31] established a nonlinear sea surface temperature anomalies prediction model using the multilayer back propagation (BP) neural network method combined with empirical mode decomposition (EMD), which proved that its correlation skills are enhanced by 0.10–0.14 compared with the linear regression model. Gupta and Malmgren [32] made a comparative study on the prediction ability of various methods relying on specific training algorithms, regression, and artificial neural networks, and showing that the RMSEP value of the neural network was 1.3 °C, which was better than other algorithms. Tripathi et al. [33] used an artificial neural network to predict sea surface temperature anomalies in a small area of the Indian Ocean and found that the model could predict sea surface temperature anomalies with considerable accuracy. Furthermore, Patil and De [34,35], Mohongo and Deo et al. [36] also used the neural network to predict SST, showing that the neural network has a certain improvement compared with traditional statistical methods. Aparna et al. [37] proposed a neural network consisting of three layers, an input layer, a linear layer, and an output layer to predict the SST of the next day at a specific location, and found that the error of the prediction is within ±0.5 °C. However, most use traditional neural networks, which have a relatively simple structure and limited learning ability. Thus, they cannot describe the complex features in the data well. At the same time, they also have shortcomings, such as low training efficiency and the inability to fully use a large amount of SST data to train prediction models [38], which are being replaced by neural networks with deeper layers.

As a typical representative of the deep neural network in long-term sequence, the long short-term memory neural network (LSTM) model has a lower computational cost and less requirement for marine expertise than the numerical model. Compared with a shallow neural network, the LSTM model has a more complex structure to extract data change rules better. Compared with the Recurrent Neural Network (RNN), which deals with time-series data, it can prevent the gradient disappearance and explosion in the backpropagation process [39]. The reason is that, under the action of the gating mechanism, LSTM can better capture long-time series data. Therefore, it is widely used in time series forecasting problems. Zhang et al. [40] used daily, weekly, and monthly SST data to forecast the Bohai Sea one day, three days, one week, and one month in advance. The results show that the LSTM model captures time-series information better than the traditional multilayer feed-forward network. To the best of our knowledge, they are the first to apply LSTM networks to SST prediction. Sarkar [41] also applied the LSTM model to SST prediction and found that the correlation coefficient (r) between the predicted value and the actual value is close to 1. Kim et al. [42] used the LSTM model to predict the SST in the coastal areas of South Korea and found that the *RMSE* of the LSTM model one day in advance is about 0.4 °C. Their prediction results are of great significance for the prevention of aquaculture. Li [43] used the LSTM model to predict the SST in the sea area where El Niño or La Niña occurred, and the correlation coefficient between the predicted value and the actual value reached 94%, which provided a noteworthy method for the monitoring and prediction of El Niño or La Niña. Of course, there are also many scholars who use the LSTM model with other methods to predict SST. For example, Xiao et al. [44,45], respectively, applied the LSTM-AdaBoost and ConvLSTM models to SST prediction in ECS and found that the LSTM-AdaBoost and ConvLSTM models have good application prospects for medium- and short-term SST prediction. Wei et al. [46] used a self-organizing mapping (SOM) algorithm to divide the entire China Sea and its adjacent areas into 130 small areas. Then, they built an LSTM model for each area to predict its SST and found that, one month in advance, the root mean square error (*RMSE*) of the prediction is 0.5 °C. Sun et al. [47] combined the graph convolutional neural network (GCN) with the LSTM neural network to create a time-series graph network (TSGN) to predict SST and found that the *RMSE* predicted 3 days in advance is 0.47 °C. Zhang et al. [48] used the gated recurrent unit (GRU) model to predict the SST in the Bohai Sea and found that it can effectively fit the actual SST, with a correlation coefficient of 0.98. However, the above studies did not explore the impact of input and prediction lengths on the accuracy of LSTM models and the model's mobility.

This paper discusses the impact of input lengths and prediction lengths of SST on the prediction performance of the LSTM model and the application of the single-point prediction model of SST in a small area, which provides a reference for the operational prediction of SST, marine pasture, aquaculture, and other industries greatly affected by sea surface temperature, especially for some aquaculture industries with simple equipment in ECS. The specific content of the experiment is as follows: (1) A set of sensitivity experiments on input and prediction lengths are designed, and the influence of input and prediction lengths on the prediction results of the LSTM model is analyzed through the results of sensitivity experiments. (2) Through a training model at a specific location to predict the SST of ECS, experiments show that more than 95% of the *RMSE* values predicted by this method 5 days in advance are within 0.4 °C. Compared with the experimental results shown by the model proposed by Zhang et al. [40], which combines the SOM algorithm with the LSTM model, the TSGCN model proposed by Sun et al. [41], and Xiao et al. [47] applied ConvLSTM to prediction of SST, the *RMSE* value of this experiment decreased by 0.1, 0.07, 0.25, respectively.

The remainder of this paper is structured as follows. Section 2 describes the satellite data and the LSTM model used in this study. Section 3 presents the experimental results and a detailed discussion. Section 4 gives the conclusion.

2. Materials and Methods

There are many sources of observation data. Compared with buoy data, high-resolution satellite data is easier to obtain. In this section, we introduce the data sources and LSTM method in detail. The details are as follows:

2.1. Materials

The high-resolution satellite remote sensing sea surface temperature data used in this study is Operational Sea Surface Temperature and Sea Ice Analysis (OSTIA). OSTIA is the operational sea temperature and sea ice analysis system [49]. Based on the data provided by Group for high resolution sea surface temperature (GHRSST), it is a daily 1/20° grid SST product made by the UK Met Office using AATSR data, SEVIRI data, AVHRR data, AMSR data, TMI data, and in situ measurements. All satellite SST data are adjusted for bias errors based on a combination of AATSR SST data and in situ SST measurements from drifting buoys. The product is generated by using an optimal algorithm, and its *RMSE* is less than 0.6 °C [50]. The spatial range of SST data used in this study is (22°N–33°N, 120°E–131°E), and the time range is 2010–2020, of which the SST data from 2010–2019 is the training data and validation data, and SST data in 2020 is the test data.

2.2. Methods

2.2.1. LSTM Neural Network

LSTM is a special form of RNN, proposed by Hochreiter and Schmidhuber in 1997 [51]. LSTM overcomes, to some extent, the most direct gradient disappearance or explosion problem caused by a traditional RNN due to an excessive number of layers in the time dimension. The main reason is that LSTM network introduces a unit state and uses a gating mechanism to save and control information flow. The cell structure is shown in Figure 1. Its first gate is the forget gate, which determines how much of the cell state C_{t-1} at the previous moment is retained to the current moment C_t. The second gate is the input gate, which determines how much of the network input X_t at the current moment is saved to the cell state C_t. The third gate is the output gate, which controls how much of the unit state C_t is exported to the current output value h_t of the LSTM. The gating mechanism and the update computation of the cell state are as follows:

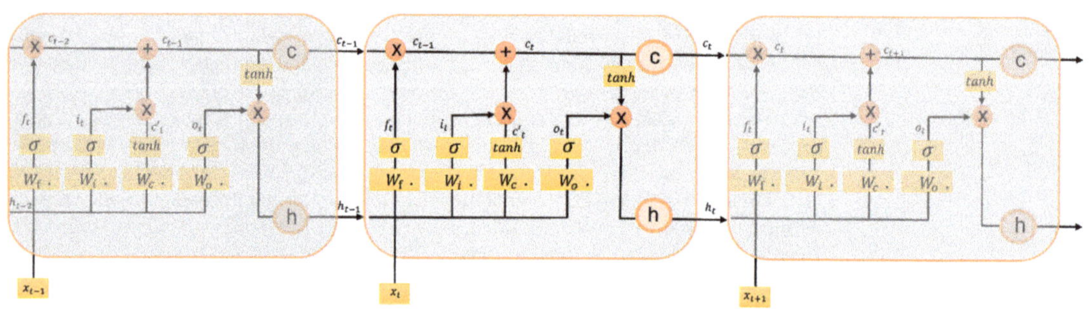

Figure 1. Structure of long short-term memory (LSTM) memory cell, including the forget, input, and output gates.

Forgotten gate:

$$f_t = \sigma(W_f.[h_{t-1}, x_t] + b_f) \tag{1}$$

Input gate:

$$i_t = \sigma(W_i.[h_{t-1}, x_t] + b_i) \tag{2}$$

$$C'_t = tanh(W_c.[h_{t-1}, x_t] + b_c) \tag{3}$$

$$C_t = f_t * C_{t-1} + i_t * C'_t \tag{4}$$

Output gate:

$$o_t = \sigma(W_o.[h_{t-1}, x_t] + b_o) \tag{5}$$

$$h_t = o_t * tanh(C_t) \tag{6}$$

where h_{t-1} represents the output value of the hidden layer at the previous moment, and x_t is the current input value. σ and tanh are activation functions, and σ represents the sigmoid function. f_t, i_t, and o_t denote forgetting gate values, enter threshold values, and output gate values. W_f, W_i, W_c, and W_o are weight matrices. b_f, b_i, b_c, and b_o are the corresponding offset terms. C_{t-1}, C'_t, and C_t represents the cell state at the previous time, the candidate state, and the cell state at the current time.

2.2.2. Model Building

This study constructed a 4-layer LSTM model based on Keras, including an input layer, two LSTM layers, and a dense layer, as shown in Figure 2. During the training process of the LSTM model, parameters, such as weight vector W and bias vector b, are updated by error back propagation. The updating methods mainly include stochastic gradient descent [52], AdaGrad, RMSProp [53], adaptive momentum estimation algorithms, and so on. Among them, the Adam optimization algorithm is an effective stochastic optimization algorithm based on gradient learning. The algorithm integrates the advantages of AdaGrad and RMSProp algorithm, has an adaptive learning rate for different parameters, and occupies fewer storage resources. Compared with other stochastic optimization algorithms, the Adam algorithm performs better in practical applications [54]. Therefore, the Adam optimization algorithm is adopted in this study.

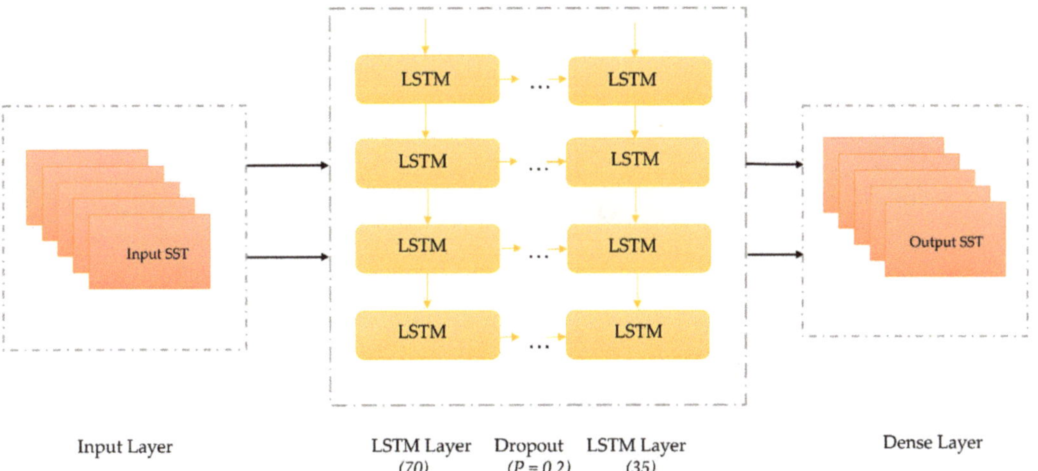

Figure 2. The architecture of the proposed LSTM deep neural network for SST prediction.

Suppose the time series of SST is expressed as $X = X_t$, $t = 1, 2, 3 \cdots n$, where X_t is the SST at time t. Given the window length of the neural network, this parameter indicates that SST at the next moment, which is described P_{t+L}, is predicted using the historical SST with the time length of L, which is represented as X_t, $X_{t+1} \cdots X_{t+L-1}$. The prediction steps of SST are as follows:

(1)   Network initialization. Weights vector W and bias vector b are randomly initialized. The initial learning rate and the maximum number of iterations are set to 0.0001 and 100, respectively, where EarlyStopping is used in the number of iterations.

(2) Data standardization. The missing values in the data are filled with the surrounding values, and the MinmaxScaler function is imported from the sklearn library to standardize the dataset X to $(-1, 1)$ to obtain the standardized dataset X.

(3) The division of dataset X. The standardized dataset X is set according to the window length L and the number of days of prediction, in which the training set and the validation set are divided into 85% and 15%, respectively.

(4) Error calculation. The error between the output of the output layer and the satellite data and the loss function are calculated using MSE.

(5) Update of weights and thresholds. Using the Adam gradient optimization algorithm, update the weights W and biases b according to the loss function.

(6) Repeat steps (3) to (5). The training ends when the training times reach the maximum number of iterations, or the value of the loss function does not change for three consecutive iterations.

2.2.3. Evaluation Indicators

To evaluate the prediction performance of SST, the predicted SST is compared with OSTIA data using the coefficient of determination (R^2), root mean square error ($RMSE$), and absolute error (AE). The formula is as follows:

$$R^2 = 1 - \frac{\sum_{i=1}^{m}\left(sst_o - sst_p\right)^2}{\sum_{i=1}^{m}\left(sst_o - \overline{sst_o}\right)^2} \tag{7}$$

$$RMSE = \sqrt{\frac{1}{m}\sum_{i=1}^{m}\left(sst_o - sst_p\right)^2} \tag{8}$$

$$AE = \left|sst_o - sst_p\right| \tag{9}$$

$$Improve\ rate = \frac{AE_{max} - AE_{min}}{AE_{max}} \tag{10}$$

where sst_o and sst_p are OSTIA value and predicted value of SST, and m is the total number of samples. The smaller the $RMSE$ and AE, the more accurate the prediction, and the closer R^2 value is to 1, the higher the fit between the predicted and true values. AE_{max} is the maximum value of each column of AE, AE_{min} is the minimum value of each column of AE.

3. Results

SST of the ECS varies greatly from nearshore to far sea, and the ocean current also greatly impacts the sea surface temperature change in this area. Therefore, according to the above reasons, the four points, L1 (31.5°N, 122°E), L2 (25.5°N, 122.5°E), L3 (24.5°N,128°E), and L4 (30.5°N, 129.2°E) as shown in Figure 3, are selected to analyze the sea surface temperature predicted by the LSTM model. The reasons for selecting these four points are as follows: (1) The seasonal variation in the Yangtze River estuary area is very obvious, and SST varies greatly, with a minimum value of approximately 7 °C and a maximum value of approximately 30 °C. Therefore, L1 is selected near the Yangtze River estuary; (2) The Kuroshio is a powerful western boundary warm current in the northwestern Pacific Ocean. It has obvious characteristics, such as fast speed, narrow flow width, large flow, high temperature, and high salinity [55,56], which have an important impact on China's climate. Therefore, to analyze the change of SST in the Kuroshio area, L2 and L3 with different water depths on both sides of the Kuroshio are selected to represent Kuroshio; (3) Compared with the other three points, the water depth value of L4 is larger than 7000 m. Thus, L3 is selected at the position shown in Figure 3.

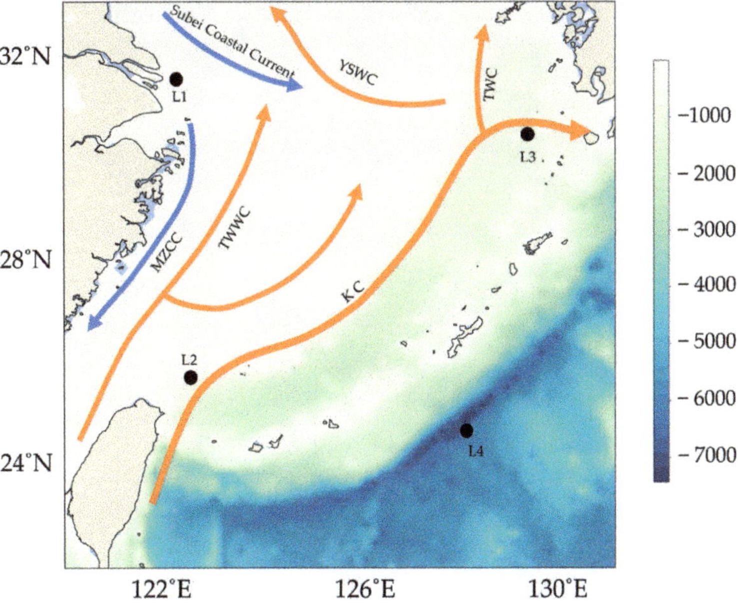

Figure 3. Location diagram of different points. The black dots represent the selected positions. MZCC, TWWC, YSWC, TWC, and KC represent the Min-Zhe Costal Current (Min is Fujian province, Zhe is Zhejiang Province), the Taiwan Warm Current, the Yellow Sea Warm Current, the Tsushima Warm Current, and the Kuroshio, respectively. Besides the blue line is the cold current, and the red line is the warm current and the unit of colorbar is meters.

3.1. The Effect of Different Parameter Settings on LSTM Prediction Performance

The prediction performance of the LSTM model is affected by parameter settings. For example, learning rate, the number of network layers, the input length, and the prediction length will all affect the prediction effect. However, this subsection mainly explores the influence of input length and prediction length on the prediction performance of LSTM models through $RMSE$, AE, and R^2.

3.1.1. The Impact of Input Length on LSTM Prediction Performance

In order to verify the influence of input length on the prediction results of the model, under the condition that hyperparameters, such as the learning rate, the number of hidden layers, and the number of neurons, do not change, the prediction length is controlled to 5, and the input length is set to 2, 5, 10, and 15 days, respectively, to discuss the impact of input length changes. The influence of input length on the prediction of the LSTM model is shown in Tables 1–3, where the bold font is the extremum value of each column. It is worth noting that, compared with other input lengths, when the input length is 2, the $RMSE$ and AE at the four positions are the maximum value and R^2 is the minimum. Then, with the increase of the input length, the $RMSE$ and AE decreases significantly and R^2 increases compared with the input length of 2. Especially when the input length is 5, the $RMSE$ and AE values of L2 and L3 positions are the smallest, and when the input length is 15, the $RMSE$ values of L1 and L4 positions are the smallest. Moreover, R^2 also becomes larger at the corresponding positions above. This proves that if the input length is too small, the LSTM model cannot capture the change law of the SST data well. Increasing the input length can improve the prediction performance of the LSTM model to a certain extent. However, no obvious positive correlation is seen between them. In fact, the improvement of the prediction performance of the LSTM model is not only related to the input length, but also related to the predicted position. The selection of appropriate input length should consider related factors, such as the predicted position.

Table 1. *RMSE* (°C) variation of different input lengths at different positions, where the bold font is the minimum value of each column.

Length of Input \ Location	L1	L2	L3	L4
2	0.3465	0.2698	0.1786	0.3331
5	0.2741	**0.0568**	**0.0458**	0.0769
10	0.2730	0.0917	0.0707	0.0764
15	**0.2461**	0.0995	0.1005	**0.0698**

Table 2. R^2 variation of different input lengths at different positions, where the bold font is the maximum value of each column.

Length of Input \ Location	L1	L2	L3	L4
2	0.9976	0.9830	0.9949	0.9884
5	0.9985	**0.9992**	**0.9996**	0.9993
10	0.9985	0.9980	0.9992	0.9994
15	**0.9988**	0.9977	0.9984	**0.9995**

Table 3. *AE* (°C) variation of different input lengths at different positions, where the bold font is the minimum value of each column.

Length of Input \ Location	L1		L2		L3		L4	
	Max	Mean	Max	Mean	Max	Mean	Max	Mean
2	1.3978	0.2454	0.9512	0.1979	0.7163	0.1356	1.1755	0.2471
5	1.1656	0.1968	**0.2773**	**0.0406**	**0.1893**	**0.0328**	0.3873	0.0574
10	1.0081	0.2003	0.5757	0.0634	0.2271	0.0540	**0.3401**	0.0551
15	**0.8816**	**0.1833**	0.5338	0.0724	0.3605	0.0773	0.3624	**0.0500**
Improve Rate	36.93%	25.31%	70.85%	79.48%	73.57%	75.81%	71.07%	79.77%

3.1.2. The Impact of Prediction Lengths on LSTM Prediction Performance

Similarly, to explore the influence of the prediction length on the prediction results of the model, when the other hyperparameters mentioned above remain unchanged, the input length is controlled to 15 in combination with Tables 1 and 2. The main reason is that, when the input length is set to 5, the input length may be short, and the data change law may not be well displayed. Furthermore, L1 has the largest difference in the extremum among the four positions, so the change law of the SST data is the most difficult to capture. When the input length is 15, the *RMSE* is the smallest and R^2 is the largest in L1, and the LSTM model has the best prediction effect. To sum up, this paper believes that it is better to set the input length to 15. Figure 4 shows the *RMSE* and R^2 values for 5 prediction steps, where the different colors of the lines represent each specific location. We can see that when the prediction length is 1, the *RMSE* value is the minimum value at any position, and the R^2 value is the maximum value. When the prediction length is 5, the *RMSE* at any position is the maximum value, and the R^2 is the minimum value, that is, the minimum value of *RMSE* and the maximum value of R^2 are obtained almost at the same time. Meanwhile, it also shows that the prediction performance of the LSTM model decreases gradually with the increase of prediction length.

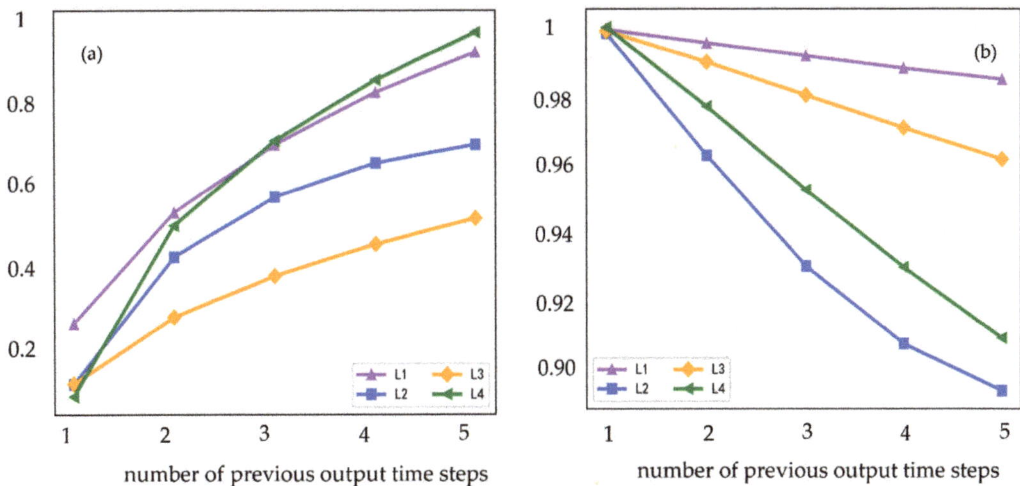

Figure 4. *RMSE* (°C) and R^2 of different prediction lengths at different positions, where lines with different colors represent different positions, (**a**) *RMSE* at different positions, (**b**) R^2 at different positions, the abscissa is the number of days predicted in advance.

3.2. Analysis of Prediction Results at Different Points

In order to analyze the variation trend of the LSTM model prediction results and error over time at different locations, Figures 5 and 6 are drawn. Meanwhile, to explore the accuracy of the LSTM model for extreme value prediction, we selected the region with the largest SST in a year, as shown in the gray rectangle in Figure 5. According to (a)–(d) of Figure 5, we found that the prediction results of the LSTM model for L1, L2, L3, and L4 are slightly different from OSTIA, and cyclical trends are represented accurately. However, from Figure 5e–g, it is found that the LSTM model is not very accurate in predicting extremum. In Figure 5h, it is found that LSTM is quite accurate for predicting extremum. The reason is that the *RMSE* value here is particularly small. That is, except that the *RMSE* value is particularly small, the LSTM model cannot predict the extremum well in most cases. According to Figure 6, it is found that the difference between the prediction results of the LSTM model and the OSTIA data changes greatly at the L1 position. The maximum value of the difference between the two is 0.7 °C and the minimum value is −0.9 °C. At the L4 position, their differences are relatively small and stable, and most of the differences are −0.1 °C. Furthermore, the maximum value of *RMSE* at L1 position is 0.2461 °C, and the minimum value of *RMSE* at L4 position is 0.0698 °C. The large difference between L1 and L4 is mainly because L1 is located at the estuary of the Yangtze River. The seasonal variation of SST at the estuary of the Yangtze River is more obvious, so that the LSTM model cannot capture the SST law of L1 position well. Moreover, the *RMSE* value of the L4 position is smaller than that of the L3 position, which may be due to the lower water depth of the L3 position than that of the L4 position.

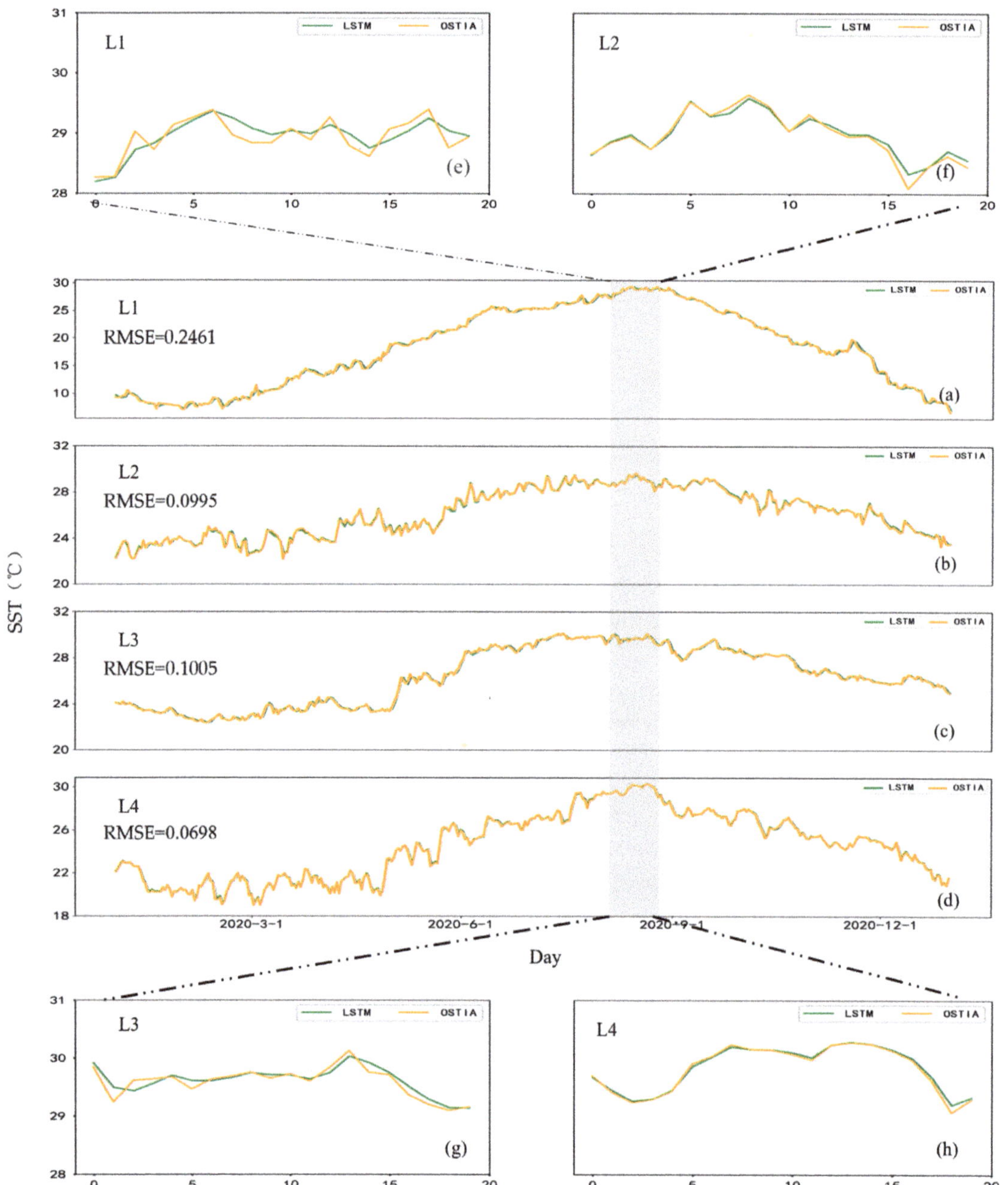

Figure 5. OSTIA data and LSTM predictions at different points, where the green line is the prediction result of LSTM, the yellow line is the OSTIA data, and the gray rectangle is the area where the maximum SST is located in 2020. (**a–d**) are in a comparison chart of the LSTM prediction results and OSTIA data in 2020. (**e–h**) are the values of the region where the maximum SST is located in 2020. The abscissa is the SST in degrees Celsius, and the ordinate is the number of days.

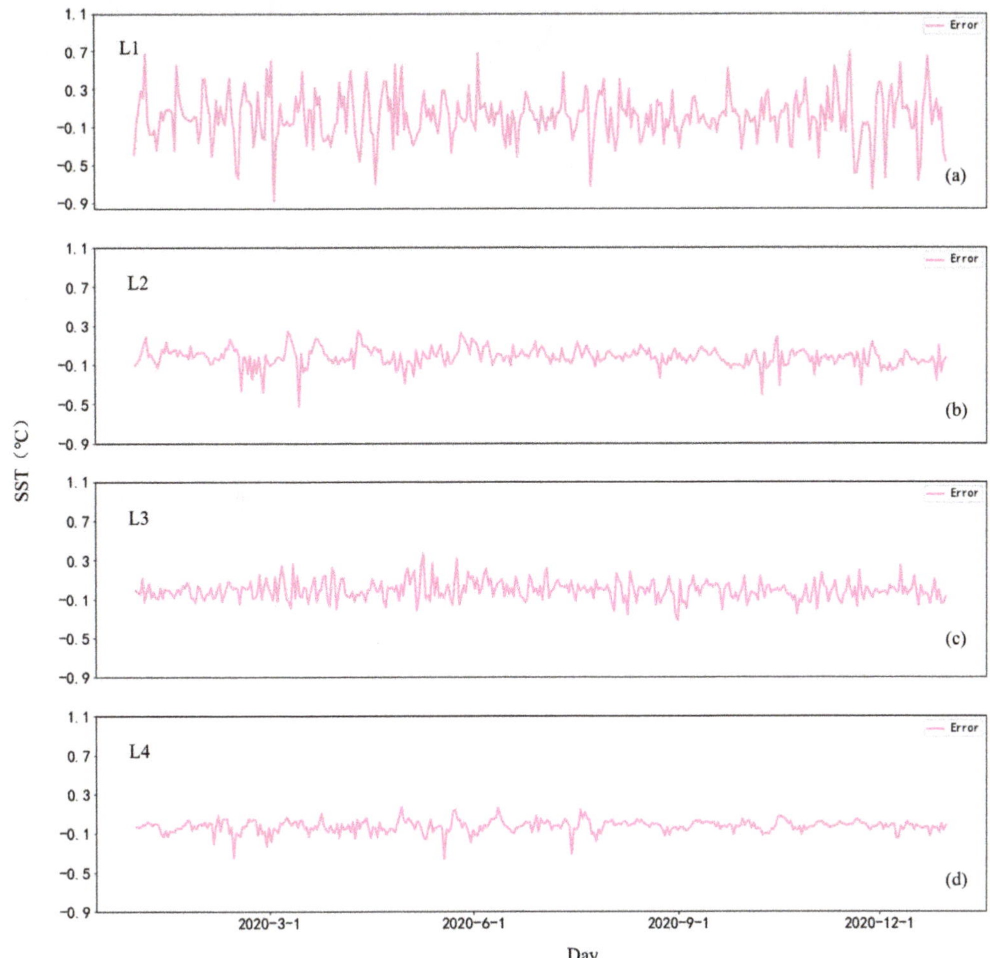

Figure 6. The error of OSTIA data and LSTM prediction results at different points, where the error is obtained by subtracting the OSTIA data and the LSTM prediction result. (**a**–**d**) are the errors of L1, L2, L3 and L4, respectively. The abscissa is SST in degrees Celsius, and the ordinate is the number of days.

3.3. Migration Analysis

This subsection mainly describes the feasibility of applying the L4 position trained model to the prediction of SST in ECS from the following two aspects: (1) Study its migration from the spatial distributions of *RMSE* and *AE* of each month as shown in Figures 7–9. (2) Due to the obvious seasonal variation of SST, the spatial distributions of *RMSE* and *AE* of four seasons are described to verify its migration as shown in Figures 10 and 11.

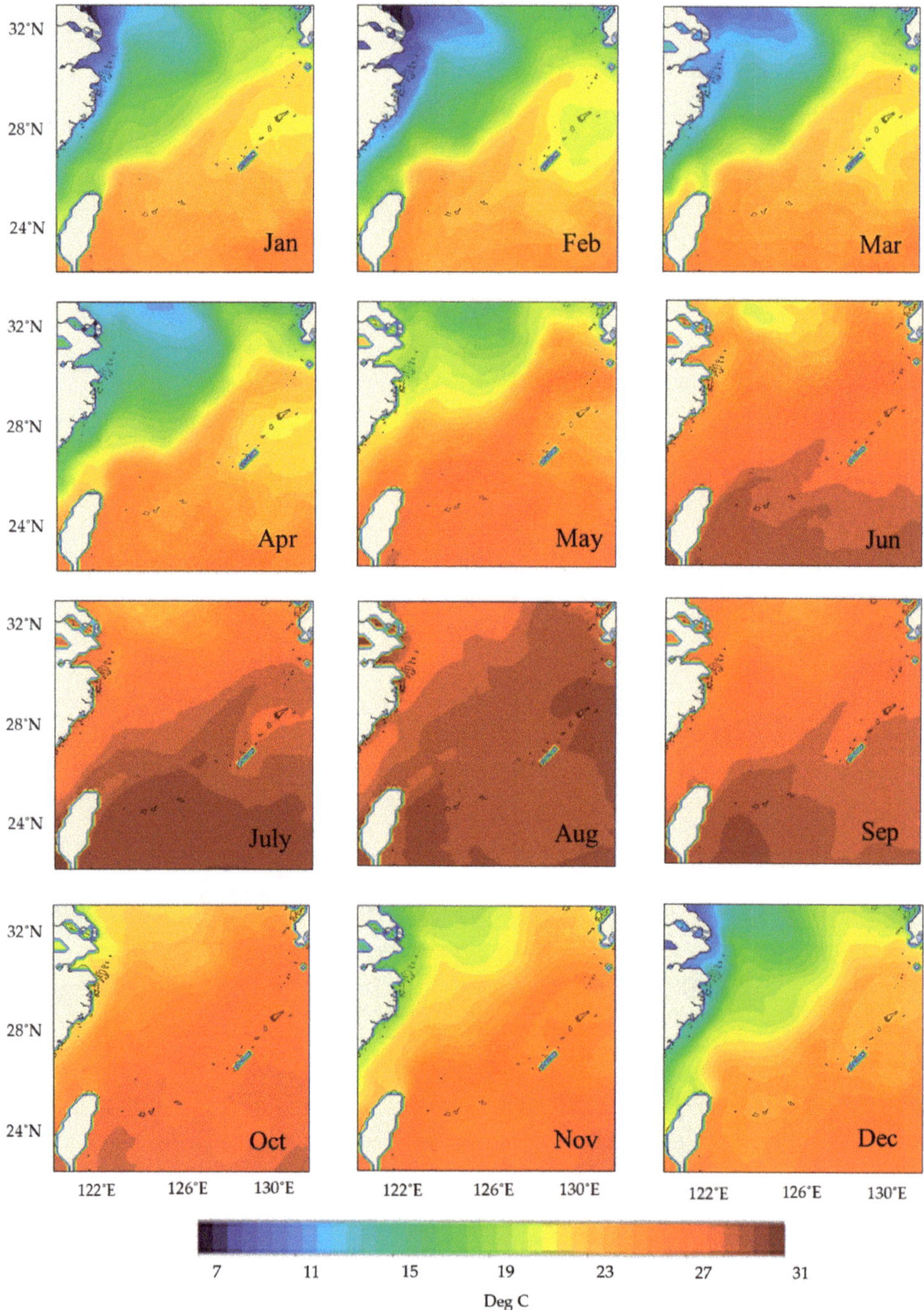

Figure 7. Spatial distribution of monthly mean sea surface temperature predicted by LSTM model in 2020, the unit of colorbar is degrees Celsius.

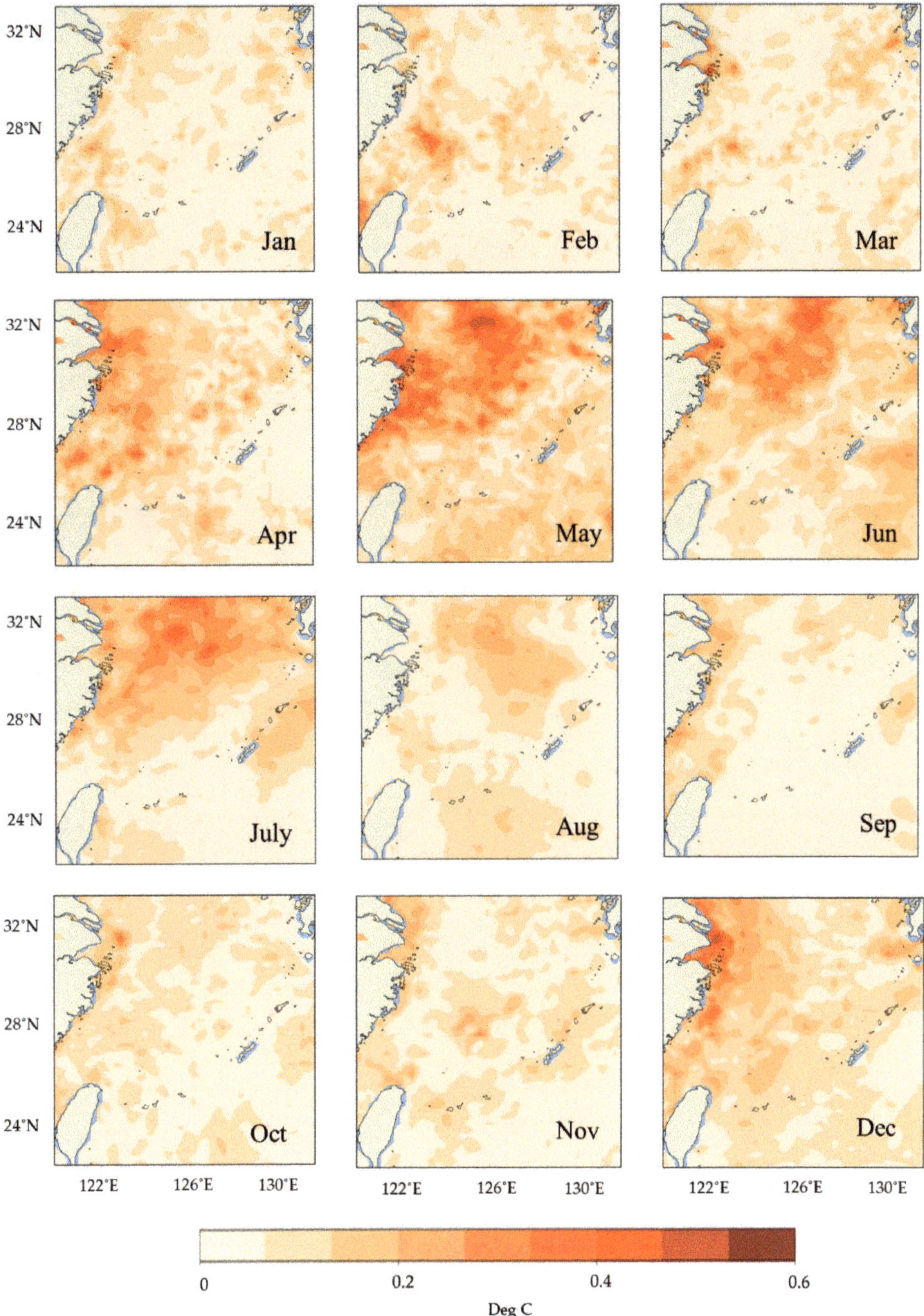

Figure 8. Spatial distribution of monthly mean *AE* of OSTIA data and LSTM predictions of sea surface temperature in 2020. The unit of colorbar is degrees Celsius.

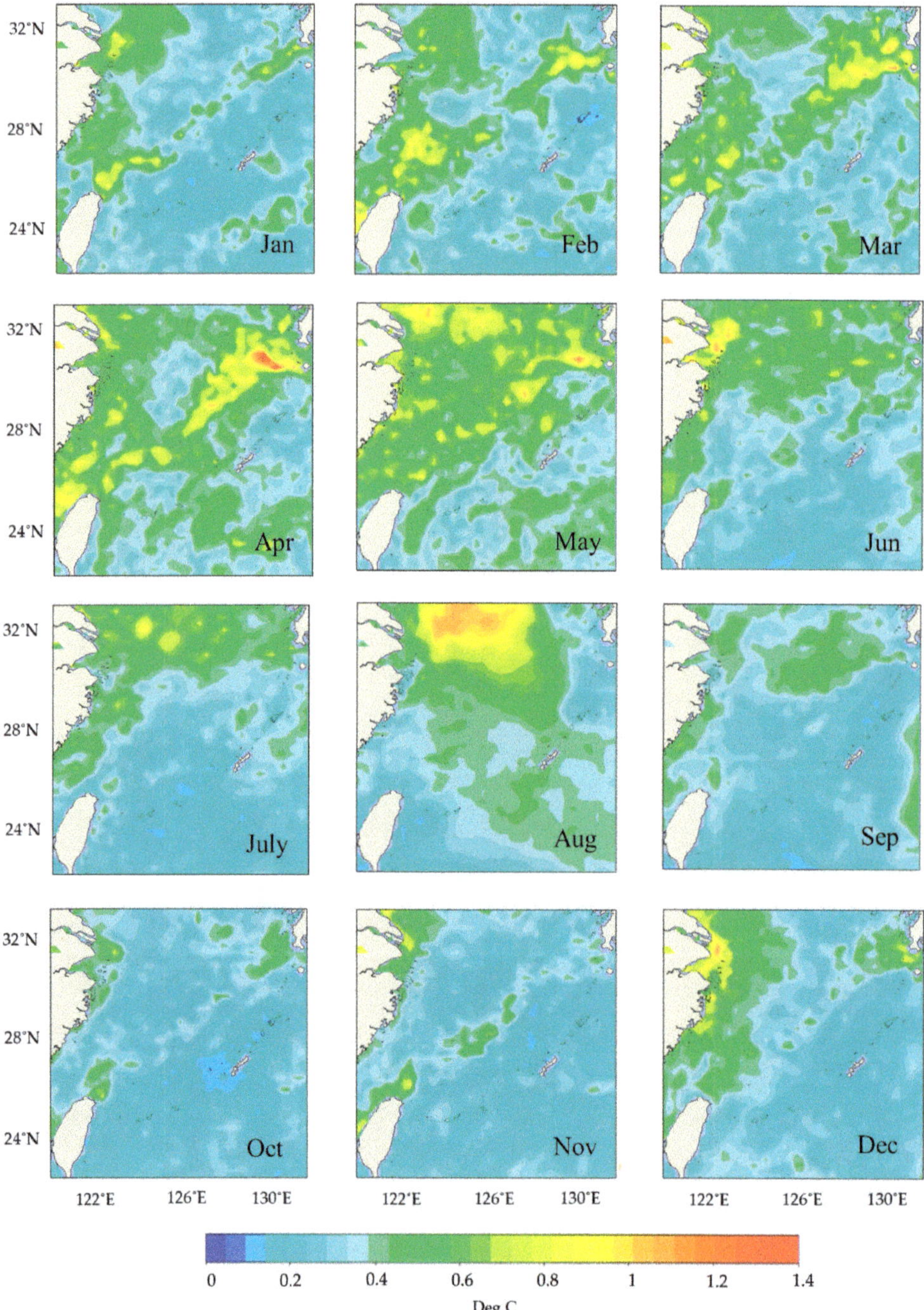

Figure 9. Spatial distribution of monthly mean *RMSE* between LSTM predicted sea surface temperature and OSTIA data in 2020. The unit of colorbar is degrees Celsius.

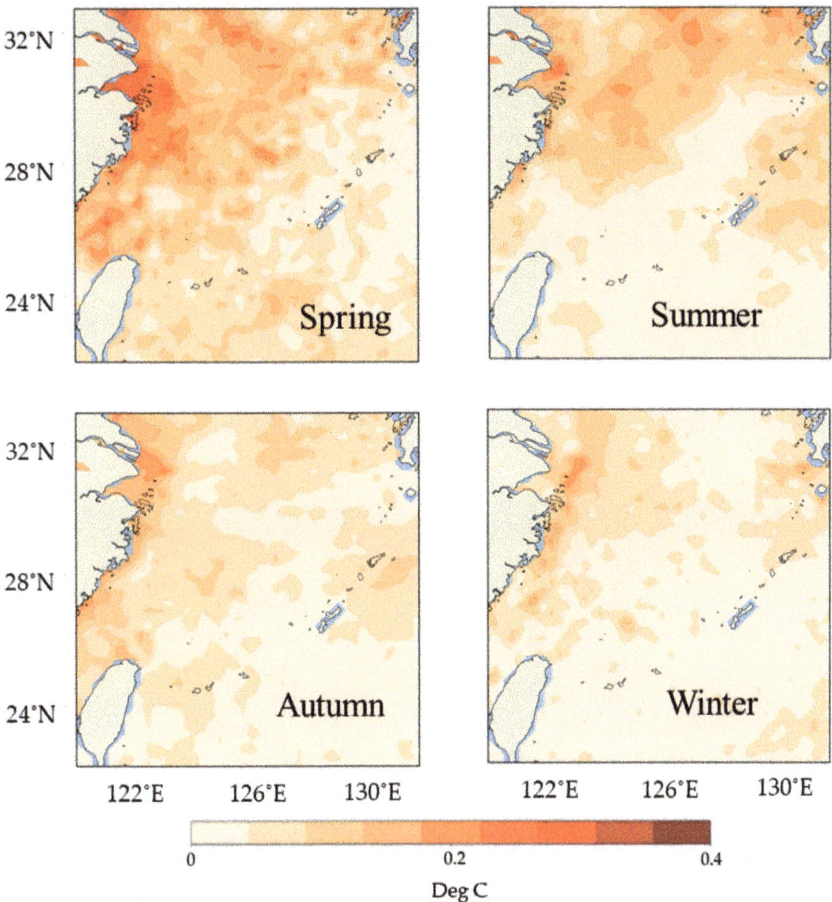

Figure 10. Spatial distribution of *AE* of OSTIA data and LSTM predictions of SST in four seasons, in which spring is from March to May, summer is from June to August, autumn is from September to November, and winter is from December to February. The unit of colorbar is degrees Celsius.

3.3.1. Migration Analysis for Monthly Changes

Given that the *RMSE* value of the L4 position in Figure 5 is the smallest, the LSTM model trained at the L4 position is selected to predict the SST of the whole study area in 2020 to prove whether the LSTM network has the characteristics of migration. The spatial distribution of SST in 2020 predicted by the LSTM model is shown in Figure 7. The characteristics of SST, such as the Kuroshio, the Min-Zhe coastal current, and the Yangtze River Diluting Water, are clearly displayed in the forecast map and show obvious seasonal changes. For the quality evaluation of LSTM prediction results, the methods of *AE* and *RMSE* are used. Figure 8 presents the *AE* between the prediction results of the model and the OSTIA satellite data in 2020. We can see that the error between the prediction results and OSTIA are mostly less than 0.4 °C, and the *AE* of the Yangtze River estuary and its northern part and Min-Zhe coastal currents is relatively large in April, May, June, July, and December. Figure 9 shows that most of the *RMSE* values in 2020 are less than 0.5 °C. In April, maximum *RMSE* is found around the Kuroshio and TWC area, that is, the dispersion of error in this region is relatively large. In August, the error dispersion in the northern part of the Yangtze River estuary is relatively high. In general, the areas with higher error dispersion in a year are located in the Yangtze River estuary and its north, the Kuroshio, and the Min-Zhe coastal current.

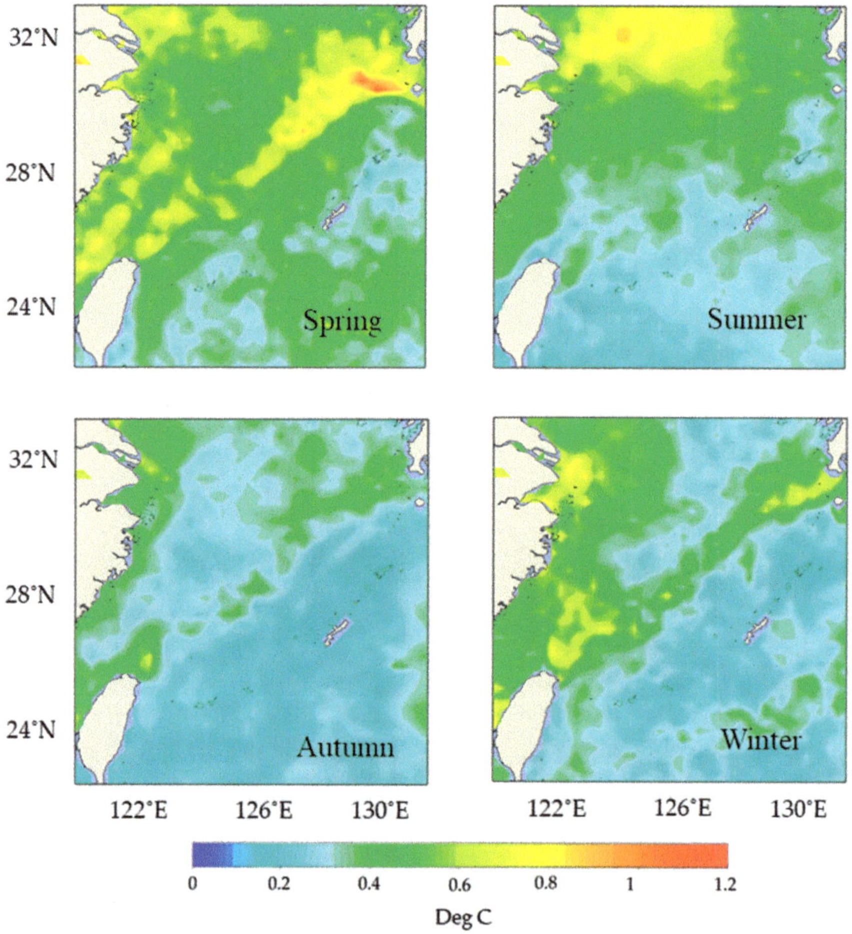

Figure 11. Spatial distribution of *RMSE* between LSTM predicted sea surface temperature and OSTIA data in four seasons, in which spring is from March to May, summer is from June to August, autumn is from September to November, and winter is from December to February. The unit of color bar is degrees Celsius.

3.3.2. Mobility Analysis of Seasonal Changes

Figure 10 is drawn for the seasonal *AE* in spring, summer, autumn, and winter to analyze whether the migration of the LSTM model in ECS can show the characteristics of seasonal changes well. *AE* between the predicted results of the LSTM model and OSTIA, shown in Figure 10, is within 0.4 °C, and the maximum value of *AE* decreases by 0.2 °C compared with Figure 8. Among the four seasons, *AE* of spring is the largest, and *AE* of autumn and winter is smaller. Moreover, the extremum of *AE* of the four seasons is mainly concentrated in the area of the Yangtze River Estuary and the Min-Zhe coastal current. Through the analysis of the dispersion degree of the error in Figure 11, the *RMSE* of the Kuroshio in spring is larger than that in other seasons, which means that the dispersion degree of errors in this area is large. In summer, the dispersion degree of error is large in the north of the Yangtze River Estuary. Through the analysis of *AE* and *RMSE* of each month and four seasons, the maximum value of *AE* and *RMSE* are 0.4 °C in four seasons, which is 0.2 °C lower than each month. However, most *AE* and *RMSE* values are relatively small, which are less than 0.4 °C and 0.5 °C, respectively. Therefore, the long-term and seasonal change law can be well represented by the migration of the LSTM model.

4. Conclusions

The past research at home and abroad has used many related SST prediction methods, such as numerical simulation, BP neural network, etc. In this study, the LSTM neural network is applied to the SST prediction, and its feasibility is discussed. The most important findings of this study are as follows:

(1) The input and prediction lengths will affect the prediction performance of the LSTM model. The increase of the input length can improve the prediction performance of the LSTM model to a certain extent, but no obvious positive correlation is seen between them. Meanwhile, the prediction performance of the LSTM model decreases with the increase of the prediction length, and an obvious negative correlation is seen between them. The effect is the best when the prediction length is 1 and the worst when it is 5.

(2) The prediction results of the LSTM model for a single site are quite accurate, but the extremum cannot be well displayed. Furthermore, affected by the seasonal variation of the Yangtze River Estuary, the prediction result of the Yangtze River Estuary site is the worst compared with other regions.

(3) By analyzing the *AE* and *RMSE* of the prediction results of the LSTM model, most of the error is found to be less than 0.4 °C and 0.5 °C, respectively, and the LSTM model has a very successful migration in the East China Sea. In addition, the *AE* and *RMSE* of the seasonal and monthly average have prominent spatial characteristics. The places with larger error are distributed in the Yangtze River estuary and its north, the Kuroshio, and the Min-Zhe coastal current.

Using the LSTM neural network to predict SST is a new prediction method, which has achieved good results in the experiment of SST prediction. Therefore, it can be a better tool and method to predict the change in SST. However, the interpretability of deep learning remains a hot issue in the computer field. Explaining the LSTM network's physical mechanism more effectively for predicting SST still needs further experimental research.

Author Contributions: Conceptualization, X.J. and Q.J.; methodology, X.J.; software, X.J.; validation, X.J., Q.J. and L.H.; formal analysis, X.J. and Q.J.; investigation, X.J. and Q.J.; resources, Q.J.; data curation, Y.L. and L.H.; writing—original draft preparation, X.J. and Q.J.; writing—review and editing, Q.J., G.H. and X.L.; visualization, X.J.; supervision, Q.J.; project administration, Q.J.; funding acquisition, Q.J. and Y.L. All authors have read and agreed to the published version of the manuscript.

Funding: This study was supported by the projects supported by Southern Marine Science and Engineering Guangdong Laboratory (Zhuhai) (SML2020SP007, 311020004), the National Natural Science Foundation of China (41806004).

Institutional Review Board Statement: Not applicable.

Informed Consent Statement: Not applicable.

Data Availability Statement: Publicly available datasets were analyzed in this study. OSTIA data sets can be found here: https://www.ncei.noaa.gov/data/oceans/ghrsst/L4/GLOB/UKMO/OSTIA/ (accessed on 1 January 2020).

Conflicts of Interest: The authors declare no conflict of interest.

References

1. Sumner, M.D.; Michael, K.J.; Bradshaw, C.J.A.; Hindell, M.A. Remote sensing of Southern Ocean sea surface temperature: Implications for marine biophysical models. *Remote Sens. Environ.* **2003**, *84*, 161–173. [CrossRef]

2. Wentz, F.J.; Gentemann, C.; Smith, D.; Chelton, D. Satellite Measurements of Sea Surface Temperature through Clouds. *Science* **2000**, *288*, 847–850. [CrossRef] [PubMed]

3.   Rauscher, S.A.; Jiang, X.; Steiner, A.; Williams, A.P.; Jiang, X. Sea Surface Temperature Warming Patterns and Future Vegetation Change. *J. Clim.* **2015**, *28*, 7943–7961. [CrossRef]
4.   Salles, R.; Mattos, P.; Iorgulescu, A.-M.D.; Bezerra, E.; Lima, L.; Ogasawara, E. Evaluating temporal aggregation for predicting the sea surface temperature of the Atlantic Ocean. *Ecol. Inform.* **2016**, *36*, 94–105. [CrossRef]
5.   Bouali, M.; Sato, O.T.; Polito, P.S. Temporal trends in sea surface temperature gradients in the South Atlantic Ocean. *Remote Sens. Environ.* **2017**, *194*, 100–114. [CrossRef]
6.   Cane, M.A.; Kaplan, A.; Clement, A.C.; Kushnir, Y. Twentieth-Century Sea Surface Temperature Trends. *Science* **1997**, *275*, 957–960. [CrossRef]
7.   Castro, S.L.; Wick, G.A.; Steele, M. Validation of satellite sea surface temperature analyses in the Beaufort Sea using UpTempO buoys. *Remote Sens. Environ.* **2016**, *187*, 458–475. [CrossRef]
8.   Chaidez, V.; Dreano, D.; Agusti, S.; Duarte, C.M.; Hoteit, I. Decadal trends in Red Sea maximum surface temperature. *Sci. Rep.* **2017**, *7*, 8144. [CrossRef]
9.   Herbert, T.D.; Peterson, L.C.; Lawrence, K.T.; Liu, Z. Tropical ocean temperatures over the past 3.5 million years. *Science* **2010**, *328*, 1530–1534. [CrossRef]
10.   Yao, S.; Luo, J.; Huang, G.; Wang, P. Distinct global warming rates tied to multiple ocean surface temperature changes. *Nat. Clim. Change* **2017**, *7*, 486–491. [CrossRef]
11.   Jiao, N.; Zhang, Y.; Zeng, Y.; Gardner, W.D. Ecological anomalies in the East China Sea: Impacts of the Three Gorges Dam? *Water Res.* **2007**, *41*, 1287–1293. [CrossRef] [PubMed]
12.   Du, B.; Yuan, X. Analysis and Forecast of Sea Surface Temperature Field for the East China Sea and the adjacent waters. *Mar. Forecast.* **1986**, *1*, 3–11.
13.   Gao, G.; Marin, M.; Feng, M.; Yin, B.; Yang, D.; Feng, X.; Ding, Y.; Song, D. Drivers of Marine Heatwaves in the East China Sea and the South Yellow Sea in Three Consecutive Summers During 2016–2018. *J. Geophys. Res. Ocean.* **2020**, *125*, e16518. [CrossRef]
14.   Tiwari, P.; Dimri, A.P.; Shenoi, S.C.; Francis, P.A.; Jithin, A.K. Impact of Surface forcing on simulating Sea Surface Temperature in the Indian Ocean—A study using Regional Ocean Modeling System (ROMS). *Dyn. Atmos. Ocean.* **2021**, *95*, 101243. [CrossRef]
15.   Gao, S.; Lv, X.; Wang, H. Sea Surface Temperature Simulation of Tropical and North Pacific Basins Using a Hybrid Coordinate Ocean Model (HYCOM). *Mar. Sci. Bull.* **2008**, *10*, 1–14.
16.   Arx, W. An Introduction to Physical Oceanography. *Am. J. Phys.* **2005**, *30*, 775–776. [CrossRef]
17.   Bell, M.J.; Schiller, A.; Le Traon, P.Y. An introduction to GODAE OceanView. *J. Oper. Oceanogr.* **2015**, *8* (Suppl. 1), s2–s11. [CrossRef]
18.   Xue, Y.; Leetmaa, A. Forecasts of tropical Pacific SST and sea level using a Markov model. *Geophys. Res. Lett.* **2000**, *27*, 2701–2704. [CrossRef]
19.   Laepple, T.; Jewson, S. Five year ahead prediction of Sea Surface Temperature in the Tropical Atlantic: A comparison between IPCC climate models and simple statistical methods. *arXiv* **2007**, arXiv:physics/0701165. [CrossRef]
20.   Collins, D.C.; Reason, C.; Tangang, F. Predictability of Indian Ocean sea surface temperature using canonical correlation analysis. *Clim. Dyn.* **2004**, *22*, 481–497. [CrossRef]
21.   Peng, Y.; Wang, Q.; Yuan, C.; Lin, K. Review of Research on Data Mining in Application of Meteorological Forecasting. *J. Arid. Meteorol.* **2015**, *33*, 9.
22.   Kusiak, A.; Zheng, H.; Song, Z. Wind farm power prediction: A data-mining approach. *Wind Energy* **2009**, *12*, 275–293. [CrossRef]
23.   Ho, H.C.; Knudby, A.; Sirovyak, P.; Xu, Y.; Hodul, M. Mapping maximum urban air temperature on hot summer days. *Remote Sens. Environ.* **2014**, *154*, 38–45. [CrossRef]
24.   Behrang, M.A.; Assareh, E.; Ghanbarzadeh, A.; Noghrehabadib, A.R. The potential of different artificial neural network (ANN) techniques in daily global solar radiation modeling based on meteorological data. *Sol. Energy* **2010**, *84*, 1468–1480. [CrossRef]
25.   Mellit, A.; Pavan, A.M.; Benghanem, M. Least squares support vector machine for short-term prediction of meteorological time series. *Theor. Appl. Climatol.* **2012**, *111*, 297–307. [CrossRef]
26.   Yue, L.; Shen, H.; Zhang, L.; Zheng, X.; Zhang, F.; Yuan, Q. High-quality seamless DEM generation blending SRTM-1, ASTER GDEM v2 and ICESat/GLAS observations. *ISPRS J. Photogramm. Remote Sens.* **2017**, *123*, 20–34. [CrossRef]
27.   Zang, L.; Mao, F.; Guo, J.; Wang, W.; Pan, Z.; Shen, H.; Zhu, B.; Wang, Z. Estimation of spatiotemporal PM 1.0 distributions in China by combining PM 2.5 observations with satellite aerosol optical depth. *Sci. Total Environ.* **2019**, *658*, 1256–1264. [CrossRef]
28.   Tangang, F.T.; Tang, B.; Monahan, A.H. Forecasting ENSO Events: A Neural Network–Extended EOF Approach. *J. Clim.* **1998**, *11*, 29–41. [CrossRef]
29.   Tangang, F.T.; Hsieh, W.W.; Tang, B.; Hsieh, W.W. Forecasting the equatorial Pacific sea surface temperatures by neural network models. *Clim. Dyn.* **1997**, *13*, 135–147. [CrossRef]
30.   Tangang, F.T.; Hsieh, W.W.; Tang, B. Forecasting regional sea surface temperatures in the tropical Pacific by neural network models, with wind stress and sea level pressure as predictors. *J. Geophys. Res. Ocean.* **1998**, *103*, 7511–7522. [CrossRef]
31.   Wu, A.; Hsieh, W.W.; Tang, B. Neural network forecasts of the tropical Pacific sea surface temperatures. *Neural Netw.* **2006**, *19*, 145–154. [CrossRef]
32.   Gupta, S.M.; Malmgren, B.A. Comparison of the accuracy of SST estimates by artificial neural networks (ANN) and other quantitative methods using radiolarian data from the Antarctic and Pacific Oceans. *Earth Sci. India* **2009**, *2*, 52–75.

33. Tripathi, K.C.; Das, I.; Sahai, A.K. Predictability of sea surface temperature anomalies in the Indian Ocean using artificial neural networks. *Indian J. Mar. Sci.* **2006**, *35*, 210–220.
34. Patil, K.; Deo, M.C.; Ghosh, S.; Ravichandran, M. Predicting Sea Surface Temperatures in the North Indian Ocean with Nonlinear Autoregressive Neural Networks. *Int. J. Oceanogr.* **2013**, *2013*, 302479. [CrossRef]
35. Patil, K.; Deo, M.C. Prediction of daily sea surface temperature using efficient neural networks. *Ocean. Dyn.* **2017**, *67*, 357–368. [CrossRef]
36. Mahongo, S.B.; Deo, M.C. Using Artificial Neural Networks to Forecast Monthly and Seasonal Sea Surface Temperature Anomalies in the Western Indian Ocean. *Int. J. Ocean. Clim. Syst.* **2013**, *4*, 133–150. [CrossRef]
37. Aparna, S.G.; D'Souza, S.; Arjun, N.B. Prediction of daily sea surface temperature using artificial neural networks. *Int. J. Remote Sens.* **2018**, *39*, 4214–4231. [CrossRef]
38. Hou, S.; Li, W.; Liu, T.; Zhou, S.; Guan, J.; Qin, R.; Wang, Z. MIMO: A Unified Spatio-Temporal Model for Multi-Scale Sea Surface Temperature Prediction. *Remote Sens.* **2022**, *14*, 2371. [CrossRef]
39. Lecun, Y.; Bengio, Y.; Hinton, G. Deep learning. *Nature* **2015**, *521*, 436. [CrossRef]
40. Zhang, Q.; Wang, H.; Dong, J.; Zhong, G.; Sun, X. Prediction of Sea Surface Temperature Using Long Short-Term Memory. *IEEE Geosci. Remote Sens. Lett.* **2017**, *14*, 1745–1749. [CrossRef]
41. Sarkar, P.P.; Janardhan, P.; Roy, P. Prediction of sea surface temperatures using deep learning neural networks. *SN Appl. Sci.* **2020**, *2*, 1458. [CrossRef]
42. Kim, M.; Yang, H.; Kim, J. Sea Surface Temperature and High Water Temperature Occurrence Prediction Using a Long Short-Term Memory Model. *Remote Sens.* **2020**, *12*, 3654. [CrossRef]
43. Li, X. Sea surface temperature prediction model based on long and short-term memory neural network. *IOP Conf. Ser. Earth Environ. Sci.* **2021**, *658*, 12040. [CrossRef]
44. Xiao, C.; Chen, N.; Hu, C.; Wang, K. Short and mid-term sea surface temperature prediction using time-series satellite data and LSTM-AdaBoost combination approach. *Remote Sens. Environ.* **2019**, *233*, 111358. [CrossRef]
45. Xiao, C.; Chen, N.; Hu, C.; Wang, K. A spatiotemporal deep learning model for sea surface temperature field prediction using time-series satellite data. *Environ. Model. Softw.* **2019**, *120*, 104501–104502. [CrossRef]
46. Wei, L.; Guan, L.; Qu, L.; Guo, D. Prediction of Sea Surface Temperature in the China Seas Based on Long Short-Term Memory Neural Networks. *Remote Sens.* **2020**, *12*, 2697. [CrossRef]
47. Sun, Y.; Yao, X.; Bi, X.; Huang, X.; Zhao, X.; Qiao, B. Time-Series Graph Network for Sea Surface Temperature Prediction. *Big Data Res.* **2021**, *25*, 100237. [CrossRef]
48. Zhang, Z.; Pan, X.; Jiang, T.; Sui, B.; Liu, C.; Sun, W. Monthly and Quarterly Sea Surface Temperature Prediction Based on Gated Recurrent Unit Neural Network. *J. Mar. Sci. Eng.* **2020**, *8*, 249. [CrossRef]
49. Donlon, C.J.; Martin, M.; Stark, J. Roberts-Jones, J.; Fiedler, E.; Wimmer, W. The Operational Sea Surface Temperature and Sea Ice Analysis (OSTIA) system. *Remote Sens. Environ.* **2012**, *116*, 140–158. [CrossRef]
50. Jiang, X.; Xi, M.; Song, Q. A comparison analysis of six sea surface temperature products. *Acta Oceanol. Sin.* **2013**, *35*, 88–97.
51. Hochreiter, S.; Schmidhuber, J. Long Short-Term Memory. *Neural Comput.* **1997**, *9*, 1735–1780. [CrossRef] [PubMed]
52. Graves, A.; Schmidhuber, J. Framewise phoneme classification with bidirectional LSTM and other neural network architectures. *Neural Netw.* **2005**, *18*, 602–610. [CrossRef] [PubMed]
53. Duchi, J.; Hazan, E.; Singer, Y. Adaptive Subgradient Methods for Online Learning and Stochastic Optimization. *J. Mach. Learn. Res.* **2011**, *12*, 2121–2159.
54. Kingma, D.P.; Ba, J. Adam: A Method for Stochastic Optimization. *arXiv* **2014**, arXiv:1412.6980. [CrossRef]
55. Su, J.L.; Guan, B.X.; Jiang, J.Z. The Kuroshio. Part I. Physical features. *Oceanogr. Mar. Biol. Annu. Rev.* **1990**, *28*, 11–71.
56. Bryden, H.L.; Roemmich, D.H.; Church, J.A. Ocean heat transport across 24°N in the Pacific. *Deep Sea Res.* **1991**, *38*, 297–324. [CrossRef]

 *remote sensing*

Article

Detection of *Sargassum* from Sentinel Satellite Sensors Using Deep Learning Approach

Marine Laval [1,2,*], Abdelbadie Belmouhcine [3], Luc Courtrai [3], Jacques Descloitres [4], Adán Salazar-Garibay [5], Léa Schamberger [6,7], Audrey Minghelli [6,7], Thierry Thibaut [2], René Dorville [1], Camille Mazoyer [2], Pascal Zongo [1] and Cristèle Chevalier [2]

1   Laboratoire des Matériaux et Molécules en Milieu Agressif (L3MA), Université des Antilles, 97275 Schoelcher, France
2   Mediterranean Institute of Oceanography (MIO), IRD, Aix Marseille Université, CNRS, Université de Toulon, 13288 Marseille, France
3   IRISA, Université de Bretagne Sud, 56000 Vannes, France
4   AERIS/ICARE Data and Services Center, University of Lille, CNRS, CNES, UMS 2877, 59000 Lille, France
5   Mexican Space Agency (AEM), Ciudad de Mexico 01020, Mexico
6   Laboratoire d'Informatique et Système (LIS), Université de Toulon, CNRS UMR 7020, 83041 Toulon, France
7   Laboratoire d'Informatique et Système (LIS), Aix Marseille Université, 13288 Marseille, France
*   Correspondence: marine.laval@mio.osupytheas.fr

Abstract: Since 2011, the proliferation of brown macro-algae of the genus *Sargassum* has considerably increased in the North Tropical Atlantic Sea, all the way from the Gulf of Guinea to the Caribbean Sea and the Gulf of Mexico. The large amount of *Sargassum* aggregations in that area cause major beaching events, which have a significant impact on the local economy and the environment and are starting to present a real threat to public health. In such a context, it is crucial to collect spatial and temporal data of *Sargassum* aggregations to understand their dynamics and predict stranding. Lately, indexes based on satellite imagery such as the Maximum Chlorophyll Index (MCI) or the Alternative Floating Algae Index (AFAI), have been developed and used to detect these *Sargassum* aggregations. However, their accuracy is questionable as they tend to detect various non-*Sargassum* features. To overcome false positive detection biases encountered by the index-thresholding methods, we developed two new deep learning models specific for *Sargassum* detection based on an encoder–decoder convolutional neural network (CNN). One was tuned to spectral bands from the multispectral instrument (MSI) onboard Sentinel-2 satellites and the other to the Ocean and Land Colour Instrument (OLCI) onboard Sentinel-3 satellites. This specific new approach outperformed previous generalist deep learning models, such as ErisNet, UNet, and SegNet, in the detection of *Sargassum* from satellite images with the same training, with an F1-score of 0.88 using MSI images, and 0.76 using OLCI images. Indeed, the proposed CNN considered neighbor pixels, unlike ErisNet, and had fewer reduction levels than UNet and SegNet, allowing filiform objects such as *Sargassum* aggregations to be detected. Using both spectral and spatial features, it also yielded a better detection performance compared to algal index-based techniques. The CNN method proposed here recognizes new small aggregations that were previously undetected, provides more complete structures, and has a lower false-positive detection rate.

Keywords: ocean color; *Sargassum*; MODIS; MSI; OLCI; Sentinel-2; Sentinel-3; convolutional neural network; deep learning

Citation: Laval, M.; Belmouhcine, A.; Courtrai, L.; Descloitres, J.; Salazar-Garibay, A.; Schamberger, L.; Minghelli, A.; Thibaut, T.; Dorville, R.; Mazoyer, C.; et al. Detection of *Sargassum* from Sentinel Satellite Sensors Using Deep Learning Approach. *Remote Sens.* **2023**, *15*, 1104. https://doi.org/10.3390/rs15041104

Academic Editors: Ana B. Ruescas, Veronica Nieves and Raphaëlle Sauzède

Received: 19 December 2022
Revised: 7 February 2023
Accepted: 11 February 2023
Published: 17 February 2023

1. Introduction

The brown macro-algae, *Sargassum,* from the genus *Sargassum* C. Agardh (Phaeophyceae, Fucales) is a type of algae commonly found in the Sargasso Sea (North Atlantic Ocean). However, since 2011, these *Sargassum* have started proliferating outside the Sargasso Sea, expanding to the Gulf of Guinea, the Caribbean Sea, and the Gulf of Mexico, forming the Great Atlantic *Sargassum* Belt [1]. Three morphotypes of the same taxon are currently recognized within the Great Atlantic *Sargassum* Belt [2–4]: *Sargassum natans* I Parr, *S. natans* VIII Parra, and *S. fluitans* III Parr. *Sargassum* aggregate and form windrows or mats up to a length of hundred meters [5]. The significant amount of *Sargassum* aggregations in the area causes sizeable beaching events on the coast [6,7], consequently impacting the local economy [8,9], the environment [10–12] and causing major sanitary issues due to the emanation of harmful gas during the decomposition [13–15].

In order to predict and thus anticipate beaching events, it is crucial to collect spatio-temporal data about *Sargassum* aggregations. Several algal indexes were developed to detect *Sargassum* from satellite imagery, such as the Maximum Chlorophyll Index (MCI) [16] or the Floating Algae Index (FAI) [17]. Recently, the FAI was improved by the Alternative Floating Algae Index, AFAI [18], which is less sensitive to clouds. Using the MCI, Gower et al. [19] initiated the detection of *Sargassum* aggregations in the Gulf of Mexico using satellite images from the medium-resolution imaging spectrometer (MERIS) on board ESA's Envisat satellite and the moderate-resolution imaging spectro-radiometer (MODIS) on board NASA's Terra and Aqua satellites. It was then extended to the Ocean and Land Colour Instrument (OLCI) [20] on board Copernicus's Sentinel-3 satellites. The AFAI proved to be performant on MODIS [18] or the multispectral instrument (MSI) [5,7,21,22] on board Copernicus's Sentinel-2 satellites. These indexes can also be used to determine the coverage of *Sargassum* per pixel [18] and then the biomass quantity of *Sargassum* as empirically defined by Wang and Hu [23].

However, the standard index-based detection (ID) method, from for example MCI or AFAI, cannot discards all radiometric noise and other non-*Sargassum* factors (i.e., residual clouds and cloud shadow, land, coastal water, surface waves, and sunglint [18,24]), which interfere with the detection of low-coverage *Sargassum*. The standard method is based on the calculation of the background index and the use of a threshold on the difference between local and background indexes to determine whether a pixel contains Sargassum (above the threshold) or not (under the threshold). However, the background calculation itself relies on the accurate screening of clouds, cloud shadows, coastal areas, land and sunglint [18,21] that may yield false positive values. Finally, uncertainties in the calculation of the background index may induce non-significant positive values. As a result, the lowest index values are not significant enough to ensure the presence of *Sargassum* and must be discarded, inevitably yielding false negatives. In order to retain most *Sargassum* while rejecting most false detections, the threshold applied on indexes must be optimally determined [18,19]. While some false detections can be removed by mathematical morphology post-treatments (e.g., erosion-dilation [18,25–28]) false detections remain, linked to residual clouds, cloud shadows, sunglint and turbidity. Another limitation encountered by the ID method is the high cloud cover in some images, because in such cases the background index, hence the index, cannot be accurately estimated. Wang and Hu [18,21] estimated false positive detections for their retrieval to be 20% of the total *Sargassum* pixels for MODIS and 6% for MSI. Recently, a machine learning model revealed a higher rate, namely 50%, of false positive detections by the ID method applied to MODIS images [24]. The remaining false positive detections cannot be discriminated on their sole index value, since their index distribution overlaps with that of true *Sargassum*. The ID method also produces false negative detections (i.e., *Sargassum* pixels not detected); it has been estimated to be 7% for MODIS [18] and 20% for MSI [21]. Therefore, the ID method has two important limitations: the need for complex pre-processing and the unavoidable yield of false detections (false positive and false negative).

An alternative to the ID method can be neural networks. Indeed, in the last few years neural networks have made significant progress in computer vision and are widely used in several domains, including imagery in oceanology [29–33]. Furthermore, they have already been successfully used for other macro-algae detections as an alternative to the ID method [34–36]. *Sargassum* in near-shore water and *Sargassum* stranding on the Atlantic coast have also been investigated using classic classifier algorithms (random forest,

K-means) and deep neural networks (UNet, AlexNet, VGG . . .) on the Mexican-Caribbean and Caribbean coasts using smartphone camera pictures [37,38] and fixed cameras [39]. On coastal and off-shore waters, machine learning has also been used for *Sargassum* detection from satellite images, e.g., random forest for false detection extractions [24] or a combination of several indexes [40] to improve *Sargassum* detection in the Atlantic Ocean. In the Yellow Sea, next to China and Korea, *Sargassum* detection was also studied with several machine learning models (i.e., fine tree, support vector machine, etc.) [41,42].

Among neural network methods, deep neural networks, such as convolutional neural networks (CNN), are efficient for image processing in object segmentation and classification [43–47]. Contrary to the previous machine learning algorithms, these types of neural networks could improve open sea *Sargassum* detection in the Atlantic Ocean by automatically learning the relationship between the reflectance and the spatial context of *Sargassum* aggregations. Arellano-Verdejo et al. [48] proposed a first deep learning method for *Sargassum* detection on the Mexican Caribbean coast, ErisNet, i.e., a pixel classifier with convolutional and recurrent layers, applied to a selection of MODIS spectral bands. ErisNet weakly takes into account neighbor pixels, thus it does not learn the structure of the *Sargassum* aggregations composed of several pixels. Recently, Wang and Hu [49] used the UNet model [47] using FAI and RGB images from high resolution satellite data such as MSI, Operational Land Imager, WorldView-II, and PlanetScope/Dove to segment *Sargassum* in off-shore waters of the Lesser Antilles and the Gulf of Mexico. UNet is a CNN using an encoder-decoder style network and this architecture allows *Sargassum* aggregation structures to be learnt. However, the spectral information can be reduced by using only three spectral bands when the information provided by other available spectral bands might be used.

To avoid all those problems and increase the performance of detection compared to previous methods, we propose two new specialized models for *Sargassum* detection based on deep neural learning in the Caribbean and Atlantic areas. Those models used all the available spectral bands provided by a sensor contrary to ID methods and UNet [49]. Furthermore, deep neural networks, and, more specifically, CNN, have a higher abstraction level than simple machine learning algorithms, and are better structured to take into account neighbor pixels at a higher scale. Thus, the CNN models are based on an encoder–decoder architecture, contrary to ErisNet. MSI and OLCI sensors were chosen due to the difference of their spatial and spectral resolutions, but CNNs can be applied to other similar sensors.

This paper is organized as follows: first, the satellite data, the study area, the previous detection methods and our CNN architecture, training and validation methods are presented in Section 2. Second, the performance of our CNN models is quantitatively analyzed in Section 3. Third, the *Sargassum* detection obtained using our CNN models and ID methods are compared from the literature images in Section 4. Finally, a conclusion is proposed in Section 5.

2. Materials and Methods

2.1. Satellite Data

This study used images from several satellite sensors with different spatial resolutions to detect *Sargassum* from space: MSI onboard Sentinel-2, OLCI onboard Sentinel-3 and MODIS onboard Terra satellites.

The Sentinel-2 and Sentinel-3 missions are carried out by the European Union's Copernicus Earth Observation programme, and each mission is currently composed of a constellation of two satellites.

The Sentinel-2A and Sentinel-2B satellites were launched in 2015 and 2017, respectively. These satellites carry the multi-spectral instrument (MSI), which acquires data in 13 multispectral bands in the visible, near-infrared, and short-wave infrared ranges of the solar spectrum (0.44 μm to 2.19 μm), with a high spatial resolution (10 m, 20 m, and 60 m, depending on the bands). In this study, images have a resolution of 20 m. Each MSI sensor has a revisit period of five days.

Onboard the two Sentinel-3 satellites (Sentinel-3A and Sentinel-3B, launched in 2016 and 2018, respectively), the Ocean Land Color Instrument (OLCI) has similar spectral bands than the MERIS sensor (2002–2012). OLCI images are composed of 21 spectral bands from the visible to the near infrared (from 0.4 µm to 1.02 µm), with a spatial resolution of 300 m. Each OLCI sensor has a revisit period of two days.

The Moderate Resolution Imaging Spectroradiometer (MODIS) instrument flies on board the Terra and Aqua satellites launched by NASA in 1999 and 2002 respectively. MODIS images have 36 spectral bands (0.4 µm to 14.4 µm) at three spatial resolutions (250 m, 500 m, and 1 km). Each MODIS sensor has a revisit period of one to two days. Only the MODIS/Terra sensor was used in this study because the overpass time is closer to the Sentinel satellites than MODIS/Aqua.

2.2. Study Area and Data Set

The west coast of the Atlantic is particularly prone to massive arrivals of *Sargassum* aggregations from the Atlantic Ocean since 2011 [1,50]. The study area is located in the Lesser Antilles, in the East Caribbean Sea and the Mexican coast (Figure 1), from 2018 to 2022.

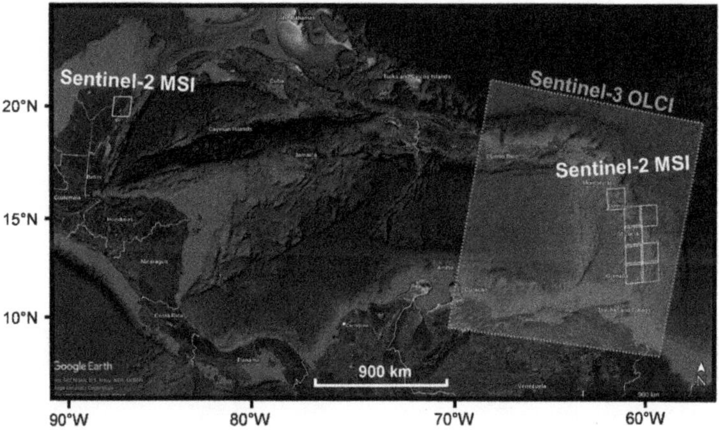

Figure 1. Location of Sentinel-3 OLCI images (white dotted boxes) and Sentinel-2 MSI tiles (white boxes) used for the learning and focus sets.

This study focused on two products: level-2A MSI products and Level-2 OLCI products from Copernicus Open Access Hub [51]. All images were corrected for Rayleigh scattering, gaseous absorption, and aerosols and have their corresponding ground truth. Images of each spectral band were normalized using their mean and standard deviations. Then, those images were divided into two types of datasets: a learning dataset and a focus dataset.

The learning dataset targeted CNN learning (training, validation, and test) and the quantitative validation process for our proposed CNNs. This dataset was composed of 14 images of size 10,980 × 10,980 pixels from MSI (Table S1) and 16 images of size 4865 × 4091 pixels from OLCI images (Table S2) from 2017 to 2022 distributed across all the study areas (Lesser Antilles and the Mexican Caribbean coast). All spectral bands were used as inputs to the model. To increase the learning dataset and avoid overfitting, we generated more images by randomly applying a horizontal flip, a vertical flip, or a rotation of 90 or 270 degrees. Additionally, a random color augmentation was applied only to OLCI images because they have more significant pixel variations than MSI images, due to the variation in the sunlight intensity from east to west. Indeed, at most, three spectral bands were picked randomly and adjusted by random values selected from [−0.1,0.1]. *Sargassum* represented, on average, only 1% of the pixels of an image. Thus, 128 × 128 sub-images centered on random positions were taken while conserving the proportion of *Sargassum* in the original images. We ensured that 66% of the sub-images had no *Sargassum* to provide a representative sample of different backgrounds. Therefore, the dataset comprised 75,000 sub-images from MSI and 50,000 sub-images from OLCI. The learning dataset was split into three subsets: a training set including 80% of the total image number, a validation set comprising 10% of the images, and a test set containing the remaining images. The test set was only used to evaluate the model's performance.

A focus dataset was then built in order to: (1) quantitatively evaluate the CNN models performance; and (2) discuss the performance compared to the ID method images published and validated in the literature. That dataset was composed of complete images, (not sub-images) from MSI and OLCI sensors, detailed in Table 1. It contained four images of level-1 files from the literature (Table 1): MODIS and MSI images from Descloitres et al. [22] and OLCI images from Schamberger et al. [52]. The data also included our CNN images from the OLCI sensor at the same dates and locations and from the MSI sensor only at the same dates and locations as those of Descloitres et al. [22]. In addition, the dataset contained a supplemental image from MSI on 6 August 2022 in the Martinique area.

Table 1. Images of the focus dataset used to evaluate the performance of our CNNs compared to images from the literature, with a summary of their characteristics: location, acquisition date and time, satellite, and sensor. The 'Source' column indicates the source of each image: a reference for images taken from the literature, and 'us' for images processed with our own CNN method.

Location	Date	Time (UTC)	Satellite-Sensor	Source
		14:35	Terra-MODIS	[22]
Grenadines (Caribbean)	29 January 2019	14:37	Sentinel-2B-MSI tiles: PQU- PQV-PRU-PRV	[22], us
		13:38	Sentinel-3-OLCI	us
Lesser Antilles	9 May 2020	13:56	Sentinel-3-OLCI	[52], us
	8 July 2017	13:55	Sentinel-3-OLCI	
Martinique	6 August 2022	14:37	Sentinel-2-MSI tile PRB	us

2.3. Three Sargassum Detection Methods

2.3.1. Standard Index-Thresholding Method

Currently, the satellites cannot detect the differences between the three morphotypes of *Sargassum* living in sympatry. Therefore, they are studied as a whole. Two indexes were essentially used in this study: the Maximum Chlorophyll Index (MCI) [16] and the Alternative Floating Algae Index (AFAI) [18]. Furthermore, the Floating Algae Index (FAI) [17], the Normalized Difference Vegetative Index (NDVI) [53], and the InfraRed Color (IRC) were also used for the ground truth computation (Section 2.3.3).

The NDVI takes advantage of the red edge of the vegetation spectral reflectance. It measures the normalized difference between near-infrared (λ_2) and red range (λ_1):

$$\text{NDVI} = \frac{R(\lambda_2) - R(\lambda_1)}{R(\lambda_2) + R(\lambda_1)} \tag{1}$$

This index is often used to study vegetation ecology [54–56]. Floating algae have also been investigated using NDVI [57,58]. MCI, FAI and AFAI indexes were calculated using Equation (2), based on wavelengths summarized in Table 2. The MCI is an index based on radiance around 705 nm, more accurately, the MERIS and OLCI spectral bands located at 709 nm. MCI reveals high chlorophyll concentrations; it was first used to detect phytoplankton bloom using the MERIS sensor [16]. That index also showed its efficiency on *Sargassum* observations in the Atlantic Ocean using MERIS, MODIS and OLCI sensors [19,20]. The

FAI was developed for floating algae in the open ocean and firstly tested with the MODIS instrument to detect the benthic macroalgae *Ulva prolifera* in the Yellow Sea and pelagic *Sargassum* in the North Atlantic Ocean [17]. Later, Wang and Hu [18] improved it with the AFAI, an index more efficient in the presence of clouds. It was developed for the MODIS sensor to detect *Sargassum* and extended to the MSI sensor [22].

$$Index = R(\lambda_2) - R(\lambda_1) - [R(\lambda_3) - R(\lambda_1)] \times \frac{\lambda_2 - \lambda_1}{\lambda_3 - \lambda_1} \tag{2}$$

Table 2. Wavelength parameters of Equations (1) and (2) for NDVI, MCI, FAI and AFAI indexes for OLCI, MSI and MODIS.

Index	Sensor	λ_1 (nm)	λ_2 (nm)	λ_3 (nm)
NDVI	OLCI	665	865	-
	MSI	665	833	-
MCI	OLCI [1]	681	709	754
FAI	MODIS [2]	645	859	1240
	MSI [3]	655	855	1609
AFAI	MODIS [4]	667	748	869
	MSI [5]	665	740	865

[1] [20]; [2] [17]; [3] [21]; [4] [18]; [5] [22].

While MCI, FAI and AFAI can be directly used for *Sargassum* detection, additional processing is required to improve the detection. Indeed, while Equation (2) guarantees that the *Index* increases with the presence of *Sargassum*, it also increases with the value of the *Index* for *Sargassum*-free surrounding water, the so-called background *Index*. This background *Index* depends on local non-*Sargassum* factors, such as water spectral reflectance, potential residual sunglint contamination, and the presence of aerosols. Then, a local *Index* deviation ($\delta Index$), i.e., the difference between the background and the local *Index*, is used to discriminate between pixels with and without *Sargassum* thanks to a threshold. This latter is optimally determined [18,19], but also depends on the author's objective and specific datasets; for example, the threshold used on OLCI images by Schamberger et al. [52] is 0.0030, while the one used on MODIS images by Descloitres et al. [22] is 0.00017 [18].

Finally, the amount of *Sargassum* may be determined with the fractional coverage (FC), it is defined as the ratio between the surface area of *Sargassum* within one pixel and the total surface of that pixel [22]. The FC is deduced thanks to the linearity of Equation (2), which ensures that the local $\delta Index$ is proportional to the FC of *Sargassum* within a sensor pixel [18,22]:

$$\delta Index(\text{FC}) = K \times \text{FC} \tag{3}$$

In this study we used K = 0.0874 for MODIS, K = 0.0824 for MSI [22] and K = 0.0579 for OLCI [59].

The biomass quantity of *Sargassum* can be estimated using a linear relationship between the fractional coverage, the pixel area and a calibration constant defined empirically by Wang and Hu [23].

2.3.2. Convolutional Neural Network (CNN) for Sargassum Retrieval

State of the Art

In this study, we used three other CNNs to compare with our method, including two CNNs already used for *Sargassum* extraction in the Atlantic and Caribbean areas since 2019.

These are the ErisNet model from Arellano-Verdejo et al. [48] and the UNet of Ronneberger et al. [47], adapted by Wang and Hu [49] for *Sargassum*, and by Yan et al. [60], for harmful cyanobacterial algal blooms detection.

ErisNet [48] contains nine convolutional layers and two recurrent layers, allowing the model to keep a history of previous predictions. The network learns using a balanced number of *Sargassum* and background pixels during the training. However, during inference, it takes all pixels and predicts for each of them whether they belong to *Sargassum* or not. ErisNet is a pixel classifier that processes pixels as 1D input, unlike UNet and SegNet, which are image segmenters and treat the image as 2D input.

The encoder–decoder style network UNet [47] was used by Wang and Hu [49] for *Sargassum* retrieval. It was initialized with the weights of a VGG16 [61] trained on ImageNet. UNet contains four convolutional layers and performs four down-samplings in the encoder and four up-samplings in the decoder. The UNet encoder groups pixels of the same types together, whereas the decoder magnifies the output of the encoder. Each decoded map is concatenated with the same-sized encoded map. UNet was also applied by Yan et al. [60] on sentinel-2 MSI images of Chaohu Lake in China to identify harmful cyanobacterial algal blooms (CyanoHABs). They compared the segmentation with three FAI-based automatic CyanoHABs extraction methods: gradient mode, fixed threshold, and the Otsu method, and showed that the accuracy of UNet was better.

SegNet [43] is another encoder-decoder-style segmentation network powerful for image segmentation. Like UNet, SegNet has reduction layers connected to enlargement layers to reconstruct pixels. In addition, it uses neighboring pixels, having the same distribution during the reduction, in the upsampling reconstruction. UNet and SegNet can process small images (here from 128 × 128) in a single pass.

We built these three models using descriptions found in the literature (parameters summarized in Table 3), and trained them with the same dataset as our own models.

Table 3. Summary of the layer, block and total parameter numbers for ErisNet, UNet, SegNet and our proposed Networks for MSI and OLCI. More detailed tables of the structure of our proposed Networks are available in Supplementary Materials for MSI (Table S3) and OLCI (Table S4).

	ErisNet	UNet	SegNet	Our Proposed Network	
				MSI	OLCI
Layers	44	83	91	32	75
Blocks	7	10	10	9	18
Total Parameters	455,554	13,400,578	29,449,350	226,762	973,145

Model Description for MSI and OLCI

We proposed two convolutional neural networks (CNN) with an encoder–decoder architecture (parameters summarized in Table 3), one targeting *Sargassum* semantic segmentation and taking Sentinel-2/MSI images as input, while the other focused on Sentinel-3/OLCI images and returning the δMCI (post-processed without clouds, land, and coasts). Indeed, with the ID method, OLCI images have more diffuse *Sargassum* aggregations than MSI. Thus, it was more relevant to train the CNN on the δMCI than on semantic segmentation. For MSI and OLCI images, input channels were composed of all the image spectral bands listed in Tables S1 and S2. The code of the two models was written using Pytorch [62].

For the CNN with MSI images, the encoder reduces the image scale, and the decoder performs a semantic segmentation. Indeed, we made only a single reduction because *Sargassum* aggregations have a width of one to three pixels. Hence, too much reduction is useless since target objects are thin. The encoder is composed of residual blocks issued from ResNet [63]. Figure 2 shows the architecture of the network; each residual block conserves its input by adding it to its output. Since there is only a need to segment *Sargassum*, a weighted binary cross-entropy loss function (WBCEWithLogitsLoss) was used. This loss is appropriate for binary problems and helped to deal with the unbalanced problem of *Sargassum* because the ratio of *Sargassum* pixels to the background pixels was meager. Let

$y_{i,j}$ be a pixel of the ground truth segmentation, $\hat{y}_{i,j}$ be the output of the network, which corresponds to the logits (log-odds function) of the semantic segmentation, and $N \times M$ be the resolution. The weighted BCEWithLogitsLoss is defined by Equation (4):

$$\text{WBCEWithLogitsLoss}(y, \hat{y}) = -\frac{1}{N \times M} \sum_{i,j} \left[w_1 \times y_{i,j} \times \log\left(\sigma\left(\hat{y}_{i,j}\right)\right) + w_2 \times \left(1 - y_{i,j} \times \log\left(1 - \sigma\left(\hat{y}_{i,j}\right)\right)\right) \right] \quad (4)$$

where $\sigma(x) = \frac{1}{1+e^{-x}}$ is the sigmoid function, w_1 is the weight of the foreground, w_2 is the weight of the background.

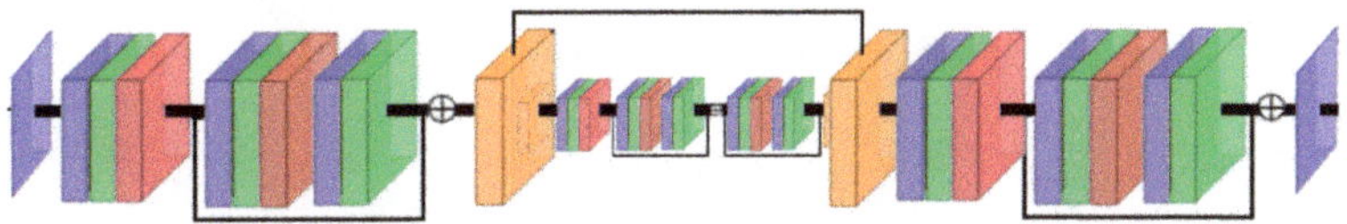

Figure 2. The neural network architecture for semantic segmentation on MSI; blue layers are convolutions, green ones represent batch normalizations, ReLU is in red, and the orange layers correspond to max pooling/unpooling layers.

Regarding OLCI, the segmentation map is substituted with the δMCI. The architecture of the network is similar to the one of MSI, except that it adds one reduction level and more channels in the residual blocks. The network contains 128, 256, and 512 blocks against 32 and 64 for MSI (Figure 3). Furthermore, a soft attention block (Attention UNet; [64]) is added to each reduction level. These blocks allow the connection between outputs of unpooling layers and inputs of reduction layers corresponding to the same scale. The goal is to use the distribution of the same scale-reduction input.

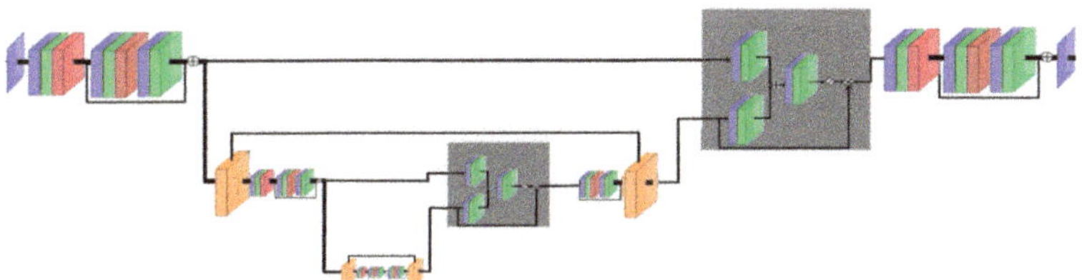

Figure 3. The neural network architecture for OLCI images; blue layers are convolutions, green ones represent batch normalizations, ReLU is in red, and yellow layers correspond to max pooling/unpooling layers. Gray blocks represent soft attention.

Because the task is a regression, the mean squared error (squared L2 norm) was used as a loss (MSELoss). This loss function compares the output δMCI image with the reference one. Let δMCI_{ij} be a pixel of the reference masked δMCI, $\widehat{\delta\text{MCI}}_{i,j}$ be a pixel of the network's output filter, and $N \times M$ be the resolution. The MSELoss is written as follows:

$$\text{MSELoss}(\delta\text{MCI}, \widehat{\delta\text{MCI}}) = -\frac{1}{N \times M} \Sigma_{i,j}(\delta\text{MCI}_{i,j} - \widehat{\delta\text{MCI}}_{i,j})^2 \quad (5)$$

Training Process

For the training process, we used the training and the validation subsets from the learning dataset described in Section 2.2. The training lasted for 1000 epochs and was done using a batch size of 92 and a stochastic gradient descent (SGD) with a learning rate of 10^{-3} and a momentum of 0.9. Finally, the model giving the best F1-score on the validation set was saved. This validation was done every ten epochs.

2.3.3. Visual Analysis to Establish the Ground Truth

The ground truth was determined for each image of the learning set (i.e., training, validation and test datasets) and the focus dataset described in Section 2.2. In the absence of in situ observations of the *Sargassum* distribution, the ground truth was provided by an "expert" who manually annotated *Sargassum* aggregations using different standard indexes (i.e., FAI, AFAI, NDVI, MCI, IRC; Table 1) as indicators from level-2 OLCI and MSI images. These indexes were chosen due to their different response to *Sargassum* signals, and the sensor used.

The ground truth of the MSI was a binary mask: 1 for pixels containing *Sargassum* and 0 for *Sargassum*-free pixels. For OLCI, the ground truth for the training process (training and validation datasets) was obtained using a binary mask built manually using several indexes to identify *Sargassum*, then applied to the δMCI (Figure 4). For the model evaluation, we only compared OLCI detections to the binary ground truth (i.e., the accuracy of the retrieved δMCI values was not evaluated).

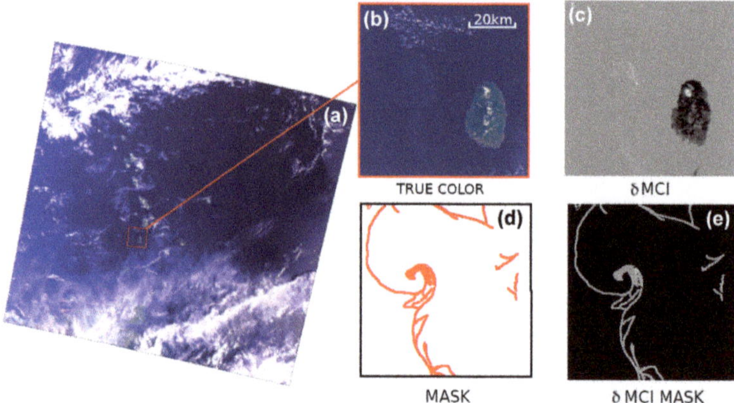

Figure 4. (**a**) OLCI image on 23 February 2017 with True color composite (band 9:674 nm:Red; band 7:560 nm:Green; Band 2:413 nm:Blue). (**b**) Sub-image example (red square); (**c**) δMCI; (**d**) binary mask manually made using several indexes and (**e**) masked δMCI built using the binary mask.

2.4. Performance Evaluation

2.4.1. Performance Metrics Using the Ground Truth

Three types of metrics were used to evaluate the performance of the ID or CNN methods by comparing the test dataset to the ground truth: the recall, the precision, and the F1-score. The latter was calculated from the two other metrics (the recall and the precision). These metrics depend on three parameters: true positive (TP), false positive (FP) and false negative (FN). TP, FP and FN were determined using the ground truth and *Sargassum* pixels detected by the retrieval technique. TPs are *Sargassum* pixels successfully recognized by the retrieval technique. FPs are *Sargassum* pixels only identified by the retrieval technique but not present in the ground truth. Finally, FNs are *Sargassum* pixels not detected by the retrieval method and present in the ground truth.

The recall metric is the ratio of the number of TP detections and the number of *Sargassum* pixels of the ground truth (Equation (6)). The precision metric is the ratio of the number of TP detections and the number of pixels detected as *Sargassum* (Equation (7)). To evaluate the model's performance, these two metrics were combined to form the F1-score metric (Equation (8)). The better the performance; the closer the F1-score is to 1. *Sargassum* detection was considered a "true" detection when its distance from an annotation on the ground truth is below three pixels.

$$Recall = \frac{TP}{TP + FN} \tag{6}$$

$$Precision = \frac{TP}{TP + FP} \tag{7}$$

$$F1\_score = 2 \times \frac{Recall \times Precision}{Recall + Precision} \tag{8}$$

2.4.2. Comparison of CNN and ID Approaches

Another evaluation was performed on the focus dataset (described in Section 2.2) based on the comparison of CNN with ID method images in order to estimate their respective performance. During this analysis, we no longer took into account the ground truth, which was not without errors, due to a degree of subjectivity present in the process of annotating *Sargassum* aggregations, especially for aggregation edges and low-FC aggregations. Indeed, the ground truth can introduce biases during the model performance evaluation with the performance metrics (Section 2.4.1).

Before the evaluation, a threshold set to 0.01 was applied to the δMCI of the CNN on OLCI images in order to optimally discard δMCI not linked to *Sargassum* aggregations.

To compare *Sargassum* aggregations obtained using CNN and ID approaches, the aggregations were extracted with a polygon function from Matlab (bwlabel function). *Sargassum* aggregation features such as their area, major axis size (i.e., length), minor axis size (i.e., width) and major/minor axis ratio (i.e., length/width ratio) were then measured.

To study the accuracy of the aggregations extracted by the CNN, the main aggregations detected by the ID method were compared. The main ID-detected aggregations were selected to: (1) have a length/width ratio higher than three [49]; and (2) to be in the 90th percentile of the length distribution. Thus, these main aggregations were larger than 3000 m and 140 m length for OLCI and MSI, respectively. In addition, *Sargassum* aggregations near coasts were discarded because of the FC high values in such areas due to the turbidity and shallow water [7,24,65,66]. The distance was set to 15 km away from coasts for OLCI and 200 m for MSI images.

3. Results

3.1. Model Performances on the Test Dataset

The model performance metrics (precision, F1-score) were calculated. They were both relatively high for MSI and OLCI. *Sargassum* pixels were accurately detected, and there were few false positive detections. Consequently, the F1-score is significant for the two proposed CNNs (Table 4), with an F1-score higher for MSI (88% of *Sargassum* pixels accurately detected versus 76% for OLCI). The lower F1-score of OLCI can be explained by a lower precision (79% versus 94% for MSI). OLCI has sparser aggregations than the ground truth. As a result, it detects more *Sargassum* pixels and has a lower precision than MSI.

Table 4. Recall, precision and F1-score of each method (indexes and Neural Networks) for Sentinel-2/MSI and Sentinel-3/OLCI. For Neural Network these metrics are calculated from the test data set (from the learning set of Tables S1 and S2).

	Sentinel-2/MSI			Sentinel-3/OLCI		
	Recall	Precision	F1-Score	Recall	Precision	F1-Score
NDVI	0.835	0.085	0.154	0.910	0.105	0.188
FAI(MSI)/MCI(OLCI)	0.587	0.120	0.200	0.619	0.167	0.320
ErisNet	0.958	0.179	0.302	0.896	0.314	0.465
UNet	0.618	0.818	0.704	0.961	0.452	0.615
SegNet	0.599	0.840	0.699	0.931	0.493	0.645
Our network	0.819	0.942	0.876	0.735	0.785	0.760

3.2. Comparison with the ID Method

The performance metrics were also calculated for the three indexes (NDVI and AFAI for MSI or MCI for OLCI; Table 4). The indexes are very sensitive to land contamination, which can bias their performance. The land was therefore masked out before computing the indexes. The recall was better than with the CNN, especially for the NDVI. This is not surprising because these indexes were used to create the ground truth. However, their precision was low, below 17%, leading to a low F1-score (lower than CNNs) for all of them. Indeed, regarding the ID method, many false detections occurred around clouds, and many isolated detections seemed not to be *Sargassum* aggregations.

3.3. Comparison with Existing Networks

For comparison purposes, the other networks (ErisNet, UNet, SegNet) were trained in the same conditions as the proposed CNN models. Note that the precision and the recall were computed using a zero threshold, and all non-zero outputs were considered positive.

Compared to other CNN models, ErisNet network had the lowest F1-score (0.30). *Sargassum* pixels were well detected (recall around 90% for OLCI and MSI). However, it had a substantial number of false positive detections (a precision of 18% and 31% for MSI and OLCI, respectively, (Table 4)) around cloud edges and in the open ocean; the network principle can explain this. ErisNet is not a segmentation network like the other, in the sense that it processes pixels independently as 1D inputs and learns one background pixel for one *Sargassum* pixel. It does not leverage information in all background pixels since *Sargassum* pixels represent just a tiny portion of all pixels. On the contrary, segmentation networks (UNet, SegNet, and ours) consider the neighboring pixels.

Among encoder–decoder networks, our proposed network had the highest F1-score with 0.88 versus ~0.70 for UNet and SegNet on MSI images, and 0.76 versus ~0.62 for OLCI images (Table 4). On MSI images, our proposed network had fewer false positive and false negative detections than SegNet and UNet, with a recall of 0.82 versus ~0.60, and a precision of 0.94 versus ~0.83. Regarding OLCI images, SegNet and UNet were more efficient at detecting *Sargassum* pixels (recall around ~0.94 for both SegNet and UNet, versus 0.74 for our proposed CNN). However, the proposed approach had significantly fewer false positive detections (precision of 0.79 versus ~0.47 for SegNet and UNet), which resulted in a higher F1-score. Indeed, the multiple levels of reduction in UNet and SegNet makes the detection of filiform objects such as *Sargassum* aggregations difficult.

3.4. Results with the Focus Image Dataset

The performance metrics were also computed for images of the focus dataset to evaluate their validity. For Sentinel-3, the two images tested had a F1-score of 0.72 and 0.79 (Table 5), i.e., a score around the value of the corresponding tested CNN (0.76; Table 4). For Sentinel-2, the F1-score was 0.90, 0.88, 0.87 and 0.81, respectively, for the PQV, PRU, and PRV tiles of 29 January 2019 and the PRB tiles of 6 August 2022. For those tiles, the F1-score was in the same range as the one found during the performance evaluation of the corresponding model (0.88; Table 4). Only the PQU tile was out of range, with an F1-score of 0.65. The results shown proved the consistency of our technique.

Table 5. Recall, Precision and F1-score from our models for some images of the focus dataset. Note that the results from 29 January 2019 images (MSI) are biased due to the use of their presence in the training set.

Satellite-Sensor	Date	Tile	Recall	Precision	F1-Score
Sentinel-3-OLCI	29 January 2019	-	0.566	0.970	0.715
	8 July 2017	-	0.851	0.739	0.791
Sentinel-2B-MSI	29 January 2019	PQU	0.495	0.951	0.651
		PQV	0.833	0.980	0.900
		PRU	0.786	0.999	0.880
		PRV	0.832	0.904	0.867
	6 August 2022	PRB	0.817	0.896	0.854

4. Discussion

The focus dataset was used here to discuss the CNN images in relation to the ID method. First, the main aggregations were analyzed to verify the reliability of CNN. Thereafter, we compare all *Sargassum* pixels of the CNN with the ID method to explain their differences and evaluate the suspected false positive and false negative detections. Finally, we compared OLCI and MSI images from the CNN method.

4.1. The CNN Reliability on Sargassum Aggregations

The main *Sargassum* aggregations of the ID method were detected by our proposed CNN on OLCI images (Figure 5a) and MSI images (Figure 6a). The CNN detected more than 70% of the FC of main aggregations detected by the ID method (70% for OLCI and 80% for MSI). This corresponds, respectively, to 90% and 48% of *Sargassum* pixels belonging to the main aggregations of OLCI and MSI images. Note that the main aggregations are also composed of low FC. In that respect, the CNNs are robust for main aggregations.

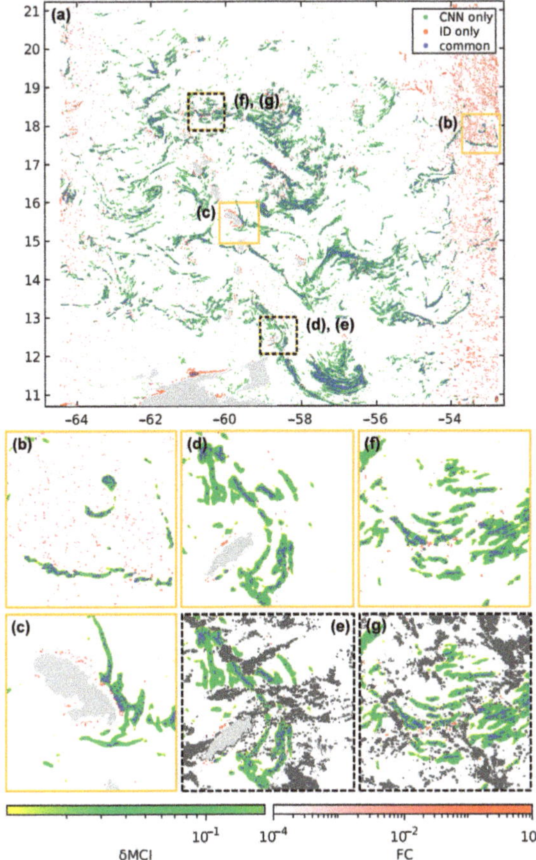

Figure 5. (a) *Sargassum* detections from OLCI with the CNN and ID methods: detected by both methods (blue), detected by the CNN but not by the ID method (green), and detected by the ID method but not by the CNN (red); (b–g) sub-images from (a) with a color scale for the ID and CNN detections; and (e,g) are respectively (d,f) with a cloud mask. MCI data for the calculation of the FC, cloud (dark gray) and land masks (light gray) come from Schamberger et al. [52]. OLCI image from 8 July 2017.

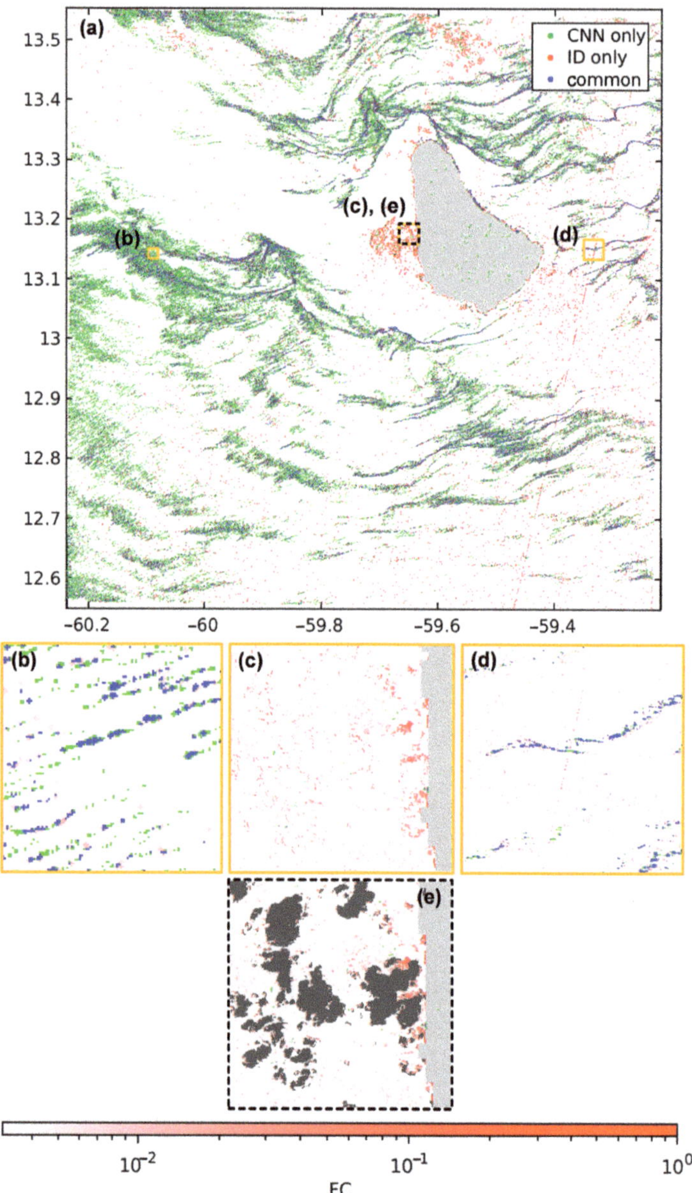

Figure 6. (**a**) *Sargassum* detections from MSI with the CNN and ID methods: detected by both methods (blue), detected by the CNN but not by the ID method (green) and detected by the ID method but not the CNN (red). Only FCs above 1.16×10^{-2} are represented in (**a**), in order to clarify the figure. Below this threshold the image is too contaminated by a large number of isolated pixels without spatial consistency (discussed in Section 4.2), which start to be visible below this threshold; and (**b**–**e**) sub-images from (**a**) with a color scale for all IDs. FC, cloud (dark gray) and land masks (light gray) come from Descloitres et al. [22]. MSI image, PRV tile from 29 January 2019.

In addition to the main aggregations, the whole FC distribution for all *Sargassum* pixels was similar for the ID and the CNN methods from Sentinel-3 images (ranging from 7.8×10^{-6} for low FC to 0.7 for high FC; Figure 7a). A slight and rather constant discrepancy (around 1 to 0.5% of pixels) can be observed.

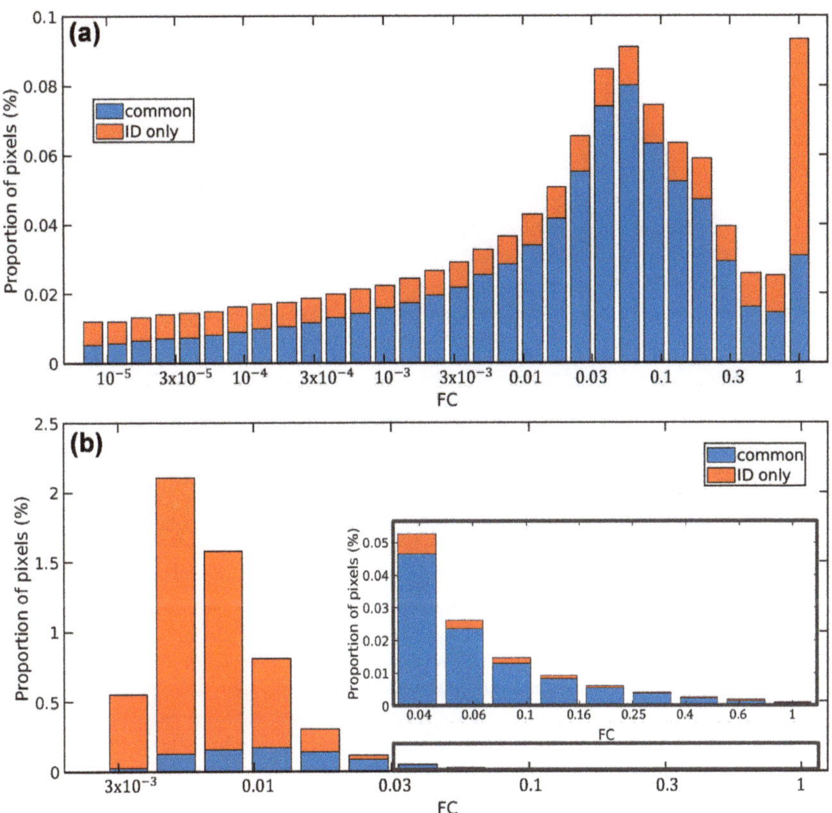

Figure 7. Distribution of the proportion of *Sargassum* pixels detected by both the CNN and the ID methods (blue) and detected by the ID method but not the CNN (red) for (**a**) OLCI; and (**b**) MSI.

Regarding MSI, the proportion of common detection depends on the FC (Figure 7b), 35% of the FC detected by the ID method was also detected by CNN. Indeed, only few IDs were detected by the CNN with low FC (Figure 8). Less than 10% of pixels by the ID method were detected as *Sargassum* pixels by the CNN under a *Sargassum* coverage of 0.006 (2.4 m^2/pixel). However, this proportion increased with the FC. This detection ability follows a Gompertz curve represented by Equation (9) and shown in Figure 8. About 90% of FC pixels above 0.026 (10.4 m^2/pixel) were detected. This curve allows the expected accuracy of *Sargassum* detections to be predicted using the CNN method as a function of FC.

$$Y = c + (d - c) \times \exp(-\exp(b \times (FC - e)))$$ (9)

where $b = -170$, $c = 4.6$, $d = 97$ and $e = 1.25 \times 10^{-2}$.

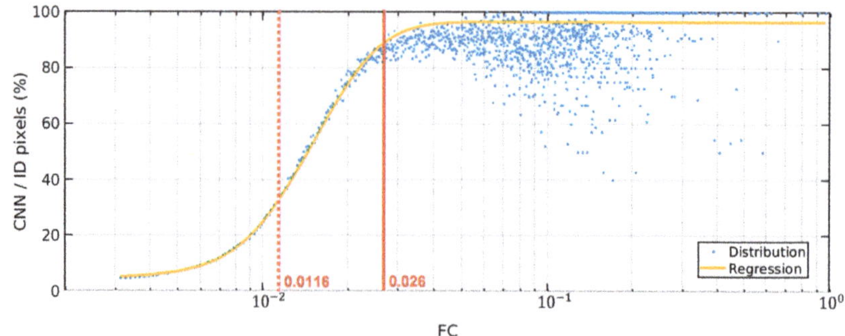

Figure 8. Proportion of MSI detections common to CNN and ID methods on the total IDs (in %) as a function of FC intervals (blue dots), each interval has a bin width of 2.79×10^{-4}. Intervals with less than 10 values are not represented. A Gompertz equation (Equation (9); yellow line) fits the blue points distribution and predicts the CNN detections accuracy. The red plain line represents the FC threshold from at least the CNN detects 90% of the IDs, and the red dotted line is the FC threshold used in Figure 6a.

Nonetheless, a large amount of low FC detections (with an FC under 0.006) with the ID method can either be associated with "true" or "false" detections [21,48] and it may be difficult to estimate the "true" *Sargassum* detection missed by the CNN. Going forward, the CNN can be improved on low index pixels to increase its performance on false positive and true positive pixels of this particular class.

4.2. Less False Detections Which Improve Sargassum Coverage Estimation

In the comparison between the two methods, some pixels from the ID method were not detected by the CNN. On MSI images, we only focused on *Sargassum* pixels associated with a FC above 0.026 (see Section 4.1). As presented before, those *Sargassum* pixels seemed to be accurately recognized by the CNN, with more than 90% detections common to the CNN and ID. However, for a few FC classes in that range, the CNN method seemed to underperform compared to the ID. (Figure 8). We focused on these ID extra detections for the MSI, whereas we analyzed all extra detections from the OLCI. Regarding OLCI, on the two OLCI images analyzed, (respectively, for 8 July 2017 and 9 May 2020) about 35% (respectively, 23% and 46%) of the total pixels of the ID method, which correspond to 50% (respectively, 45% and 55%) of the total FC were not detected by the CNN.

As we can easily identify through visual inspection, a large part of those non-detections in OLCI and MSI images should be attributed to false detections from ID methods, which the CNN discards. For instance, haloed false detections around land masses are present on MODIS images using the ID method (Figure 9), and to a lesser extent, on OLCI (Figure 5c) and MSI images (Figure 6c). The ID method also yields very small unrealistic isolated detections, unexpected aggregations in cloudy areas (Figures 5d–g and 6c,e), or artifact patterns linked to the index (satellite swath edges, radiometric noises, isolated detections on satellite image edges) (Figures 5b and 6d). In addition, on OLCI images, the high detection rate for the FC classes closest to one by the ID method (Figure 7a) was poorly recognized by the CNN (only 33% of the class pixels). These extra ID pixels are associated with turbid water that the ID method used here as an undiscarded reference [59]. All the false detection origins presented here are also confirmed by different authors [7,23,24], namely the extra-detection rate in OLCI images is close to the false detection rate observed by Podlejski et al. [24] using the ID method on MODIS images. Such estimations confirm that a large part of the extra-detections is linked to false detection and leads to an overestimation of the coverage of *Sargassum* (FC) by 50%. On MSI images, extra ID detections above a FC of 0.026 are mainly, as for OLCI, false detections obtained by the ID method and discarded by the CNN. Hence, ID methods for the 0.026 FC threshold seem to induce an overestimation of 14% of the FC for that range. Furthermore, the overestimation can increase by considering the false detections from lower FC pixels (below 0.026).

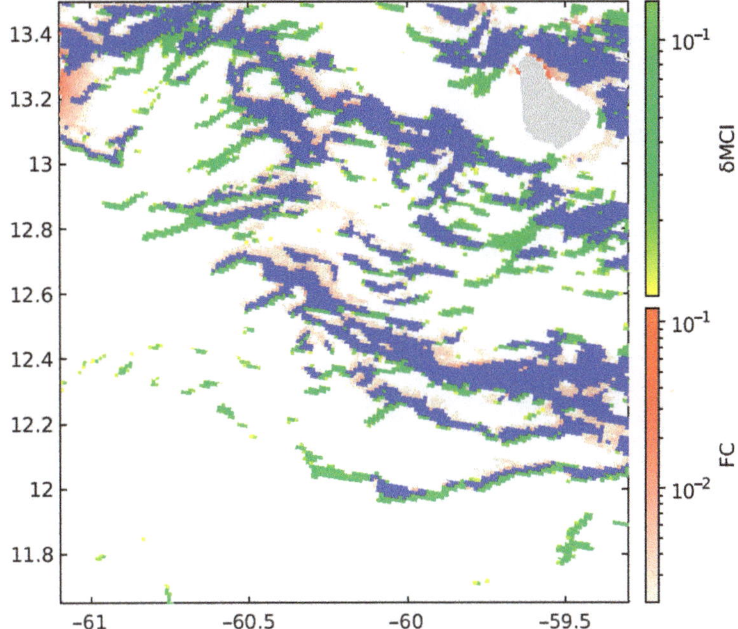

Figure 9. Superposition of *Sargassum* detections from MODIS with the ID method and OLCI with the CNN method (δMCI): detected with both MODIS and OLCI (blue), detected with OLCI but not with MODIS (green) and detected with MODIS but not with OLCI (red). Terra and OLCI images from 29 January 2019 at 14:35 UTC and 13:38 UTC respectively.

Therefore, our proposed CNNs discarded false positive detections through an understanding of the spatial context and features (e.g., shape) of *Sargassum* aggregations [24,36,42], leading to a more accurate estimation to the *Sargassum* coverage than the ID method.

4.3. Better Estimation of the Aggregation Shape

The CNN generated *Sargassum* pixels not detected by the ID method, which constituted 85% and 56% of *Sargassum* pixels of the CNN dataset for OLCI and MSI, respectively. These extra detections were mainly located in the continuation or along the edges of *Sargassum* aggregations, thus forming bigger aggregations than the ID method. In most cases, these bigger aggregations also included several smaller ID-detected aggregations (Figures 5b–d,f and 6b). The last part of CNN extra detections represented *Sargassum* aggregations not found by the ID method (Figures 5c,d and 6b). The coverage of all CNN extra detections on OLCI were confirmed by the ones found by the ID method on MODIS (Figure 9).

Regarding OLCI images, the CNN method detected four times fewer aggregations than the ID method (around 3000 for the CNN against 11,000 for the ID method). Indeed, the CNN identified larger and longer aggregations that were erroneously detected as several aggregations (Figure 5b–d,f). For instance, on July 8, 2017 (Figure 5a), CNN *Sargassum* aggregations measured, on average 9 km length, versus 2 km for the ID method and covered 50 km^2 (CNN) versus 2 km^2 (ID). Furthermore, in the case of a blocked signal, such as tiny clouds, the CNN was able to reconnect aggregations with each other and reconstruct the whole aggregations (Figure 5d,e; lower part). However, this only worked with small clouds (Figure 5d,e; upper part).

Considering MSI images, similarly to OLCI, the CNN detected fewer aggregations than the ID method: ~60,000 versus ~200,000. The CNN aggregations were slightly bigger than the ID-detected aggregations, 5600 m^2 and 3600 m^2, respectively.

As a result, using our proposed CNNs, the aggregations' shape was more realistic and less discontinuous; more like those observed from satellites of better resolution [49] or from in situ observations [5].

4.4. Complementarity of MSI and OLCI Images

Sargassum features retrieved by the CNN method using MSI and OLCI differed. The highest-resolution sensor (MSI) had more detailed *Sargassum* aggregations with a shape clearly defined, close to in situ observations [5]. Moreover, it detected *Sargassum* aggregations near the coasts and small aggregations not present in the OLCI images (Figure 10c,d). For instance, in Figure 10a,b, within 1.5 km from the coasts of Barbados Island, around 210 MSI *Sargassum* pixels on the OLCI grid were detected (3200 MSI *Sargassum* pixels), whereas only 80 *Sargassum* pixels are detected using OLCI.

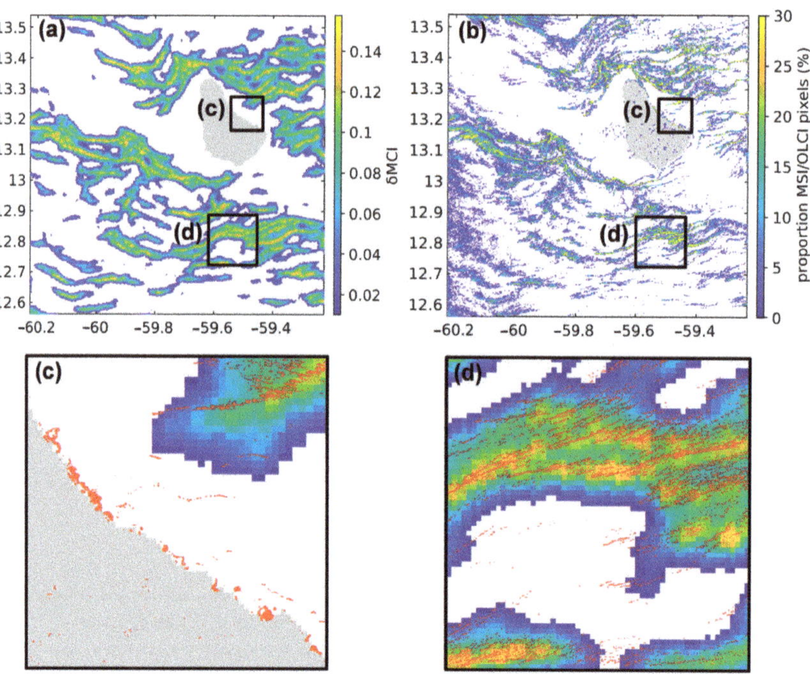

Figure 10. (a) *Sargassum* detections from CNN with OLCI; (b) proportion of MSI *Sargassum* pixels inside 1 OLCI pixel; and (c,d) Superposition of *Sargassum* detections from CNN with MSI (red) and with OLCI (color scale); (c,d) sub-images from (a,b). MSI image, PRV tile and OLCI image from 29 January 2019 at 14:37 and 13:38 respectively.

On the other hand, overall, OLCI has a higher percentage of *Sargassum* pixels than MSI. For instance, in Figure 8, 38% of OLCI pixels are *Sargassum* versus 27% for MSI extrapolated on OLCI pixels (corresponding to 1.89% MSI *Sargassum* pixels). Hence, on average, the CNN detected larger aggregations using OLCI than using MSI (see Section 4.3). Moreover, most MSI detections coincided with higher *Sargassum* signals from the OLCI (Figure S1).

When visually comparing the *Sargassum* aggregations detected from OLCI images with the CNN and MODIS images with the ID method (Figure 9), we retrieved overall the same *Sargassum* aggregations, with a slight shift due to the difference of 1 h between the two sensor acquisition times. The CNN conserved the same proportion of *Sargassum* pixels as MODIS, for instance in Figure 9, 15.7% for the CNN method versus 15.3% for the ID method are *Sargassum* pixels.

OLCI and MSI sensors are complementary. The high resolution MSI sensor detects smaller *Sargassum* aggregations, aggregations nearer to the coasts, and morphology of aggregations with better accuracy. On the other hand, OLCI provides daily data and detects pixels with lower FC.

5. Conclusions

The increase in the number of methods and the improvement of the quality of *Sargassum* detection by satellite is crucial for the prediction, and therefore, the management of the future standing of the algae along the coasts. The coupling between satellite detections, ground truth, and modeling remains the best way to understand the dynamics of *Sargassum* along the Great *Sargassum* Atlantic Belt.

In this study, we proposed a new encoder–decoder to detect floating pelagic *Sargassum*. The proposed CNNs were trained using two types of satellite images (OLCI, MSI) with different resolutions. This new model appeared to be more efficient than existing CNNs, such as ErisNet, UNet and SegNet, for *Sargassum* detection, with fewer false positive detections and more accurate *Sargassum* detections. Indeed, the consideration of neighboring pixels avoided some of the false detections made by ErisNet, and fewer reductions improved the performance of UNet.

Our proposed CNNs detected the same large *Sargassum* aggregations detected by the ID method, but with a more accurate estimation of the *Sargassum* coverage. Indeed, with the use of all the spectral bands in the images and the *Sargassum* spatial context, the CNN more efficiently discarded false positive detections as it detected more realistic *Sargassum* aggregations. The *Sargassum* fractional coverage corresponding to the discarded false positive detections was estimated to be 50% for OLCI, and 14% of high FC for MSI.

Furthermore, the CNNs need fewer supplementary post- and pre-treatments than the ID method, and once the model is trained, the use of indexes is not required anymore.

Finally, the study also considered satellite scale characteristics. With MSI, our proposed CNN provided more detailed and distinct aggregations than with OLCI, and was able to detect *Sargassum* aggregations in coastal water with higher confidence thanks to the higher resolution of MSI. The combination of a regional-scale sensor (MSI) and a large-scale sensor (OLCI) may be relevant for the Antilles area, which contains a mix of islands and open sea.

Supplementary Materials: The following supporting information can be downloaded at: https://www.mdpi.com/article/10.3390/rs15041104/s1, Table S1: Sentinel-2 MSI images used for the CNN training between 2018 and 2022; Table S2: Sentinel-3 OLCI images of the Lesser Antilles used for the CNN training between 2017 and 2022; Table S3: Detail of our proposed network architecture for MSI images; Table S4: Detail of our proposed network architecture for OLCI images; Figure S1: Distribution of δMCI computed by the CNN.

Author Contributions: Conceptualization, M.L., A.B., L.C., J.D., A.S.-G. and C.C.; Data curation, J.D. and C.M.; Formal analysis, M.L., A.B., L.C., J.D. and C.C.; Funding acquisition, R.D. and P.Z.; Methodology, M.L., A.B., L.C., J.D., A.S.-G. and C.C.; Resources, M.L., A.B., L.C., J.D., L.S., A.M.; Software, A.B., L.C. and J.D.; Supervision, C.C.; Validation, M.L., A.B., L.C., J.D. and C.C.; Writing—original draft, M.L., A.B., L.C. and C.C.; Writing—review & editing, A.B., L.C., J.D., L.S., A.M., T.T., R.D., P.Z. and C.C. All authors have read and agreed to the published version of the manuscript.

Funding: This research was funded by the National Research Agency (ANR) and by the Territorial Authority of Martinique (CTM) as part of the collaborative FORESEA project (ANR grant number: ANR-19-SARG-0007-07 and CTM grant number: MQ0027405).

Data Availability Statement: Not applicable.

Acknowledgments: The authors would like to thank Malik Chami for his contribution of the dataset of comparison for OLCI and his relevant suggestions.

Conflicts of Interest: The authors declare no conflict of interest.

References

1. Wang, M.; Hu, C.; Barnes, B.B.; Mitchum, G.; Lapointe, B.; Montoya, J.P. The Great Atlantic Sargassum Belt. *Science* **2019**, *365*, 83–87. [CrossRef] [PubMed]
2. Schell, J.M.; Goodwin, D.S.; Siuda, A.N.S. Recent Sargassum Inundation Events in the Caribbean: Shipboard Observations Reveal Dominance of a Previously Rare Form. *Oceanography* **2015**, *28*, 8–11. [CrossRef]
3. Amaral-Zettler, L.A.; Dragone, N.B.; Schell, J.; Slikas, B.; Murphy, L.G.; Morrall, C.E.; Zettler, E.R. Comparative Mitochondrial and Chloroplast Genomics of a Genetically Distinct Form of Sargassum Contributing to Recent "Golden Tides" in the Western Atlantic. *Ecol. Evol.* **2017**, *7*, 516–525. [CrossRef] [PubMed]
4. Dibner, S.; Martin, L.; Thibaut, T.; Aurelle, D.; Blanfuné, A.; Whittaker, K.; Cooney, L.; Schell, J.M.; Goodwin, D.S.; Siuda, A.N.S. Consistent Genetic Divergence Observed among Pelagic Sargassum Morphotypes in the Western North Atlantic. *Mar. Ecol.* **2022**, *43*, e12691. [CrossRef]
5. Ody, A.; Thibaut, T.; Berline, L.; Changeux, T.; André, J.-M.; Chevalier, C.; Blanfuné, A.; Blanchot, J.; Ruitton, S.; Stiger-Pouvreau, V.; et al. From In Situ to Satellite Observations of Pelagic Sargassum Distribution and Aggregation in the Tropical North Atlantic Ocean. *PLoS ONE* **2019**, *14*, e0222584. [CrossRef]
6. Széchy, M.T.; Baeta-Neves, M.; Guedes, P.; Oliveira, E. Verification of Sargassum Natans (Linnaeus) Gaillon (Heterokontophyta: Phaeophyceae) from the Sargasso Sea off the Coast of Brazil, Western Atlantic Ocean. *Check List* **2012**, *8*, 638–641. [CrossRef]
7. Maréchal, J.-P.; Hellio, C.; Hu, C. A Simple, Fast, and Reliable Method to Predict Sargassum Washing Ashore in the Lesser Antilles. *Remote Sens. Appl. Soc. Environ.* **2017**, *5*, 54–63. [CrossRef]
8. Chávez, V.; Uribe-Martínez, A.; Cuevas, E.; Rodríguez-Martínez, R.E.; van Tussenbroek, B.I.; Francisco, V.; Estévez, M.; Celis, L.B.; Monroy-Velázquez, L.V.; Leal-Bautista, R.; et al. Massive Influx of *Pelagic Sargassum* Spp. on the Coasts of the Mexican Caribbean 2014–2020: Challenges and Opportunities. *Water* **2020**, *12*, 2908. [CrossRef]
9. Schling, M.; Guerrero Compeán, R.; Pazos, N.; Bailey, A.; Arkema, K.; Ruckelshaus, M. *The Economic Impact of Sargassum: Evidence from the Mexican Coast*; Inter-American Development Bank: Washington, DC, USA, 2022. [CrossRef]
10. van Tussenbroek, B.I.; Hernández Arana, H.A.; Rodríguez-Martínez, R.E.; Espinoza-Avalos, J.; Canizales-Flores, H.M.; González-Godoy, C.E.; Barba-Santos, M.G.; Vega-Zepeda, A.; Collado-Vides, L. Severe Impacts of Brown Tides Caused by Sargassum Spp. on near-Shore Caribbean Seagrass Communities. *Mar. Pollut. Bull.* **2017**, *122*, 272–281. [CrossRef]
11. Rodríguez-Martínez, R.E.; Medina-Valmaseda, A.E.; Blanchon, P.; Monroy-Velázquez, L.V.; Almazán-Becerril, A.; Delgado-Pech, B.; Vásquez-Yeomans, L.; Francisco, V.; García-Rivas, M.C. Faunal Mortality Associated with Massive Beaching and Decomposition of Pelagic Sargassum. *Mar. Pollut. Bull.* **2019**, *146*, 201–205. [CrossRef]
12. Maurer, A.S.; Stapleton, S.P.; Layman, C.A.; Burford Reiskind, M.O. The Atlantic Sargassum Invasion Impedes Beach Access for Nesting Sea Turtles. *Clim. Chang. Ecol.* **2021**, *2*, 100034. [CrossRef]
13. Resiere, D.; Mehdaoui, H.; Névière, R.; Mégarbane, B. Sargassum Invasion in the Caribbean: The Role of Medical and Scientific Cooperation. *Rev. Panam. Salud Pública* **2019**, *43*, e52. [CrossRef]
14. Resiere, D.; Mehdaoui, H.; Banydeen, R.; Florentin, J.; Kallel, H.; Nevière, R.; Mégarbane, B. Effets sanitaires de la décomposition des algues sargasses échouées sur les rivages des Antilles françaises. *Toxicol. Anal. Clin.* **2021**, *33*, 216–221. [CrossRef]
15. de Lanlay, D.B.; Monthieux, A.; Banydeen, R.; Jean-Laurent, M.; Resiere, D.; Drame, M.; Neviere, R. Risk of Preeclampsia among Women Living in Coastal Areas Impacted by Sargassum Strandings on the French Caribbean Island of Martinique. *Environ. Toxicol. Pharmacol.* **2022**, *94*, 103894. [CrossRef]
16. Gower, J.; King, S.; Borstad, G.; Brown, L. Detection of Intense Plankton Blooms Using the 709 Nm Band of the MERIS Imaging Spectrometer. *Int. J. Remote Sens.* **2005**, *26*, 2005–2012. [CrossRef]
17. Hu, C. A Novel Ocean Color Index to Detect Floating Algae in the Global Oceans. *Remote Sens. Environ.* **2009**, *113*, 2118–2129. [CrossRef]
18. Wang, M.; Hu, C. Mapping and Quantifying Sargassum Distribution and Coverage in the Central West Atlantic Using MODIS Observations. *Remote Sens. Environ.* **2016**, *183*, 350–367. [CrossRef]
19. Gower, J.; Hu, C.; Borstad, G.; King, S. Ocean Color Satellites Show Extensive Lines of Floating Sargassum in the Gulf of Mexico. *IEEE Trans. Geosci. Remote Sens.* **2006**, *44*, 3619–3625. [CrossRef]
20. Gower, J.; King, S. The Distribution of Pelagic Sargassum Observed with OLCI. *Int. J. Remote Sens.* **2020**, *41*, 5669–5679. [CrossRef]
21. Wang, M.; Hu, C. Automatic Extraction of Sargassum Features from Sentinel-2 MSI Images. *IEEE Trans. Geosci. Remote Sens.* **2021**, *59*, 2579–2597. [CrossRef]
22. Descloitres, J.; Minghelli, A.; Steinmetz, F.; Chevalier, C.; Chami, M.; Berline, L. Revisited Estimation of Moderate Resolution Sargassum Fractional Coverage Using Decametric Satellite Data (S2-MSI). *Remote Sens.* **2021**, *13*, 5106. [CrossRef]
23. Wang, M.; Hu, C. On the Continuity of Quantifying Floating Algae of the Central West Atlantic between MODIS and VIIRS. *Int. J. Remote Sens.* **2018**, *39*, 3852–3869. [CrossRef]

24. Podlejski, W.; Descloitres, J.; Chevalier, C.; Minghelli, A.; Lett, C.; Berline, L. Filtering out False Sargassum Detections Using Context Features. *Front. Mar. Sci.* **2022**, *9*, 1–15. [CrossRef]

25. Chen, C.; Qin, Q.; Zhang, N.; Li, J.; Chen, L.; Wang, J.; Qin, X.; Yang, X. Extraction of Bridges over Water from High-Resolution Optical Remote-Sensing Images Based on Mathematical Morphology. *Int. J. Remote Sens.* **2014**, *35*, 3664–3682. [CrossRef]

26. Kaur, B.; Garg, A. Mathematical Morphological Edge Detection for Remote Sensing Images. In Proceedings of the 2011 3rd International Conference on Electronics Computer Technology, Kanyakumari, India, 8–10 April 2011; Volume 5, pp. 324–327.

27. Siddiqi, M.H.; Ahmad, I.; Sulaiman, S.B. Weed Recognition Based on Erosion and Dilation Segmentation Algorithm. In Proceedings of the 2009 International Conference on Education Technology and Computer, Singapore, 17–20 April 2009; Springer: Berlin/Heidelberg, Germany, 2009; pp. 224–228. [CrossRef]

28. Soille, P.; Pesaresi, M. Advances in Mathematical Morphology Applied to Geoscience and Remote Sensing. *IEEE Trans. Geosci. Remote Sens.* **2002**, *40*, 2042–2055. [CrossRef]

29. Cao, H.; Han, L.; Li, L. A Deep Learning Method for Cyanobacterial Harmful Algae Blooms Prediction in Taihu Lake, China. *Harmful Algae* **2022**, *113*, 102189. [CrossRef]

30. Li, X.; Liu, B.; Zheng, G.; Ren, Y.; Zhang, S.; Liu, Y.; Gao, L.; Liu, Y.; Zhang, B.; Wang, F. Deep-Learning-Based Information Mining from Ocean Remote-Sensing Imagery. *Natl. Sci. Rev.* **2020**, *7*, 1584–1605. [CrossRef]

31. Ren, Y.; Li, X.; Yang, X.; Xu, H. Development of a Dual-Attention U-Net Model for Sea Ice and Open Water Classification on SAR Images. *IEEE Geosci. Remote Sens. Lett.* **2022**, *19*, 1–5. [CrossRef]

32. Vasavi, S.; Divya, C.; Sarma, A.S. Detection of Solitary Ocean Internal Waves from SAR Images by Using U-Net and KDV Solver Technique. *Glob. Transit. Proc.* **2021**, *2*, 145–151. [CrossRef]

33. Zheng, G.; Li, X.; Zhang, R.-H.; Liu, B. Purely Satellite Data–Driven Deep Learning Forecast of Complicated Tropical Instability Waves. *Sci. Adv.* **2020**, *6*, eaba1482. [CrossRef]

34. Gao, L.; Li, X.; Kong, F.; Yu, R.; Guo, Y.; Ren, Y. AlgaeNet: A Deep-Learning Framework to Detect Floating Green Algae from Optical and SAR Imagery. *IEEE J. Sel. Top. Appl. Earth Obs. Remote Sens.* **2022**, *15*, 2782–2796. [CrossRef]

35. Guo, Y.; Gao, L.; Li, X. Distribution Characteristics of Green Algae in Yellow Sea Using a Deep Learning Automatic Detection Procedure. In Proceedings of the 2021 IEEE International Geoscience and Remote Sensing Symposium IGARSS, Brussels, Belgium, 11–16 July 2021; pp. 3499–3501. [CrossRef]

36. Qiu, Z.; Li, Z.; Bilal, M.; Wang, S.; Sun, D.; Chen, Y. Automatic Method to Monitor Floating Macroalgae Blooms Based on Multilayer Perceptron: Case Study of Yellow Sea Using GOCI Images. *Opt. Express* **2018**, *26*, 26810–26829. [CrossRef] [PubMed]

37. Arellano-Verdejo, J.; Lazcano-Hernández, H.E. Collective View: Mapping Sargassum Distribution along Beaches. *PeerJ Comput. Sci.* **2021**, *7*, e528. [CrossRef] [PubMed]

38. Vasquez, J.I.; Uriarte-Arcia, A.V.; Taud, H.; García-Floriano, A.; Ventura-Molina, E. Coastal Sargassum Level Estimation from Smartphone Pictures. *Appl. Sci.* **2022**, *12*, 10012. [CrossRef]

39. Portillo, J.A.L.; Casasola, I.G.; Escalante-Ramírez, B.; Olveres, J.; Arriaga, J.; Appendini, C. Sargassum Detection and Path Estimation Using Neural Networks. In Proceedings of the Optics, Photonics and Digital Technologies for Imaging Applications VII, Strasbourg, France, 6–7 April 2022; SPIE: Strasbourg, France, 2022; Volume 12138, pp. 14–25. [CrossRef]

40. Cuevas, E.; Uribe-Martínez, A.; Liceaga-Correa, M.d.l.Á. A Satellite Remote-Sensing Multi-Index Approach to Discriminate Pelagic Sargassum in the Waters of the Yucatan Peninsula, Mexico. *Int. J. Remote Sens.* **2018**, *39*, 3608–3627. [CrossRef]

41. Shin, J.; Lee, J.-S.; Jang, L.-H.; Lim, J.; Khim, B.-K.; Jo, Y.-H. Sargassum Detection Using Machine Learning Models: A Case Study with the First 6 Months of GOCI-II Imagery. *Remote Sens.* **2021**, *13*, 4844. [CrossRef]

42. Chen, Y.; Wan, J.; Zhang, J.; Zhao, J.; Ye, F.; Wang, Z.; Liu, S. Automatic Extraction Method of Sargassum Based on Spectral-Texture Features of Remote Sensing Images. In Proceedings of the IGARSS 2019—2019 IEEE International Geoscience and Remote Sensing Symposium, Yokohama, Japan, 28 July–2 August 2019; pp. 3705–3707. [CrossRef]

43. Badrinarayanan, V.; Kendall, A.; Cipolla, R. SegNet: A Deep Convolutional Encoder-Decoder Architecture for Image Segmentation. *IEEE Trans. Pattern Anal. Mach. Intell.* **2017**, *39*, 2481–2495. [CrossRef]

44. Girshick, R.; Donahue, J.; Darrell, T.; Malik, J. Rich Feature Hierarchies for Accurate Object Detection and Semantic Segmentation. In Proceedings of the 2014 IEEE Conference on Computer Vision and Pattern Recognition, Colombus, OH, USA, 23–28 June 2014; pp. 580–587. [CrossRef]

45. Krizhevsky, A.; Sutskever, I.; Hinton, G.E. ImageNet Classification with Deep Convolutional Neural Networks. *Commun. ACM* **2017**, *60*, 84–90. [CrossRef]

46. LeCun, Y.; Bengio, Y.; Hinton, G. Deep Learning. *Nature* **2015**, *521*, 436–444. [CrossRef]

47. Ronneberger, O.; Fischer, P.; Brox, T. U-Net: Convolutional Networks for Biomedical Image Segmentation. In Proceedings of the Medical Image Computing and Computer-Assisted Intervention—MICCAI 2015, Munich, Germany, 5–9 October 2015; pp. 234–241. [CrossRef]

48. Arellano-Verdejo, J.; Lazcano-Hernandez, H.E.; Cabanillas-Terán, N. ERISNet: Deep Neural Network for Sargassum Detection along the Coastline of the Mexican Caribbean. *PeerJ* **2019**, *7*, e6842. [CrossRef]

49. Wang, M.; Hu, C. Satellite Remote Sensing of Pelagic Sargassum Macroalgae: The Power of High Resolution and Deep Learning. *Remote Sens. Environ.* **2021**, *264*, 112631. [CrossRef]

50. Gower, J.; Young, E.; King, S. Satellite Images Suggest a New Sargassum Source Region in 2011. *Remote Sens. Lett.* **2013**, *4*, 764–773. [CrossRef]

51. Open Access Hub. Available online: https://scihub.copernicus.eu/ (accessed on 2 December 2022).

52. Schamberger, L.; Minghelli, A.; Chami, M.; Steinmetz, F. Improvement of Atmospheric Correction of Satellite Sentinel-3/OLCI Data for Oceanic Waters in Presence of Sargassum. *Remote Sens.* **2022**, *14*, 386. [CrossRef]

53. Rouse, J.W.; Hass, R.H.; Deering, D.W.; Schell, J.A.; Harlan, J.C. *Monitoring the Vernal Advancement and Retrogradation (Green Wave Effect) of Natural Vegetation*; E75-10354; NASA/GSFC: Greenbelt, MD, USA, 1974; Type III Final Report.

54. Gamon, J.A.; Field, C.B.; Goulden, M.L.; Griffin, K.L.; Hartley, A.E.; Joel, G.; Penuelas, J.; Valentini, R. Relationships Between NDVI, Canopy Structure, and Photosynthesis in Three Californian Vegetation Types. *Ecol. Appl.* **1995**, *5*, 28–41. [CrossRef]

55. Pettorelli, N.; Vik, J.O.; Mysterud, A.; Gaillard, J.-M.; Tucker, C.J.; Stenseth, N.C. Using the Satellite-Derived NDVI to Assess Ecological Responses to Environmental Change. *Trends Ecol. Evol.* **2005**, *20*, 503–510. [CrossRef]

56. Wang, Q.; Adiku, S.; Tenhunen, J.; Granier, A. On the Relationship of NDVI with Leaf Area Index in a Deciduous Forest Site. *Remote Sens. Environ.* **2005**, *94*, 244–255. [CrossRef]

57. Son, Y.B.; Min, J.-E.; Ryu, J.-H. Detecting Massive Green Algae (Ulva Prolifera) Blooms in the Yellow Sea and East China Sea Using Geostationary Ocean Color Imager (GOCI) Data. *Ocean Sci. J.* **2012**, *47*, 359–375. [CrossRef]

58. Garcia, R.A.; Fearns, P.; Keesing, J.K.; Liu, D. Quantification of Floating Macroalgae Blooms Using the Scaled Algae Index. *J. Geophys. Res. Oceans* **2013**, *118*, 26–42. [CrossRef]

59. Schamberger, L.; Minghelli, A.; Chami, M. Quantification of Underwater Sargassum Aggregations Based on a Semi-Analytical Approach Applied to Sentinel-3/OLCI (Copernicus) Data in the Tropical Atlantic Ocean. *Remote Sens.* **2022**, *14*, 5230. [CrossRef]

60. Yan, K.; Li, J.; Zhao, H.; Wang, C.; Hong, D.; Du, Y.; Mu, Y.; Tian, B.; Xie, Y.; Yin, Z.; et al. Deep Learning-Based Automatic Extraction of Cyanobacterial Blooms from Sentinel-2 MSI Satellite Data. *Remote Sens.* **2022**, *14*, 4763. [CrossRef]

61. Simonyan, K.; Zisserman, A. Very Deep Convolutional Networks for Large-Scale Image Recognition. In Proceedings of the 3rd International Conference on Learning Representations (ICLR 2015), San Diego, CA, USA, 7–9 May 2015; pp. 1–14.

62. PyTorch. Available online: https://www.pytorch.org (accessed on 2 December 2022).

63. He, K.; Zhang, X.; Ren, S.; Sun, J. Deep Residual Learning for Image Recognition. In Proceedings of the 2016 IEEE Conference on Computer Vision and Pattern Recognition (CVPR), Las Vegas, NV, USA, 27–30 June 2016; pp. 770–778. [CrossRef]

64. Oktay, O.; Schlemper, J.; Folgoc, L.L.; Lee, M.; Heinrich, M.; Misawa, K.; Mori, K.; McDonagh, S.; Hammerla, N.Y.; Kainz, B.; et al. Attention U-Net: Learning Where to Look for the Pancreas. *arXiv* **2018**, arXiv:1804.03999. [CrossRef]

65. Gower, J.; King, S. Satellite Images Show the Movement of Floating Sargassum in the Gulf of Mexico and Atlantic Ocean. *Nat. Preced.* **2008**. [CrossRef]

66. Hu, C.; Feng, L.; Hardy, R.F.; Hochberg, E.J. Spectral and Spatial Requirements of Remote Measurements of Pelagic Sargassum Macroalgae. *Remote Sens. Environ.* **2015**, *167*, 229–246. [CrossRef]

remote sensing

Article

End-to-End Neural Interpolation of Satellite-Derived Sea Surface Suspended Sediment Concentrations

Jean-Marie Vient [1,2,*], Ronan Fablet [2], Frédéric Jourdin [3] and Christophe Delacourt [1]

[1] UBO, Technopôle Brest-Iroise, 29238 Brest, France
[2] IMT-Atlantique Bretagne-Pays de la Loire, Technopôle Brest-Iroise, 29238 Brest, France
[3] Service Hydrographique et Océanographique de la Marine (SHOM), 29603 Brest, France
* Correspondence: jean-marie.vient@univ-brest.fr

Abstract: The characterization of suspended sediment dynamics in the coastal ocean provides key information for both scientific studies and operational challenges regarding, among others, turbidity, water transparency and the development of micro-organisms using photosynthesis, which is critical to primary production. Due to the complex interplay between natural and anthropogenic forcings, the understanding and monitoring of the dynamics of suspended sediments remain highly challenging. Numerical models still lack the capabilities to account for the variability depicted by in situ and satellite-derived datasets. Through the ever increasing availability of both in situ and satellite-derived observation data, data-driven schemes have naturally become relevant approaches to complement model-driven ones. Our previous work has stressed this potential within an observing system simulation experiment. Here, we further explore their application to the interpolation of sea surface sediment concentration fields from real gappy satellite-derived observation datasets. We demonstrate that end-to-end deep learning schemes—namely 4DVarNet, which relies on variational data assimilation formulation—apply to the considered real dataset where the training phase cannot rely on gap-free references but only on the available gappy data. 4DVarNet significantly outperforms other data-driven schemes such as optimal interpolation and DINEOF with a relative gain greater than 20% in terms of RMSLE and improves the high spatial resolution of patterns in the reconstruction process. Interestingly, 4DVarNet also shows a better agreement between the interpolation performance assessed for an OSSE and for real data. This result emphasizes the relevance of OSSE settings for future development calibration phases before the applications to real datasets.

Keywords: interpolation; data-driven models; neural networks; variational data assimilation; missing data; suspended particulate matter; observing system experiment; Bay of Biscay

check for updates

Citation: Vient, J.-M.; Fablet, R.; Jourdin, F.; Delacourt, C. End-to-End Neural Interpolation of Satellite Sea Surface Suspended Sediment Concentrations. *Remote Sens.* **2022**, *14*, 4024. https://doi.org/10.3390/rs14164024

Academic Editors: Ana B. Ruescas, Veronica Nieves and Raphaëlle Sauzède

Received: 13 July 2022
Accepted: 15 August 2022
Published: 18 August 2022

Publisher's Note: MDPI stays neutral with regard to jurisdictional claims in published maps and institutional affiliations.

1. Introduction

Marine sediment fluxes result from a combination of natural and anthropogenic forcing factors [1,2]. The main source of sediment load comes from land, and the resuspension of sediments occurs under the effect of waves, tides and the oceanic general circulation, but also from fish trawling and maritime development, such as harbor sediment dredging and dumping, aggregate extraction, submarine cable installation, offshore wind farm exploitation, oil and gas activities, etc. [3]. Besides these latter anthropogenic stresses, additional ones are expected in the foreseeable future through climate change, involving sea-level and waves' rise, modifying the remobilization and transport in the coastal zone and the sediment inputs from the continent by the modification in the drainage basins hydraulic regime due to modified rainfall [4].

Tracking suspended particles in shelf seas is of interest for coastal management and marine ecosystem monitoring. Yet, the assessment of sediment fluxes, especially near the bottom of the ocean, is a key issue in the investigation of coastal morphological evolution, habitat changes and pollutant dispersion and behavior [5–7]. As the turbidity induced by fine sediment suspensions, especially near the surface of the ocean, impacts the primary production by narrowing the thickness of the euphotic zone [8], the quantification of the suspended sediment concentrations and fluxes, at the scale of the continental shelf, is also a critical aspect to fulfill the boundary conditions of their fine mesh-grid coastal hydro-dynamic models [9] used for impact studies. However the quantification of suspended sediment fluxes is generally a difficult task due to the complexity of the hydrodynamic and morpho-dynamic processes in play. In absolute terms, assessing the overall sediment dynamics requires understanding of transport processes of mineral particles in the water column as well as their behavior in the seabed, with resuspension capacities and consolidation within the sediment, oftentimes under the influence of biota, impacting flocculation processes in the water column and the biochemical behavior in the sediment [10].

In this context, deterministic (physics-based) numerical models are usually computationally intensive and inaccurate when assessing sediment fluxes from their continental source to the shelf edge [11,12]. Data-driven methods have emerged as appealing approaches to benefit the available datasets coming from observations and model simulations [13–16]. Recent advances especially bridge data assimilation formulation and machine learning paradigms [17,18]. These schemes are particularly relevant to addressing the irregular space–time sampling of satellite-derived sea surface dynamics. Following our previous study within an OSSE (Observing System Simulation Experiment) setting [19], we aim at evaluating whether such learning-based schemes apply to real satellite-derived datasets. Then, in this article, in the same way we designed real data experiments. They are typically called OSE (Observing System Experiment). Our contribution is two-fold: (i) we develop a novel application of 4DVarNet schemes [18] for satellite-derived sea surface suspended sediment concentrations (SSSC); (ii) we propose an evaluation framework based on real MODIS satellite image series to benchmark data-driven and learning-based schemes for the reconstruction of satellite-derived SSSC fields. We further assess how OSSE benchmarking experiments based on hydrosedimentary numerical simulations [20] inform performance metrics for real datasets in the OSE experiments.

The remainder is organized as follows. Section 2 details the considered datasets. Section 3 details the processes defined to interpolate observation data, which are the Optimal Interpolation and the new 4DVarNet scheme. Section 4 shows the global and specific performance for each method. And finally Section 5 compares OSSE and OSE configurations and characterizes the limits of the 4DVarNet interpolator.

2. Data

The area of study, presented in the first subsection, is located in the Bay of Biscay. The main geophysical parameter of study is the sea surface suspended sediment concentration (SSSC), which relates to the sediment dynamics. Two sets of SSSC data were used. We present the datasets considered in this study, namely numerical simulation data (Section 2.2) and real satellite-derived MODIS data (Section 2.3). These two sets of data have been used to perform two different kind of experiments carried out in parallel: the simulated data are dedicated to OSSE (Observing System Simulation Experiment) while the real data are dedicated to OSE (Observing System Experiment). Later, Sections 4 and 5 will compare the results obtained by these two different kinds of experiments.

2.1. Area of Interest

The study area encompasses a major part of the northern region of the Bay of Biscay (BoB), located on the west coast of France (North-East Atlantic). In this area the bathymetry extends from the shallow waters of the coast to the great depths of the abyssal plain. The continental shelf is wide (Figure 1). The shelf break, dotted with canyons, crosses the area like a transverse line from its north-west corner to the south-east one. The bottom sedimentology of the BoB can be divided into three main seafloor patterns: a large muddy area located in the middle of the shelf and referred to as the "Grande Vasière" (e.g., [21], coastal areas characterized by rocky and sandy seabeds, and the shelf break with a predominance of rocks. The water column experiences a variety of physical forcings and processes: tides, internal waves (especially from the shelf break), trapped waves, density gradients and seasonal winds driven circulations (with winter storms notably), mixing and stratification, eddies, fronts, filaments, upwelling/downwelling and discharges from rivers [22]. Concerning the latter, the Gironde and Loire rivers are the main sources of water and sediment suspended loads [23]. Their estuaries are located at the northern latitudes of 45.6° and 47.2° respectively. The particle dynamics in the surface layers of this oceanic area is driven by the hydrodynamics superimposed with biologic cycles, which notably are well characterised by phytoplankton blooms appearing along the Armorican shelf break, especially in spring.

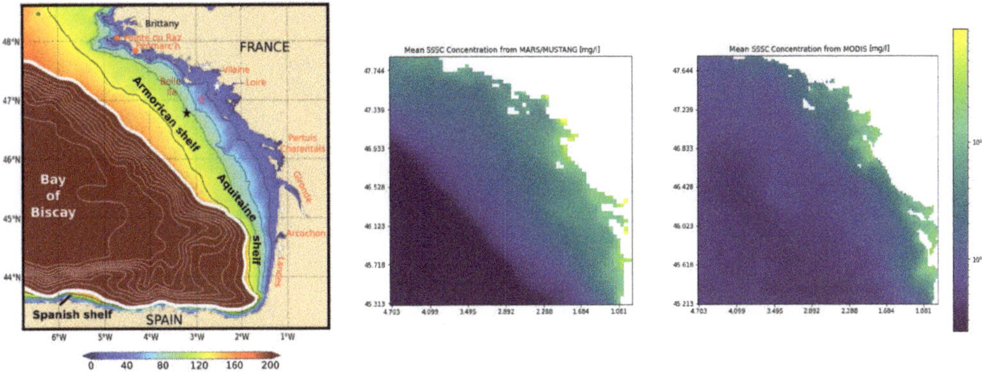

Figure 1. (**Left**) Bathymetry of the Bay of Biscay. Black lines represent isobaths 40, 70, 100, and 130 m. The thick white line (corresponding to the 180 m isobath) approximately delimits the shelf edge. Stars are validation points in [21] (**Middle**) Mean spatial distribution of SSSC (in mg/L) from the MARS-MUSTANG hydrosedimentary model. (**Right**) Mean spatial distribution of SSSC (in mg/L) from MODIS observations.

2.2. MARS Model Simulations (for OSSE)

The simulations dataset comes from a hydro-sedimentary model called MARS-MUSTANG [21]. MARS is the hydrodynamic module. MUSTANG is the sedimentary module. MARS computes the general circulation according to meteo forcings from meteo models ARPEGE and WaveWatchIII (atmospheric pressure, wind and waves), and density gradients bring by boundary conditions upon climatological dataset. MUSTANG computes sediment resuspension and settling through erosion and deposition laws followed with advection-dispersion and settling velocity equations, in connection with a dynamic seabed sediment layer model, initiated with a nature of seabed chart based on in situ measurements [21]. The main output of this couple model is values of suspended sediment concentrations in the whole water column, from its surface to depths. Since MARS-MUSTANG is designed to model the fate of terrigenous sediments only, the suspended sediment obtained is of mineral (or inorganic) origin, and should appropriately be termed Suspended Particulate Inorganic Matter (SPIM). In particular, SPIM does not include detrital particles because MARS-MUSTANG is not a biological model of primary production with a detrital compartment. Such information on the nature of particles modelled here is necessary when MARS-MUSTANG results are compared with satellite images that "mixes up" more components: especially SPIM and detritus (see following Section 2.3).

Only surface values of suspended sediment concentrations obtained with MARS-MUSTANG will be exploited. In Section 4 (Results), these concentrations will be called using the general term SSSC (surface suspended sediment concentration). For information, these SSSC values were already exploited by our team in order to obtain the present OSSE results in a previous article [19]. This latter article also provides condensed information on the validation of the MARS-MUSTANG simulations with the satellite data, which shows consistent behavior. In particular, the corresponding configuration of the model takes into account the discharges from the two main rivers (Gironde and Loire) and also the Vilaine river (the estuary of which is located slightly north of the Loire one). Figure 1 Middle displays the mean SSSC obtained with the model at the ocean surface and well shows that higher SSSC occurs preferably in the vicinity of the coast. This is due to wave exposure and tidal range in combination with the higher terrigenous sediment loads. Above the abyssal plain, suspended sediments are nearly absent. The threshold of 0.1 mg/L is well correlated with the isobath of 180 m corresponding to the shelf edge. The Figure exhibits area where SSSC values are greater than 10 mg/L near estuaries of the main rivers (Gironde, Loire, Vilaine).

The model is configured with a spatial mesh grid resolution of 2.5 km over a wider area than our area of interest, extending from latitude 41°N to 55°N, and longitude 18°W to 9°30′E [21]. Outputs from MARS-MUSTANG were then extracted in our area of interest, for our present OSSE experiments, and lead to images having a size corresponding to a spatial grid of 128 × 128. In terms of time data frame, the overall sea surface field values extracted from the MARS simulations represent a time series of 1430 daily images spanning from 1 January 2007 to 8 December 2010. The MARS-MUSTANG model and simulations will simply be referred to as MARS hereafter for short.

2.3. MODIS Real Satellite Data (for OSE)

Our satellite-derived dataset is based on the MODIS sensor images acquired on board both Aqua and Terra satellites. The MODIS sensor is part of the 1991 NASA-initiated Earth observation system. It aims at monitoring, among others, the ocean dynamics. Here we exploit the Level-2 geophysical variable called Non-Algal Particles (NAP) that is processed using Francis Gohin et al. bio-optical algorithm [24] applied to the MODIS normalized remote-sensing reflectances. All clouds and cloud shadows in raw satellite images were flagged with a low detection threshold so as to remove all questionable signals. Also, atmospheric over-corrections are taken into account using the reflectance at 412 nm [24]. Their algorithm was specifically calibrated for the Bay of Biscay using dedicated in situ measurements from 20 field cruises that took place over the shelf, and which represent a total amount of about 1000 in situ data points (see Table 1 and Figure 2 of [24]). The NAP concentration (in mg/L) is computed as the difference between the total suspended matter concentration (deduced from the remote-sensing radiances at 550 and 670 nm) and the phytoplankton biomass (derived from their Chlorophyll-*a* specific algorithm). All products (NAP and Chlorophyll-*a*) were validated according to additional in situ measurements [25,26]. In particular products accuracy have been extensively validated against coastal in situ measurements from 15 stations located along the French Atlantic coast and 3 stations along the Mediterranean coast, all stations recording the turbidity every 15 days between 1 January 2003 and 31 December 2009. The results show a confidence of 95% (see Figure 13 of [27] in French language) between yearly mean and percentile 90 of the turbidity (in NTU) recorded at all stations and the total suspended matter measured by the MODIS sensor (converted in NTU according to [28]). Part of those results can also be found in English language in [29], including VIIRS (Visible Infrared Imaging Radiometer Suite) and OLCI (Ocean Land color Instrument) satellite sensors, along with MODIS.

For a fine comparison between satellite images and outputs from the MARS model, one should know that the NAP particles observed by the satellite comprise not only mineral particles (labeled as SPIM, see the previous Section 2.2) but also detrital particles, yielding to (e.g., [30]):

$$\text{NAP} = \text{SPIM} + det, \tag{1}$$

where *det* represents the amount of detrital particles. For instance, Figure 1 Right displays the mean SSSC obtained with the satellite (based on the NAP algorithm) at the ocean surface. In particular, it shows that, contrary to the MARS model, there is a weak but significant SSSC mean signal (of the order of 1 mg/L) above the abyssal plain. This signal obviously corresponds to the detrital particles linked to the open sea primary production. Nonetheless, for simplification purposes, in Section 4 (Results) the suspended matter will be called using the general term SSSC (Surface Suspended Sediment Concentration).

In the OSE experiments, our dataset is comprised of daily MODIS images spanning from 1 January 2003 to 31 December 2009. During that period, the mean cloud cover amounted to about 75% of the whole oceanic surface of the imaged area. Each MODIS image has a 1 km spatial resolution, which leads to a 256 × 256 grid for our case-study region. The area extends from latitude 45°17′N to 47°50′N, and longitude 4°55′W to 1°5′W. In terms of validation, as mentioned in Section 2.1, this area covers the main part of the Bay of Biscay (BoB) which experiences various physical and biological forcings. This bay is a well-known fine testing ground in terms of spatial and temporal variability of the turbidity (e.g., [31]).

3. Methods

This section details the proposed space–time interpolation of satellite-derived SSSC fields based on 4DVarNet scheme [32] in Section 3.1, along with the considered evaluation framework, in Section 3.2, performance metrics in Section 3.3 and benchmarked approaches in Section 3.4.

3.1. 4DVarNet Scheme

Deep learning schemes have rapidly become the state-of-the-art approaches for a wide range of pattern recognition and image processing applications, including in geoscience [33]. This also includes neural network approaches dedicated to interpolation issues. Recent studies [17,19,32,34] have stressed the relevance of end-to-end deep learning architectures to address space–time interpolation issues with large missing data rates. Especially, 4DVarNet schemes, which rely on variational data assimilation formulation, have been shown to significantly outperform zero-filling learning-based strategies for interpolation problems [35]. Applications to sea surface height mapping from satellite altimetry [32,34] further support their relevance over other data-driven approaches to better retrieve fine-scale patterns. This study presents an application of 4DVarNet schemes to SSC interpolation. We provide below a short introduction to 4DVarNet schemes. We refer the reader to [18,35] for a detailed presentation.

4DVarNet framework relies on the formulation of the interpolation problem as a variational minimization issue:

$$\hat{x} = \arg\min_{x} \|y - x\|_{\Omega}^2 + \lambda \|x - \Phi(x)\|^2. \tag{2}$$

Ω refers to the space–time subdomain where observation y are sampled. Let us point out that we consider a matrix formulation where x and y refer to the space–time process, represented by 2D + t tensors. Operator Φ states the space–time prior to state x. Φ may refer to the flow operator when considering a dynamical ODE or PDE prior. Φ can also derive from a covariance-based prior as in the optimal interpolation framework. Here, following [32], we consider a state-of-the-art neural architecture, namely a UNet [36], such that Φ can be regarded as a projection operator. λ states the relative importance of the observation term of the prior in the minimization problem.

Given minimization problem (2), the 4DVarNet framework implements a trainable iterative gradient-based solver with a predefined number of iterations. As sketched by Figure 2, it delivers an end-to-end architecture which exploits as inputs gappy observation data and as outputs a gap-free state. The trainable solver combines the evaluation of the gradient of variational cost (2) using automatic differentiation tools embedded in deep learning framework with a recurrent network, namely a convolutional LSTM. A more detailed description of 4DVarNet schemes including experiments with different parameterizations of operator Φ and of the trainable solver can be found in [32].

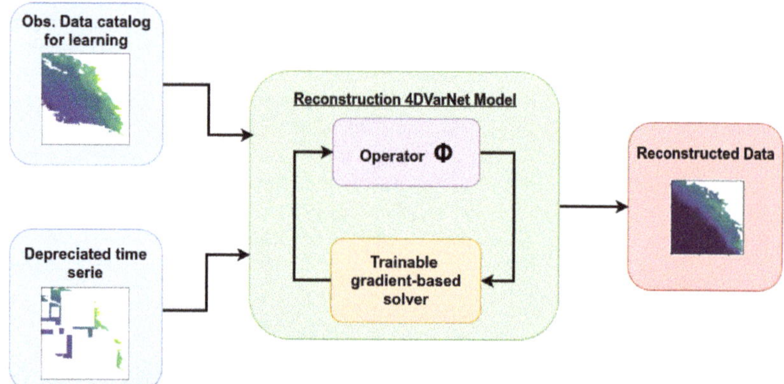

Figure 2. Workflow of the proposed framework for the reconstruction NAP field with the 4DVarNet. A given observation dataset catalogue is provide for the learning phase of the model. The interpolator Φ is trained and optimised by the solver. Then the model is able to reconstruct the irregularly sampled time series input into the reconstructed data.

Regarding learning issues, we consider here training losses evaluated on gappy data, as no gap-free reference data are available. For the trainable solver, we consider a convolution LSTM-based solver with a 35-dimensional hidden state.

3.2. Training and Evaluation Framework

For training and evaluation purposes, the whole set of data is divided in two parts. All the available data except those of the year 2011 are dedicated to the training while the year 2011 is reserved for the evaluation.

In OSSEs, the "nature run" (or model simulation) provides the reference gap-free field used as ground truth to asses the performance of the interpolation. This helps in evaluating the performance of the associated interpolation methods. For OSEs with real satellite data, no such gap-free reference field is available. We exploit a random sampling strategy as follows. For the considered dataset, we randomly sample a binary mask applied to the real satellite observation patterns. As such, we withhold some observed data from the input data provided to the interpolation methods and use them as reference data to assess reconstruction performance metrics. We may emphasize that this dataset is not an actual groundtruthed dataset as real satellite data are noisy. Available in situ datasets are too scarce to provide a relevant alternative. In the reported experiments, we subsample 50% of available satellite-derived observations for dates at which at least 500 observation points are available (i.e., 3% of the total pixels located above the ocean). We exploit two random strategies: the "pixel-wise" strategy randomly samples 50% of the observed pixels; the "patch-wise" strategy randomly samples $H \times W$ patches, width W and height H being randomly sampled according to a uniform distribution between 0 and 100. We report in (Figure 3) examples of randomly sampled patterns. Contrary to the "pixel-wise" strategy, the patch-wise one better matches the expected independence between the training and test datasets.

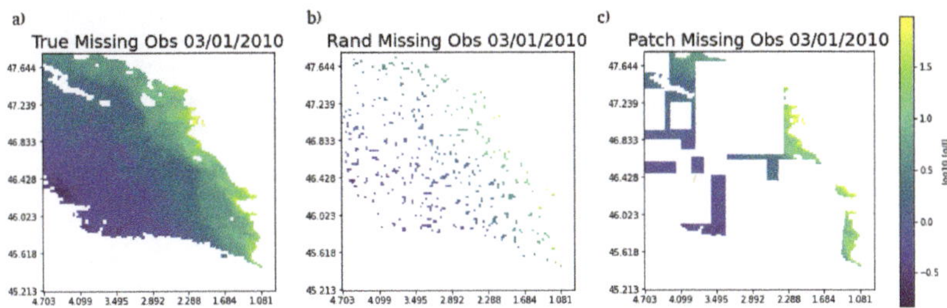

Figure 3. Different sampling strategy for validation by using missing observations: (**a**) True observation from MODIS L2 image dataset (**b**) Random points sampling strategy for observation sampling (**c**) Patch sampling strategy for observation sampling.

3.3. Performance Metrics

In terms of performance metrics, we exploit the explained variance (R-score) and the global root mean square error (RMSE) applied to the log_{10} of concentration values. The latter leads in fact to the RMSLE (root mean square log error) of concentrations, which can be expressed as follows:

$$RMSLE = \sqrt{\frac{1}{n}\sum_{i=1}^{n}\left(log_{10}(C_i^{Pred}) - log_{10}(C_i^{Obs})\right)^2},$$ (3)

where C_i^{Pred} are predicted values of SSSC concentrations and C_i^{Obs} are the observed values, i stands for the index of evaluated data, n refers to the amount of available observations points. RMSLE and R-Score in all the following Tables are evaluated with a total amount of points of the order of 10^6. We also evaluate these metrics for the gradient of the log_{10} of SSSC fields. This validation of results with the standard RMSE on log_{10} values (i.e., RMSLE) has been chosen for two main reasons:

- First, the statistical distribution of particle concentrations typically follows a lognormal probability distribution [37] so that log_{10} values follow a Gaussian distribution. Then, providing bias is negligible (all biases in all experiments were found equal or inferior to 0.01 in absolute values), the RMSE is comparable to a standard deviation and then completely characterizes the statistical distribution;
- Second, the evaluation on log_{10} of concentrations emphasizes the validation of low concentrations, which are important in the determination of water transparency, which is a main goal in our studies.

3.4. Reference Methods for Comparison

For benchmarking purposes, we consider two state-of-the-art approaches, an optimal interpolation [38,39] and DINEOF scheme [40]. The Optimal Interpolation (OI), also referred to as kriging, is a method widely applied in ocean remote sensing and geoscience. Numerous operational satellite-derived products in earth science rely on OI. We refer the reader to [41] for a detailed review. In our experiments, we implement an OI with a Gaussian covariance model empirically tuned through cross-validation experiments. DINEOF (Data Interpolating Empirical Orthogonal Functions) is an EOF-based technique for the reconstruction of gap-free fields from irregularly-sampled observations. It has been successfully applied to satellite-derived sea surface products [42], including sea surface turbidity [19,43]. DINEOF iterates a projection–reconstruction step using the EOF basis, while observed variables are kept unchanged after each iteration. Here, we select the first 56 EOF modes to account for $\tilde{9}7\%$ of the Variance of the considered datasets and apply a 10-iteration DINEOF. We may point out that the proposed 4DVarNet framework can be regarded as a generalization of DINEOF with a state-dependent covariance model and a gradient-based solver instead of the fixed-point solver implemented by DINEOF [32].

4. Results

We report below the synthesis of the considered numerical experiments. First, a global analysis of all benchmarking experiments is reported in Section 4.1. From these results, Section 4.2 focuses on the comparison between OSSE and OSE performance metrics and Section 4.3 on the performance of our specifically proposed 4DVarNet method.

4.1. Global Performance

Tables 1 and 2 summarize the performance metrics evaluated with all configurations and interpolation methods. The 4DvarNet method clearly outperformed the other methods tested here for all scores and experiments. For instance, in OSE, when real satellite data are used, 4DvarNet improves R-scores by about 30% w.r.t. OI and 23% w.r.t. DINEOF (slightly depending on the chosen subsampling strategy). We effectively note that reported performance metrics are consistent for the two random sampling strategies used to compute these metrics for the real satellite-derived datasets. The slightly better performance observed with the pixel-wise strategy relates to a lower independence between the input data and the evaluation dataset. A greater stability between the metrics computed according to these two strategies then indicates better generalization properties as exhibited for the 4DVarNet scheme. In OSSE, when simulated data are used, the improvement in R-score with 4DVarNet appears much lower (about 6% w.r.t. OI and 5% w.r.t. DINEOF) but its final R-score is much higher (nearly 97% for OSSE instead of nearly 90% for OSEs). Globally speaking, it appears that R-scores related to OSSE are pretty different to those related to OSEs. This simply reflects the different content in terms of data and data errors between simulated and real satellite images (see the discussion about this subject in Section 5.1. Finally, it is interesting to note the really poor performance of the standard OI method when applied to real data. Indeed, we can see that the correlation between interpolated and real satellite values only amounts to about 60%. Concerning the RMSLE values, Table 2 clearly shows consistent RMSLE values with regard to the R-scores. In particular, OI appears to have low accuracy when applied to real data (around 0.32) and 4DVarNet proved to have the highest accuracy with either simulated or real data (around 0.16 with real data and 0.10 with simulated ones).

Table 1. R-score performance in % for the considered methods and validation configurations. OSE refers to the real data (MODIS) interpolating process. The sub-sampling strategy is described in Section 3.2. OSSE dataset refers to the previous work [19] based on the MARS results.

Experiment	Dataset	Sub-Sampling	OI	DINEOF	4DVarNet
OSE	MODIS	Random	60.5	76.4	**89.5**
OSE	MODIS	Patch	56.5	73.8	**87.3**
OSSE	MARS	-	90.4	91.3	**96.6**

Table 2. RMSLE performance in log_{10}[mg/L] for the considered methods and validation configurations. OSE refers to the real data (MODIS) interpolating process. The sub-sampling strategy is described in Section 3.2. OSSE dataset refers to the previous work [19] based on the MARS results.

Experiment	Dataset	Sub-Sampling	OI	DINEOF	4DVarNet
OSE	MODIS	Random	0.304	0.237	**0.156**
OSE	MODIS	Patch	0.346	0.253	**0.168**
OSSE	MARS	-	0.176	0.167	**0.104**

4.2. OSSE versus OSE Comparison

Table 3 allows a further analysis on how performance metrics for simulation datasets (OSSE) inform the interpolation performance for real satellite-derived datasets (OSE). The minus signs in front of all values show that, when applied to real satellite data, the accuracy of all tested interpolation methods worsens. Of all interpolation methods,

OI loses accuracy the most. This could be due to a misrepresentation of the satellite data noise using this method (see Section 5.1). Concerning the other methods, we point out that OSEs involve an additional complexity at two levels for the training procedure: the reference data are noisy, the reference dataset is gappy. This explains why we also report a lower performance for these methods for OSE settings compared with the OSSE baseline. However, both DINEOF and 4DVarNet lose less accuracy (around 50%) than OI (between 73% and 97%). Thus, OSSE performance metrics provide a sensible assessment of the performance for the real satellite-derived dataset. This is less true for OI, the performance of which is degraded by either 73% or 97% depending on the applied subsampling strategy (Random or Patch). The latter likely relates to the spatial correlation length of the considered covariance model, such that the interpolation capability degrades at a distance greater than the correlation length.

Table 3. Evolution of accuracy, from OSSE to OSE, in the form of a performance rate according to the formula $1 - \text{RMSLE(OSE)}/\text{RMSLE(OSSE)}$ expressed in percentage, using the RMSLE reported in Table 2.

Sub-Sampling	OI	DINEOF	4DVarNet
Random	−73%	−42%	−50%
Patch	−97%	−51%	−62%

4.3. 4DVarNet Performance

We further analyze the clear improvement reported for 4DVarNet. Table 4 reports the relative performance gains with regard to OI and DINEOF. When dealing with real data, it shows a great improvement of around 50% over OI and a little less, 34%, over DINEOF. These two values quantifying the improvement also appear to be almost insensitive to the subsampling strategy used (Random or Patch). The OSSE is able to quantify a similar amount of improvement over the two methods (OI and DINEOF) with a value of around 40%, but does not see much difference between OI and DINEOF (41% and 38% respectively). This discrepancy between OSSE and OSEs could be due to a different representation of the satellite data noise using OI and DINEOF methods (see Section 5.1).

Table 4. Evolution of accuracy, from OI or DINEOF to 4DVarNet, in the form of a performance rate according to the formula $1 - \text{RMSLE(4DVarNet)}/\text{RMSLE}(\cdot)$ expressed in percentage, when using the RMSLE reported in Table 2.

Experiment	Dataset	Sub-Sampling	OI	DINEOF
OSE	MODIS	Random	49%	34%
OSE	MODIS	Patch	51%	34%
OSSE	MARS	-	41%	38%

The evaluation of the interpolation metrics for the gradient of the SSSC fields in Table 5 supports the hypothesis that the improvement obtained with 4DVarNet relates to a better reconstruction of fine-scale patterns. Previous work with similar 4DVar based architecture shows a significant improvement of a high resolution spatial pattern [44]. Surprisingly, the metrics are much better for the OSE. We interpret this aspect as a consequence of the lower spatial variability observed in numerical simulations compared with satellite-derived data, as supported by Figure 4. The mean gradient norm of 4DVarNet nicely recovers the main front structures of the true field compared with the other approaches. More specifically, these gradient fields depict a clearly visible contour offshore. This contour broadly follows the 50 m isobath, which borders the "Grande Vasière". The OI clearly overestimates the spatial gradient and does not succeed in capturing the finer scale. Though not as bad, DINEOF (based on EOF decomposition) may be limited by the explained variance rate at 97% of the selected EOF components. Besides, for real satellite-derived data, the mean gradient field involves local artifacts. By contrast, 4DVarNet retrieves mean gradient fields which are close to the reference in coastal areas without an overestimation pattern. We can also note a spatial smoothing, which may partially relate to the observation noise of real satellite-derived measurements.

Table 5. Gradient norm reconstruction performance R-score evaluation for different methods and validation configurations in %. OSSE dataset refers to the previous work ([19]) based on the MARS/MUSTANG results in an OSSE application. MODIS dataset refers to the real data interpolating process, with learning based only on observations. The sub-sampling strategy is described in Section 3.2.

Experiment	Dataset	Sub-Sampling	OI	DINEOF	4DVarNet
OSE	MODIS	Random	58.3	72.5	**88.9**
OSE	MODIS	Patch	56.6	67.4	**91.2**
OSSE	MARS	-	16.0	40.6	**63.7**

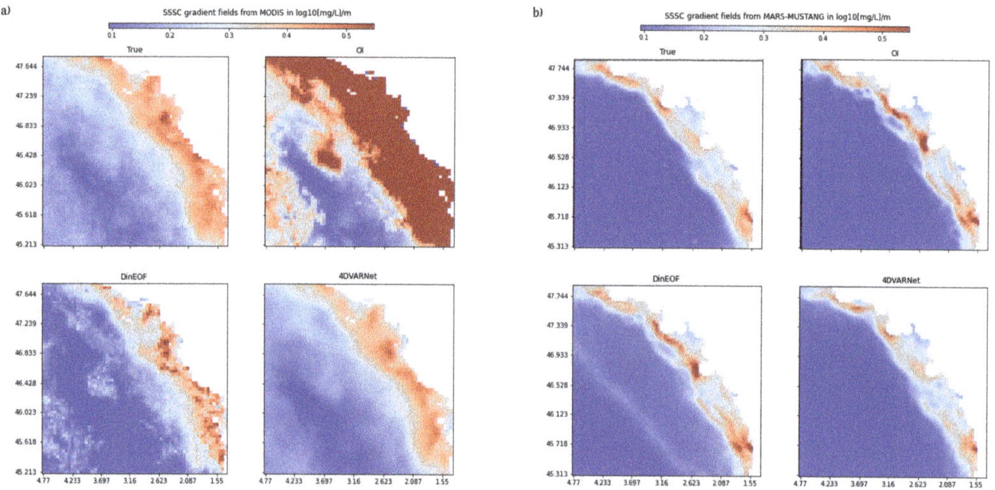

Figure 4. Reconstruction norm of SSSC gradient fields for the OSE (MODIS) and OSSE (MARS) application for the whole validation period: (**a**) OSE configuration (**b**) OSSE configuration.

5. Discussion

This study presents and evaluates a novel learning-based interpolation approach referred to as 4DVarNet for the space–time interpolation of satellite-derived sea surface suspended sediment concentrations. Numerical experiments on real and synthetic dataset support the relevance of this scheme compared with state-of-the-art approaches. We further discuss below our main contribution, namely the extent to which OSSEs can inform performance metrics for real datasets (Section 5.1), the relevance of neural network schemes for operational applications (Section 5.2) and the ability to retrieve fine-scale patterns (Section 5.3).

5.1. From OSSE to OSE

A first aim of this study was to compare the performance of interpolation methods when applied to simulated data (through OSSE) and real satellite data (through OSE). At first, the comparison shows that all the methods lost accuracy by a significant amount (more than 40%) when applied to real data. The observation noise in real satellite data may be the driving factor for this pattern. Indeed, in our OSSE, simulated data involve a simple (white) Gaussian noise evenly distributed throughout the whole dataset. By contrast, errors in real satellite data are expected to be much more complex. OI and DINEOF methods may be more impacted as they explicitly or implicitly hypothesize Gaussian noise models. In this retrospect, numerous studies have stressed the ability of neural network approaches to address denoising problems with non-Gaussian noise patterns [33,45].

Beyond the observation noise, other effects may support the difference in performance between OSSE and OSE. For instance, due to the random subsampling strategy, OSSEs involve a slightly lower missing data rate and the reference training data is noise-free. We may also note that we consider a 16-year training time series for OSEs compared with a 4-year one for OSSEs such that the pixel-wise number of training data is the same between OSSEs and OSEs. We may, however, expect a greater intrinsic variability over 16 years which may in turn result in a more complex interpolation problem. We could account for these different aspects in the design of OSSE under the hypothesis that we are provided with longer numerical simulations.

Nevertheless, the numerical simulations used in the OSSEs cannot reveal all the complexity of real satellite data. The hydrosedimentary model does not take into account all the processes we observe in a satellite image. This is of course a major and typically well known drawback of OSSE (e.g., [46]). Here, for instance, a typical discrepancy we can see between the model and a satellite image relates to the biogenic detritus, which follows the dynamics of the oceanic primary production, and which is obviously not modelled by the MARS model (see Equation (1) and the discussion about the difference between SPIM and NAP in Sections 2.2 and 2.3). Detrital particles can be well observed by satellite remote sensing of ocean color, especially beyond the shelf break (i.e., above the abyssal plain) where the MARS model always shows concentrations (off the shelf) close to zero. Coccolith-derived turbidity is especially amongst the most intense signals detected by a satellite in these areas [47]. Hence, in an OSE, interpolation methods have to address not only the dynamics of purely mineral SPIM particles but also dynamical patterns of detrital particles driven by the primary production.

Finally, we must not forget that the resulting errors in the OSE are biased in the sense that the results are compared with already noisy satellite data, contrary to the OSSE where results are compared with a "true ocean" (the modelled ocean) assumed to be a strict exact solution. Therefore, OSE validation errors are significantly larger than those of the OSSE. However, given these different aspects, the worsening of the interpolation performance of 4DVarNet from the OSSE to the OSE remains fairly limited. This supports the relevance of OSSEs as an initial testbed for the development and evaluation of interpolation algorithms, especially to rank new schemes with respect to previously benchmarked ones within a fully-controlled environment.

5.2. Comparison of Interpolation Methods

Our numerical experiments clearly stress that state-of-the-art data-driven interpolation methods can significantly improve the retrieval of operational gap-free satellite-derived products, which are often based on OI schemes. Both 4DVarNet and DINEOF can account for more complex covariance than the one used by the OI, including anisotropic ones (see in particular [48], for they developed an anisotropic OI method for that case). We believe, however, that a key issue of OI for our case study is its poorer ability to deal with the noise patterns of real satellite-derived observations.

DINEOF performs much better than OI, particularly with real satellite data. This is what our present article demonstrates with the results obtained with the OSE experiments (using real data) compared with what was expected (similar performance between OI and DINEOF) after the OSSE experiments (using simulated data). Furthermore, it has been well demonstrated [17] that DINEOF is well suited for complex areas comprising at the same time coastal and open sea domains. Given that DINEOF is quite simple to implement and does not require strong expertise, this method should definitely be considered as a baseline scheme for routine operational ocean color products similar to those studied in this article. For instance, at a higher spatial resolution, it has recently been applied by the HIGHROC consortium to deliver Sentinel 2 ocean color L4 products corresponding to the CMEMS product OCEANcolor_IBI_BGC_HR_L4_NRT_009_210 (https://doi.org/10.48670/moi-00 108, accessed on 12 July 2022) for the IBI area.

4DVarNet further stresses the greater potential of physics-informed learning schemes, which combine some physics-aware representation (here, a variational data assimilation formulation) with the computational efficiency of deep learning. Our results are in line with recent studies dedicated to other satellite-derived products, such as ocean color [49] and satellite altimetry [35]. Compared with [49], 4DVarNet involves an additional modeling flexibility through the learning of a trainable gradient-solver for the variational cost. This also opens the floor to interpolation schemes using multi-source input data through additional trainable observation operators [18]. As such, this work will likely serve as a baseline for future work aiming at further improving the space–time reconstruction of SSSC dynamics.

5.3. Retrieval of Fine-Scale Turbidity Patterns from Satellite Data

Given the irregular sampling of satellite-derived observations, the retrieval of fine-scale patterns is a critical issue for fulfilling operational needs such as the identification of specific areas where waters are highly transparent, which threatens Navy submarines over possible airborne visual detection. Our experiments stress significant differences in the ability of data-driven approaches to retrieve fine-scale patterns. By construction, OI schemes with Gaussian covariance models cannot reconstruct scales smaller than the a priori correlation distance. DINEOF schemes also relate to covariance-based models, but result in more complex covariance models learnt from data. This may actually improve the reconstruction of space–time dynamics. 4DVarNet schemes may be regarded as moving a step further with non-linear quadratic priors through operator Φ in (2), rather than linear-quadratic ones. We regard the combination of such a prior and of a trainable solver as the key features which support the improvement reported for the proposed 4DVarNet schemes.

In our experiments, we exploit RMSE metrics computed for the norm of the gradient of SSSC fields to assess the retrieval of the fine-scale patters. We may note that OSSE scores in Table 5 are significantly lower than the OSE ones, which may be surprising. Numerical simulations involve lower gradient values, whereas real satellite data depict much sharper spatial gradients as illustrated in Figure 4. Given the spatial grid resolution of the MARS model (2.5 km), we expect numerical simulations to resolve spatial scales from 20 km. By contrast, the spatial resolution of MODIS satellite observations is close to 1 km. Besides, as mentioned in Section 5.1, numerical simulations do not include all the processes in play in real satellite observations such as the turbulent behavior in detrital NAP processes. Overall, this results in lower mean gradient norm values for the OSSE, which in turn leads to a lower R-score as this score is normalized by the mean gradient norm. Despite these differences, OSSE and OSE metrics share the same ranking of the benchmarked methods for gradient-related scores. Future developments of hydrosedimentary simulations may improve the ability to reproduce the actual variability of SSSC fields.

Contrary to OI and DINEOF approaches, the proposed 4DVarNet scheme leads to better gradient-based metrics when considering the patch-based random sampling strategy. A similar behavior is observed when applying this method (4DVarNet) to sea surface satellite heights and sea surface temperatures from satellites [44]. In our case, this likely relates to a larger number of training examples with truly observed pixel-level gradients as, by construction, the pixel-level random sampling increases the likelihood that two neighboring pixels are not observed. This further highlights the ability of 4DVarNet schemes to exploit fine-scale patterns in real observation datasets. It also suggests further exploring these random sampling strategies in future work to make the most of available gappy observation datasets.

6. Conclusions

This study presented a novel end-to-end neural scheme for the space–time interpolation of remote sensed sea surface suspended sediment fields, referred to as 4DVarNet. We assessed its reconstruction performance for both real and simulation-based datasets. We reported a clear improvement with respect to the state-of-the-art schemes, namely OI and DINEOF, in terms of global interpolation error of the retrieval of SSSC gradients. To our knowledge, this study is among the few which demonstrate the readiness of end-to-end neural schemes for the processing of L4 gap-free satellite products.

Besides its integration in operational processing pipelines for satellite-derived products, future work could further exploit the variational formulation that 4DVarNet relies on. Through new forcing terms in this variational formulation, it provides a well-posed basis towards the exploitation of forcing variables, for instance associated with wave and barotropic current processes, to further improve the reconstruction of SSSC fields. Similarly, the proposed scheme naturally extends to short-term forecasting applications, which are also of key interest. Overall, we expect this study to serve as a basis for the development of physics-informed deep learning frameworks for ocean remote sensing.

Author Contributions: Conceptualization: R.F., C.D. and F.J.; methodology: R.F.; software: J.-M.V.; validation: J.-M.V. and C.D.; formal analysis: C.D.; writing original draft preparation: J.-M.V.; writing review and editing: F.J., J.-M.V., R.F. and C.D. All authors have read and agreed to the published version of the manuscript.

Funding: This research was funded by the AID (French Agency of Defense Innovation through a PhD scholarship and the city of Brest (BMO). It was supported by ANR Project OceaniX; it was also supported by CNES (French Space Agency) under Project ML4SECCHI (Machine Learning for Secchi visibility). It benefited from HPC and GPU resources from Azure (Microsoft EU Ocean awards).

Data Availability Statement: The computer code and dataset for this study can be found in the https://github.com/Jvient/4DVarNet-OSSE, accessed on 12 July 2022.

Abbreviations

The following abbreviations are used in this manuscript:

ARPEGE	Action de Recherche Petite Echelle Grande Echelle
BoB	Bay of Biscay
CMEMS	Copernicus Marine Environment Monitoring Service
DINEOF	Data INterpolating Empirical Orthogonal function
EOF	Empirical Orthogonal function
HIGHROC	HIGH spatial and temporal Resolution Ocean color products and services
IBI	Iberian-Biscay-Ireland
LSTM	Long Short Term Memory
MARS	Model for Applications at Regional Scales
MODIS	Moderate-Resolution Imaging Spectroradiometer
MUSTANG	MUd and Sand TrAnsport modelliNG
NAP	Non-Algal Particles
NN	Neural Network
NTU	Nephelometric Turbidity Unit
OI	Optimal Interpolation
OLCI	Ocean and Land color Instrument
OSE	Observing System Experiment (real data)
OSSE	Observing System Simulation Experiment (simulated data)
RMSE	Root Mean Square Error
RMSLE	Root Mean Square Logarithm Error
SPIM	Suspended Particulate Inorganic Matter

SSSC	(sea) Surface Suspended Sediment Concentration
VIIRS	Visible Infrared Imaging Radiometer Suite
4DVar	Four-Dimensional Variational data assimilation (model-driven)
4DVarNet	Four-Dimensional Variational (neural) Network data assimilation (data-driven)

References

1. Owens, P.N. Soil erosion and sediment dynamics in the Anthropocene: A review of human impacts during a period of rapid global environmental change. *J. Soils Sediments* **2020**, *20*, 4115–4143. [CrossRef]
2. Irabien, M.J.; Cearreta, A.; Gómez-Arozamena, J.; Gardoki, J.; Martín-Consuegra, A.F. Recent coastal anthropogenic impact recorded in the Basque mud patch (southern Bay of Biscay shelf). *Quat. Int.* **2020**, *566–567*, 357–367. [CrossRef]
3. Borja, Á.; Elliott, M.; Carstensen, J.; Heiskanen, A.S.; van de Bund, W. Marine management - Towards an integrated implementation of the European marine strategy framework and the water framework directives. *Mar. Pollut. Bull.* **2010**, *60*, 2175–2186. [CrossRef]
4. Elliott, M.; Borja, Á.; McQuatters-Gollop, A.; Mazik, K.; Birchenough, S.; Andersen, J.H.; Painting, S.; Peck, M. Force majeure: Will climate change affect our ability to attain Good Environmental Status for marine biodiversity? *Mar. Pollut. Bull.* **2015**, *95*, 7–27. [CrossRef] [PubMed]
5. Larcombe, P.; Morrison-Saunders, A. Managing marine environments and decision-making requires better application of the physical sedimentary sciences. *Australas. J. Environ. Manag.* **2017**, *24*, 200–221. [CrossRef]
6. Tecchiato, S.; Collins, L.; Parnum, I.; Stevens, A. The influence of geomorphology and sedimentary processes on benthic habitat distribution and littoral sediment dynamics: Geraldton, Western Australia. *Mar. Geol.* **2015**, *359*, 148–162. [CrossRef]
7. James, I.D. Modelling pollution dispersion, the ecosystem and water quality in coastal waters: A review. *Environ. Model. Softw.* **2002**, *17*, 363–385. [CrossRef]
8. Mitchell, C.; Cunningham, A. Remote sensing of spatio-temporal relationships between the partitioned absorption coefficients of phytoplankton cells and mineral particles and euphotic zone depths in a partially mixed shelf sea. *Remote Sens. Environ.* **2015**, *160*, 193–205. [CrossRef]
9. Saulnier, J.B.; Escobar-Valencia, E.; Grognet, M.; Waeles, B. 3D Modelling for the Dispersion of Sediments Dredged in the Port of La Rochelle with Open TELEMAC-MASCARET. In Proceedings of the Papers Submitted to the 2020 TELEMAC-MASCARET User Conference, Antwerp, Belgium, 14–15 October 2021; pp. 129–137.
10. Diaz, M.; Grasso, F.; Le Hir, P.; Sottolichio, A.; Caillaud, M.; Thouvenin, B. Modeling Mud and Sand Transfers Between a Macrotidal Estuary and the Continental Shelf: Influence of the Sediment Transport Parameterization. *J. Geophys. Res. Ocean.* **2020**, *125*, 1–37. [CrossRef]
11. Le Hir, P.; Monbet, Y.; Orvain, F. Sediment erodability in sediment transport modelling: Can we account for biota effects? *Cont. Shelf Res.* **2007**, *27*, 1116–1142. [CrossRef]
12. Wang, Y.P.; Voulgaris, G.; Li, Y.; Yang, Y.; Gao, J.; Chen, J.; Gao, S. Sediment resuspension, flocculation, and settling in a macrotidal estuary. *J. Geophys. Res. Ocean.* **2013**, *118*, 5591–5608. [CrossRef]
13. Renosh, P.R.; Jourdin, F.; Charantonis, A.A.; Yala, K.; Rivier, A.; Badran, F.; Thiria, S.; Guillou, N.; Leckler, F.; Gohin, F.; et al. Construction of multi-year time-series profiles of suspended particulate inorganic matter concentrations using machine learning approach. *Remote Sens.* **2017**, *9*, 1320. [CrossRef]
14. Moore, A.M.; Martin, M.J.; Akella, S.; Arango, H.G.; Balmaseda, M.; Bertino, L.; Ciavatta, S.; Cornuelle, B.; Cummings, J.; Frolov, S.; et al. Synthesis of ocean observations using data assimilation for operational, real-time and reanalysis systems: A more complete picture of the state of the ocean. *Front. Mar. Sci.* **2019**, *6*, 90. [CrossRef]
15. Nazeer, M.; Bilal, M.; Alsahli, M.; Shahzad, M.; Waqas, A. Evaluation of Empirical and Machine Learning Algorithms for Estimation of Coastal Water Quality Parameters. *ISPRS Int. J. Geo-Inf.* **2017**, *6*, 360. [CrossRef]
16. Jin, D.; Lee, E.; Kwon, K.; Kim, T. A Deep Learning Model Using Satellite Ocean Color and Hydrodynamic Model to Estimate Chlorophyll- a Concentration. *Remote Sens.* **2021**, *13*, 2003. [CrossRef]
17. Barth, A.; Alvera-Azcárate, A.; Troupin, C.; Beckers, J.M. DINCAE 2.0: A convolutional neural network with error estimates to reconstruct sea surface temperature satellite observations. *Geosci. Model Dev.* **2022**, *15*, 2183–2196. [CrossRef]
18. Fablet, R.; Beauchamp, M.; Drumetz, L.; Rousseau, F. Joint Interpolation and Representation Learning for Irregularly Sampled. *Front. Appl. Math. Stat.* **2021**, *7*, 655224. [CrossRef]
19. Vient, J.M.; Jourdin, F.; Fablet, R.; Mengual, B.; Lafosse, L.; Delacourt, C. Data-driven interpolation of sea surface suspended concentrations derived from ocean color remote sensing data. *Remote Sens.* **2021**, *13*, 3537. [CrossRef]
20. Mengual, B.; Hir, P.L.; Cayocca, F.; Garlan, T. Modelling fine sediment dynamics: Towards a common erosion law for fine sand, mud and mixtures. *Water* **2017**, *9*, 564. [CrossRef]
21. Mengual, B.; Le Hir, P.; Cayocca, F.; Garlan, T. Bottom trawling contribution to the spatio-temporal variability of sediment fluxes on the continental shelf of the Bay of Biscay (France). *Mar. Geol.* **2019**, *414*, 77–91. [CrossRef]
22. Huthnance, J.; Hopkins, J.; Berx, B.; Dale, A.; Holt, J.; Hosegood, P.; Inall, M.; Jones, S.; Loveday, B.R.; Miller, P.I.; et al. Ocean shelf exchange, NW European shelf seas: Measurements, estimates and comparisons. *Prog. Oceanogr.* **2022**, *202*, 102760. [CrossRef]

23. Castaing, P.; Froidefond, J.M.; Lazure, P.; Weber, O.; Prud'Homme, R.; Jouanneau, J.M. Relationship between hydrology and seasonal distribution of suspended sediments on the continental shelf of the Bay of Biscay. *Deep-Sea Res. Part II Top. Stud. Oceanogr.* **1999**, *46*, 1979–2001. [CrossRef]
24. Gohin, F.; Loyer, S.; Lunven, M.; Labry, C.; Froidefond, J.M.; Delmas, D.; Huret, M.; Herbland, A. Satellite-derived parameters for biological modelling in coastal waters: Illustration over the eastern continental shelf of the Bay of Biscay. *Remote Sens. Environ.* **2005**, *95*, 29–46. [CrossRef]
25. Gohin, F.; Saulquin, B.; Oger-Jeanneret, H.; Lozac'h, L.; Lampert, L.; Lefebvre, A.; Riou, P.; Bruchon, F. Towards a better assessment of the ecological status of coastal waters using satellite-derived chlorophyll-a concentrations. *Remote Sens. Environ.* **2008**, *112*, 3329–3340. [CrossRef]
26. Petus, C.; Chust, G.; Gohin, F.; Doxaran, D.; Froidefond, J.M.; Sagarminaga, Y. Estimating turbidity and total suspended matter in the Adour River plume (South Bay of Biscay) using MODIS 250-m imagery. *Cont. Shelf Res.* **2010**, *30*, 379–392. [CrossRef]
27. Gohin, F.; Saulquin, B.; Bryere, P. Atlas de la Température, de la Concentration en Chlorophylle et de la Turbidité de Surface du Plateau Continental Français et de ses Abords de L'Ouest Européen; 2010. p. 53. Available online: https://archimer.ifremer.fr/doc/00057/16840/14306.pdf (accessed on 12 July 2022).
28. Nechad, B.; Ruddick, K.G.; Park, Y. Calibration and validation of a generic multisensor algorithm for mapping of total suspended matter in turbid waters. *Remote Sens. Environ.* **2010**, *114*, 854–866. [CrossRef]
29. Gohin, F.; Bryère, P.; Lefebvre, A.; Sauriau, P.G.; Savoye, N.; Vantrepotte, V.; Bozec, Y.; Cariou, T.; Conan, P.; Coudray, S.; et al. Satellite and in situ monitoring of chl-a, turbidity, and total suspended matter in coastal waters: Experience of the year 2017 along the french coasts. *J. Mar. Sci. Eng.* **2020**, *8*, 665. [CrossRef]
30. Bellacicco, M.; Cornec, M.; Organelli, E.; Brewin, R.J.; Neukermans, G.; Volpe, G.; Barbieux, M.; Poteau, A.; Schmechtig, C.; D'Ortenzio, F.; et al. Global Variability of Optical Backscattering by Non-algal particles From a Biogeochemical-Argo Data Set. *Geophys. Res. Lett.* **2019**, *46*, 9767–9776. [CrossRef]
31. Borja, A.; Amouroux, D.; Anschutz, P.; Gómez-Gesteira, M.; Uyarra, M.C.; Valdés, L. *The Bay of Biscay*, 2nd ed.; Elsevier Ltd.: Amsterdam, The Netherlands, 2018; pp. 113–152. [CrossRef]
32. Fablet, R.; Drumetz, L.; Rousseau, F.; Fablet, R.; Drumetz, L.; End-to end, F.R. End-to-End Learning of Variational Models and Solvers for the Resolution of Interpolation Problems. 2021. Available online: https://hal-imt-atlantique.archives-ouvertes.fr/hal-03139133/document (accessed on 12 July 2022).
33. Zhang, L.; Zhang, L.; Du, B. Deep learning for remote sensing data: A technical tutorial on the state of the art. *IEEE Geosci. Remote Sens. Mag.* **2016**, *4*, 22–40. [CrossRef]
34. Beauchamp, M.; Fablet, R.; Ubelmann, C.; Ballarotta, M.; Chapron, B. Intercomparison of data-driven and learning-based interpolations of along-track nadir and wide-swath swot altimetry observations. *Remote Sens.* **2020**, *12*, 3806. [CrossRef]
35. Fablet, R.; Chapron, B.; Drumetz, L.; Memin, E.; Pannekoucke, O.; Rousseau, F. Learning Variational Data Assimilation Models and Solvers. *arXiv* **2020**, arXiv:2007.12941.
36. Gao, Y.; Guan, J.; Zhang, F.; Wang, X.; Long, Z. Attention-Unet-Based Near-Real-Time Precipitation Estimation from Fengyun-4A Satellite Imageries. *Remote Sens.* **2022**, *14*, 2925. [CrossRef]
37. Eleveld, M.A.; Pasterkamp, R.; van der Woerd, H.J.; Pietrzak, J.D. Remotely sensed seasonality in the spatial distribution of sea-surface suspended particulate matter in the southern North Sea. *Estuar. Coast. Shelf Sci.* **2008**, *80*, 103–113. [CrossRef]
38. Daley, R. Atmospheric data Assimilation. *J. Meterolog. Soc. Jpn.* **1997**, *75*, 319–329. [CrossRef]
39. Cressie, N.A.C.; Wikle, C.K. *Statistics for Spatio-Temporal Data*; Wiley: Hoboken, NJ, USA, 2011; p. 588.
40. Beckers, J.M.; Rixen, M. EOF calculations and data filling from incomplete oceanographic datasets. *J. Atmos. Ocean. Technol.* **2003**, *20*, 1839–1856. [CrossRef]
41. OLIVER, M.A.; WEBSTER, R. Kriging: A method of interpolation for geographical information systems. *Int. J. Geogr. Inf. Syst.* **2007**, *4*, 313–332. [CrossRef]
42. Alvera-Azcárate, A.; Vanhellemont, Q.; Ruddick, K.; Barth, A.; Beckers, J.M. Analysis of high frequency geostationary ocean color data using DINEOF. *Estuar. Coast. Shelf Sci.* **2015**, *159*, 28–36. [CrossRef]
43. Liu, X.; Wang, M. Analysis of ocean diurnal variations from the Korean Geostationary Ocean Color Imager measurements using the DINEOF method. *Estuar. Coast. Shelf Sci.* **2016**, *180*, 230–241. [CrossRef]
44. Fablet, R.; Febvre, Q.; Chapron, B. Multimodal 4DVarNets for the reconstruction of sea surface dynamics from SST-SSH synergies. *arXiv* **2022**, arXiv:2207.01372.
45. Guo, X.; Liu, X.; Zhu, E.; Yin, J. Deep Clustering with Convolutional Autoencoders. In *Lecture Notes in Computer Science (Including Subseries Lecture Notes in Artificial Intelligence and Lecture Notes in Bioinformatics), Proceedings of the 24th International Conference, ICONIP 2017, Guangzhou, China, 14–18 November 2017*; Springer: Berlin/Heidelberg, Germany, 2017. [CrossRef]
46. Hoffman, R.N.; Atlas, R. Future observing system simulation experiments. *Bull. Am. Meteorol. Soc.* **2016**, *97*, 1601–1616. [CrossRef]
47. Perrot, L.; Gohin, F.; Ruiz-Pino, D.; Lampert, L.; Huret, M.; Dessier, A.; Malestroit, P.; Dupuy, C.; Bourriau, P. Coccolith-derived turbidity and hydrological conditions in May in the Bay of Biscay. *Prog. Oceanogr.* **2018**, *166*, 41–53. [CrossRef]

48. Saulquin, B.; Gohin, F.; Fanton d'Andon, O. Interpolated fields of satellite-derived multi-algorithm chlorophyll-a estimates at global and European scales in the frame of the European Copernicus-Marine Environment Monitoring Service. *J. Oper. Oceanogr.* **2019**, *12*, 47–57. [CrossRef]
49. Barth, A.; Alvera-Azcárate, A.; Licer, M.; Beckers, J.M. DINCAE 1.0: A convolutional neural network with error estimates to reconstruct sea surface temperature satellite observations. *Geosci. Model Dev.* **2020**, *13*, 1609–1622. [CrossRef]

remote sensing

Article

AICCA: AI-Driven Cloud Classification Atlas

Takuya Kurihana [1,2] , Elisabeth J. Moyer [2,3] and Ian T. Foster [1,4,*]

1   Department of Computer Science, University of Chicago, Chicago, IL 60637, USA
2   The Center for Robust Decision-Making on Climate and Energy Policy, Chicago, IL 60637, USA
3   Department of the Geophysical Sciences, University of Chicago, Chicago, IL 60637, USA
4   Data Science and Learning Division, Argonne National Laboratory, Lemont, IL 60439, USA
*   Correspondence: foster@uchicago.edu

Abstract: Clouds play an important role in the Earth's energy budget, and their behavior is one of the largest uncertainties in future climate projections. Satellite observations should help in understanding cloud responses, but decades and petabytes of multispectral cloud imagery have to date received only limited use. This study describes a new analysis approach that reduces the dimensionality of satellite cloud observations by grouping them via a novel automated, unsupervised cloud classification technique based on a convolutional autoencoder, an artificial intelligence (AI) method good at identifying patterns in spatial data. Our technique combines a rotation-invariant autoencoder and hierarchical agglomerative clustering to generate cloud clusters that capture meaningful distinctions among cloud textures, using only raw multispectral imagery as input. Cloud classes are therefore defined based on spectral properties and spatial textures without reliance on location, time/season, derived physical properties, or pre-designated class definitions. We use this approach to generate a unique new cloud dataset, the AI-driven cloud classification atlas (AICCA), which clusters 22 years of ocean images from the Moderate Resolution Imaging Spectroradiometer (MODIS) on NASA's Aqua and Terra instruments—198 million patches, each roughly 100 km × 100 km (128 × 128 pixels)—into 42 AI-generated cloud classes, a number determined via a newly-developed stability protocol that we use to maximize richness of information while ensuring stable groupings of patches. AICCA thereby translates 801 TB of satellite images into 54.2 GB of class labels and cloud top and optical properties, a reduction by a factor of 15,000. The 42 AICCA classes produce meaningful spatio-temporal and physical distinctions and capture a greater variety of cloud types than do the nine International Satellite Cloud Climatology Project (ISCCP) categories—for example, multiple textures in the stratocumulus decks along the West coasts of North and South America. We conclude that our methodology has explanatory power, capturing regionally unique cloud classes and providing rich but tractable information for global analysis. AICCA delivers the information from multi-spectral images in a compact form, enables data-driven diagnosis of patterns of cloud organization, provides insight into cloud evolution on timescales of hours to decades, and helps democratize climate research by facilitating access to core data.

Keywords: cloud classification; MODIS; artificial intelligence; deep learning; machine learning

Citation: Kurihana, T.; Moyer, E.J.; Foster, I.T. AICCA: AI-Driven Cloud Classification Atlas. *Remote Sens.* **2022**, *14*, 5690. https://doi.org/10.3390/rs14225690

Academic Editors: Veronica Nieves, Ana B. Ruescas and Raphaëlle Sauzède

Received: 16 September 2022
Accepted: 3 November 2022
Published: 10 November 2022

1. Introduction

Over the past several decades, advancements in satellite-borne remote sensing instruments have produced petabytes of global multispectral imagery that capture cloud structure, size distributions, and radiative properties at a near-daily cadence. While understanding trends in cloud behavior is arguably the principal challenge in climate science, these enormous datasets are underutilized because climate scientists cannot in practice manually examine them to analyze spatial-temporal patterns. Instead, some kind of automated algorithm is needed to identify physically relevant cloud types. However, the diversity of cloud morphologies and textures, and their multi-scale properties, makes classifying them into meaningful groupings a difficult task.

Existing classification schemes are necessarily simplistic. The most standard classification, the ISCCP (International Satellite Cloud Climatology Project) schema, simply defines a grid of nine global classes based on low, medium, or high values of cloud altitude (cloud top pressure) and optical thickness [1–3]. Because this classification is typically applied pixel by pixel, it cannot capture spatial structures and can produce an incoherent spatial distribution of cloud types in cloud imagery. The World Meteorological Organization's International Cloud Atlas [4], a more complex cloud classification framework, defines 28 different classes (of which 10 are considered 'basic types') with a complex coding procedure that depends on subjective judgments, such as whether a cloud has yet "become fibrous or striated." The schema is subjective and difficult to automate, and furthermore does not capture the full diversity of important cloud types. For example, it does not distinguish between open- and closed-cell stratocumulus clouds, placing them both in "stratocumulus," though the two have different circulation patterns, rain rates, and radiative effects [5]. Because the human eye serves as a sensitive tool for pattern classification, human observers can in principle group clouds into a larger set of types based on texture and shape as well as altitude and thickness. In practice, however, it has been difficult to devise a set of artificial cloud categories that encompass all cloud observations and can be applied consistently by human labelers.

These issues motivate the application of artificial intelligence (AI)-based algorithms for cloud classification. In the last several years, a number of studies have sought to develop AI-based cloud classification by using *supervised learning* [6–10]. In these approaches, ML models are trained to classify cloud images based on a training set to which humans have assigned labels. However, the difficulty of generating meaningful and consistent labels is a constant problem, and supervised learning approaches tend to succeed best when used on limited datasets containing classic examples of well-known textures. For example, Rasp et al. [7] classified just four particular patterns of stratocumulus defined and manually labeled by Stevens et al. [11]. Supervised methods cannot discover unknown cloud types that may be relevant to climate change research.

To serve the needs of climate research free from assumptions that may limit novel discoveries, the more appropriate choice is *unsupervised learning*, in which unknown patterns in data are learned without requiring predefined labels. The first demonstrations of unsupervised methods applied to cloud images were made in the 1990s [12,13]. Even with the primitive neural networks then available, Tian et al. [13] showed that cloud images from the GOES-8 satellite could be sorted automatically into ten clusters that reproduced the ten 'basic' WMO classes with 65–75% accuracy. In 2019, Denby [14] and Kurihana et al. [15] leveraged advances in deep neural network (DNN) methods to prototype unsupervised cloud classification algorithms that used convolutional neural networks (CNNs, DNNs with convolutional layers) and produced cloud classes from the resulting compact representations via hierarchical agglomerative clustering (HAC) [16]. Both works used only 12 classes and neither was rotation-invariant, but both successfully produced reasonable-seeming classifications—for Denby [14], from near-infrared images from the GOES satellite in the tropical Atlantic, and for Kurihana et al. [15], from global multispectral images from the Moderate Resolution Imaging Spectroradiometer (MODIS) instruments on NASA's Aqua and Terra satellites). Kurihana et al. [15] were the first to use an autoencoder [17], a class of unsupervised DNNs widely used for dimensionality reduction, for cloud classification, and Kurihana et al. [18] extended the work by adding a more complex loss function to the autoencoder to produce rotation-invariant cloud clustering (RICC). Kurihana et al. [18] also developed a formal evaluation protocol to ensure that the resulting cloud classes were physically meaningful.

The work described here builds on these previous results to generate a standardized science product: an AI-driven Cloud Classification Atlas (AICCA) of global-scale unsupervised classification of MODIS satellite imagery into 42 cloud classes. We first describe and apply the protocol that we have developed to determine this optimal number of clusters when applying RICC to the MODIS dataset. (The first author calculated this number before being informed of its occurrence in an unrelated context [19]). We demonstrate that the resulting classes are coherent geographically, temporally, and in altitude-optical depth space. Finally, we describe a workflow that allows us to apply the $RICC_{42}$ algorithm to the full two decades of MODIS imagery to provide a publicly available dataset. The result is an automated, unsupervised classification process that discovers classes based on both cloud morphology and physical properties to yield unbiased cloud classes free from artificial assumptions that capture the diversity of global cloud types. AICCA is intended to support studies of the response of clouds to forcing on timescales from hours to decades and to allow data-driven diagnosis of cloud organization and behavior and their evolution over time as CO_2 and temperatures increase.

We describe this dataset as follows: Section 2 describes the MODIS imagery, information used, and structure of output data. Section 3 describes the algorithm used for classification, including the training procedure on one million randomly selected ocean-cloud patches (Section 3.2). Section 4 evaluates the stability of the clustering step, and Section 5 describes the characteristics of the resulting cloud clusters: their distribution geographically, seasonally, and in altitude-optical depth space.

2. AICCA: Data and Outputs

The dataset described in this article, $AICCA_{42}$ (or simply AICCA), provides AI-generated cloud class labels for all 128×128 pixels ($\sim$100 km by 100 km) ocean cloud *patches* sampled by MODIS instruments over their 22 years of operation. (An ocean cloud patch is defined as a patch with only ocean pixels and at least 30% cloud pixels). The cloud labels are generated by the rotation-invariant cloud clustering (RICC) method of Kurihana et al. [18]. In general, clusters produced by RICC may vary according to (1) the patches used to train RICC, (2) the number of clusters chosen, and (3) the patches to which the trained RICC is applied to generate centroids. We therefore define $AICCA_{42}$ as the dataset produced by training RICC on a subset of the data described in Section 2.1, clustered into 42 classes with a set of reference centroids based on OC-Patches$_{HAC}$, as defined in Section 4.4.

The labeled output is provided in two ways: per patch, which provides the finest granularity of labels and associated physical properties, and resampled to $1° \times 1°$ grid cells, which supplies information in a daily global grid format that is familiar to climate scientists.

2.1. MODIS Data

The MODIS instruments hosted on NASA's Aqua and Terra satellites have been collecting visible to mid-infrared radiance data in 36 spectral bands from 2002 (Aqua) [20] and 2000 (Terra) [21] through 2021. The instruments collect data over an approximately 2330 km by 2030 km *swath* every five minutes, with a spatial resolution of 1 km. AICCA is based on the MODIS Level 1B calibrated radiance product (MOD02). (Note that, while NASA uses the prefixes MOD and MYD to distinguish between Terra and Aqua, respectively, for simplicity, we use MOD to refer to both throughout this article). We limit the dataset to the six spectral bands most relevant for derivation of physical properties: bands 6, 7, and 20 relate to cloud optical properties, and bands 28, 29, and 31 relate to the separation of high and low clouds and the detection of the cloud phase. For the Aqua instrument, we use band 5 as an alternative to band 6 due to a known stripe noise issue in Aqua band 6 [22]. (See also Kurihana et al. [18] for more details). The total number of swath images per band is (12 swath/h) $\times$ (12 h/day) $\times$ (365 day/year) $\times$ (20 + 22 years, for Aqua and Terra, respectively) $\approx$ 2.2 million.

MODIS multispectral data are processed by NASA to yield a variety of derived products, several of which we employ for post-processing or analysis. We take latitude and longitude from the MOD03 geolocation fields to regrid the AICCA patches, and use selected derived physical properties from the MOD06 product to evaluate the cloud classes: four physical parameters related to cloud optical properties and cloud top properties. Note that we employ the MOD06 variables only as a diagnostic, to evaluate associations between AICCA clusters and cloud physical properties. They are not included in our RICC training data, which are thus free from any assumptions made by the producers of MOD06 variables. The data used in generating AICCA, listed in Table 1, have an aggregate size of 801 terabytes. All MODIS products are accessible via the NASA Level-1 and Atmosphere Archive and Distribution System (LAADS), grouped into per-swath files.

Table 1. MODIS products used to create the AICCA dataset. Each product name *MOD0X* in the first column refers to both the Aqua (MYD0X) and Terra (MOD0X) products. Source: NASA Earthdata.

Product	Description	Band	Primary Use	Process
MOD02	Shortwave infrared (1.230–1.250 μm)	5	Land/cloud/aerosol properties	
	Shortwave infrared (1.628–1.652 μm)	6	Land/cloud/aerosol properties	
	Shortwave infrared (2.105–2.155 μm)	7	Land/cloud/aerosol properties	
	Longwave thermal infrared (3.660–3.840 μm)	20	Surface/cloud temperature	Section 3.1
	Longwave thermal infrared (7.175–7.475 μm)	28	Cirrus clouds water vapor	
	Longwave thermal infrared (8.400–8.700 μm)	29	Cloud properties	
	Longwave thermal infrared (10.780–11.280 μm)	31	Surface/cloud temperature	
MOD03	Geolocation fields		Latitude and Longitude	
MOD06	Cloud mask		Cloud pixel detection	Section 3.1
	Land/Water		Background detection	
	Cloud optical thickness		Thickness of cloud	
	Cloud top pressure		Pressure at cloud top	Section 3.3
	Cloud phase infrared		Cloud particle phase	
	Cloud effective radius		Radius of cloud droplet	

2.2. AICCA Patch-Level Data

The AICCA dataset uses all patches from Aqua and Terra MODIS image data during 2000–2021, subject to the constraints that they (1) are disjoint in space and/or time; (2) include no non-ocean pixels, and 3) each includes at least 30% cloud pixels. The resulting set comprises about 198,676,800 individual 128×128 pixel ($\sim$100 km by 100 km) ocean-cloud patches, for each of which $AICCA_{42}$ provides the following information (and see Table 2):

- Source is either Aqua or Terra;
- Swath, Location, and Timestamp locate the patch in time and space;
- Training indicates whether the patch was used for training;
- Label is an integer in the range 1..42, generated by the rotation-invariant cloud clustering system configured for 42 clusters, $RICC_{42}$ (see Section 4 for the stability protocol used to select this number of clusters);
- COT_patch, CTP_patch, and CER_patch, the mean and standard deviation, across all pixels in the patch, for three MOD06 physical values: cloud optical thickness (COT), cloud top pressure (CTP), and cloud effective radius (CER); and
- CPI_patch, cloud phase information (CPI), four numbers representing the number of the 128×128 pixels in the patch that are estimated as clear-sky, liquid, ice, or undefined, respectively.

The resulting 146 bytes per patch represents a $16,159 \times$ reduction in size relative to the raw multispectral imagery.

The additional information shown in Table 2 to assist users in understanding individual patches is extracted from MOD06 by using the patch's geolocation index and timestamp (Location and Timestamp in Table 2) to locate the patch's data in the appropriate MOD06 file. These mean values summarize the patch's average physical characteristics; the standard deviations provide some indication as to the existence of multiple clouds (especially low- and high-altitude clouds). We do not use the MOD06 multilayered cloud flag.

Output is provided as NetCDF [23] files that combine patches from each MODIS swath into a single file. While AICCA contains no raw satellite data, it includes for each patch an identifier for the source MODIS swath and a geolocation index; thus, users can easily link AICCA results with the original MOD02 satellite imagery and other MODIS products. The complete `OC-Patches` set contains around $(20 + 22 \text{ years}) \times (365 \text{ days/year}) \times (26{,}000 \text{ patches}) \times 146 \text{ B} \approx 54.2$ gigabytes.

Table 2. Information provided in AICCA for each 128×128 pixel ocean-cloud patch: metadata that locate the patch in space and time, and indicate whether the patch was used to train RICC; a cloud class label computed by RICC; and a set of diagnostic quantities obtained by aggregating MODIS data over all pixels in the patch. A quotation mark indicates a repetition.

Variables	Description	Values	Type
Swath	Identifier for source MODIS swath	1	float32
Location	Geolocation index for the upper left corner of patch	2	float32
Timestamp	Time of observation	1	float32
Training	Whether patch used for training	1	binary
Label	Class label assigned by RICC: integer in range $1..k^*$	1	int32
COT_patch	Mean and standard deviation of pixel values in patch	2	float32
CTP_patch	"	"	"
CER_patch	"	"	"
CPI_patch	Number of pixels in patch in {clear-sky, liquid, ice, undefined}	4	int32

2.3. AICCA Grid Cell-Level Data

In addition to providing per-patch data, we follow common practice in climate datasets by also providing data organized on a per-latitude/longitude grid cell basis. The second element of the AICCA_{42} dataset spatially aggregates the patch-level class label and diagnostic values at a resolution of $1° \times 1°$, a total of 181×360 *grid cells* over the globe. For each grid cell, AICCA_{42} provides the information listed in Table 3, a total of 32 bytes:

- `Source` is either Aqua or Terra;
- `Cell` gives a latitude and longitude for the grid cell;
- `Timestamp` locates the grid cell in time;
- `Label_1deg` represents the most frequent class label in the grid cell (an integer in the range 1..42); and
- `COT_1deg`, `CTP_1deg`, `CER_1deg`, and `CPI_1deg` aggregate values for four diagnostic variables, as described in Section 2.3.

The aggregation process uses values from individual days from the Aqua and Terra satellites, a reasonable choice since the swaths taken by each satellite's MODIS instrument generally do not overlap in a daily period. Since a single 2330 km by 2030 km MODIS swath extends across multiple 1 degree by 1 degree grid cells, we extract the latitude and longitude at the center of each `OC-Patch` by using MOD03, and aggregate the information listed in Table 2 to each $1° \times 1°$ grid cell (i.e., the area extending from $-0.5°$ to $+0.5°$ from the grid cell center). To assign a class label to each grid cell on each day, we use the class of the single ocean-cloud patch with the largest overlap with the grid cell. To provide physical properties for each grid cell, we implement one simplification to reduce the use of computing memory: instead of averaging pixel values within each grid cell, we identify *all* ocean-cloud patches that overlap with the cell, and simply average those patches' mean COT, CTP, and CER values. To assign a cloud particle phase (clear–sky, liquid, ice, or undefined), we use the most frequent phase in the overlapping patches. Grid cells with no clouds are labeled as a missing value.

In some cases, especially at high latitudes, swaths may overlap within a single day. When this occurs, patches with different timestamps will overlap a given grid cell on the same day. In these cases, we discard one timestamp, to avoid inconsistent values between grid cells. That is, when accumulating the most frequent label and aggregating values on the overlapping cell, we use only those patches with a timestamp close to that of the neighboring grid cells. This neighboring selection mitigates the problem of inconsistent values between nearly grid cells due only to timing. Finally, we accumulate the aggregated grid-cell values to create the daily files. Given the MODIS orbital coverage, the complete OC-Gridcell set contains around $(20 + 22 \text{ years}) \times (365 \text{ days/year}) \times (65{,}160 \text{ grid cells}) \times 32 \text{ B} \approx 29.8$ gigabytes.

Table 3. AICCA information for each $1° \times 1°$ grid cell: a cloud class label computed by RICC and diagnostic quantities obtained by aggregating MODIS data over all patch pixels for that grid cell.

Variables	Description	Values	Type
Cell	(lat, long) for grid cell	2	float32
Timestamp	Time of observation	1	float32
Label	Most frequent class label in grid cell	1	int32
COT_1deg	Mean of pixel values in grid cell	1	float32
CTP_1deg	"	"	"
CER_1deg	"	"	"
CPI_1deg	Most frequent particle phase in grid cell	1	int32

3. Constructing AICCA

The AICCA production workflow, shown in Figure 1, consists of four principal stages: (1) download, archive, and prepare MODIS satellite data; (2) train the RICC unsupervised learning algorithm, and cluster cloud patterns and textures; (3) evaluate the reasonableness of the resulting clusters and determine an optimal cluster number; and (4) assign clusters produced by RICC to other MODIS data unseen during RICC training. We describe each stage in turn. The RICC code and Jupyter notebook [24] used in the analysis are available online [25], and the trained RI autoencoder used for this study is archived at the Data and Learning Hub for science (DLHub) [26], a scalable and low-latency model repository to share and publish machine learning models to facilitate reuse and reproduction.

3.1. Stage 1: Download, Archive, and Prepare MODIS Data

Download and archive. As noted in Section 2.1, we use subsets of three MODIS products in this work, a total of 801 terabytes for 2000–2021. In order to employ high-performance computing resources at Argonne National Laboratory for AI model training and inference, we copied all files to Argonne storage. Transferring the files from NASA archives is rapid for the subset that are accessible on a Globus endpoint at the NASA Center for Climate Simulation, which can be transferred via the automated Globus transfer system [27]. The remaining files were transferred from NASA LAADS via the more labor-intensive option of wget commands, which we accelerated by using the funcX [28] distributed function-as-a-service platform to trigger concurrent downloads on multiple machines.

Prepare. The next step involves preparing the *patches* used for ML model training and inference. We extract from each swath multiple 128 pixel by 128 pixel (roughly 100 km $\times$ 100 km) non-overlapping patches, for a total of $\sim$331 million patches. We then eliminate those patches that include any non-ocean pixels as indicated by the MOD06 land/water indicator, since, in these cases, radiances depend in part on underlying topography and reflectance. (Note that even ocean-only pixels may involve surface-related artifacts in cases when the ocean is covered in sea ice). We also eliminate those with less than 30% cloud pixels, as indicated by the MOD06 cloud mask. The result is a set of 198,676,800 *ocean-cloud patches*, which we refer to in the following as OC-Patches. For each ocean-cloud patch, we take from the MOD02 product six bands (out of 36 total) for use in training and testing the rotation-invariant (RI) autoencoder. We also extract the MOD04 and MOD06 data used for location and cluster evaluation, as described in Section 2. For an in-depth discussion of data selection, see Kurihana et al. [18].

We also construct a training set OC-Patches_{AE} by selecting one million patches at random from the entirety of OC-Patches. Because we do not expect our unsupervised RI autoencoder to be robust to the MODIS data used for training, we collect the 1M patches that they are not overly imbalanced among seasons or locations.

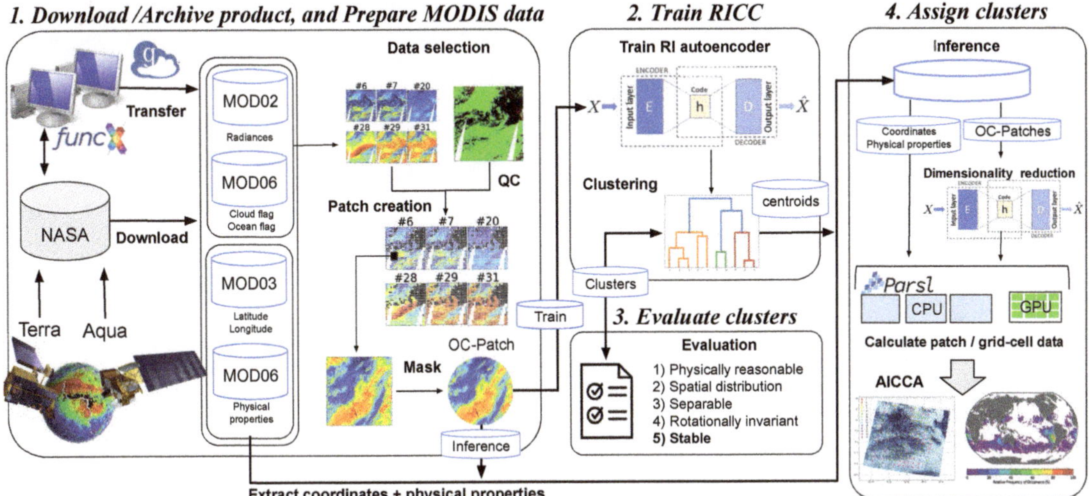

Figure 1. The AICCA production workflow comprises four principal stages. **(1) Download/Archive and Prepare MODIS data**: Download calibrated and retrieved MODIS products from the NASA Level-1 and Atmosphere Archive and Distribution System (LAADS), using FuncX and Globus for rapid and reliable retrieval of 801 terabytes of three different MODIS products between 2000–2021. Store downloaded data at Argonne National Laboratory. Select six near-infrared to thermal bands related to clouds and subdivide each swath into non-overlapping 128 × 128 pixel patches by six bands. Select patches with >30% cloud pixels over ocean regions, and apply a circular mask for optimal training of our rotation-invariant autoencoder, yielding OC-Patches. **(2) Train RICC**: Train an autoencoder on 1 M randomly selected patches to generate latent representations, and cluster those latent representations to determine cluster centroids [18]. **(3) Evaluate clusters**: Apply five protocols to evaluate whether the clusters produced are meaningful and useful. **(4) Assign clusters**: Use trained autoencoder and centroids to assign cloud labels to unseen data. We use the Parsl parallel Python library to scale the inference process to hundreds of CPU nodes plus a single GPU, and to generate the AICCA dataset in NetCDF format. We then calculate physical properties and other metadata information for each patch and for each 1° × 1° grid cell.

3.2. Stage 2: Train the RICC Autoencoder and Cluster Cloud Patterns

In this stage, we first train the RI autoencoder and then define cloud categories by clustering the compact latent representations produced by the trained autoencoder.

Train RICC. The goal of training is to produce an RI autoencoder capable of generating latent representations (a lower-dimensional embedding as the intermediate layer of the autoencoder) that explicitly capture the variety of input textures among ocean clouds and also map to differences in physical properties. We introduce general principles briefly here; see Kurihana et al. [18] for further details of the RI autoencoder architecture and training protocol.

An autoencoder [17,29] is a widely used unsupervised learning method that leverages dimensionality reduction as a preprocessing tool prior to image processing tasks such as clustering, regression, anomaly detection, and inpainting. An autoencoder comprises an encoder, used to map input images into a compact lower-dimensional latent representation, followed by a decoder, used to map that representation to output images. During training, a loss function minimizes the difference between input and output. The resulting latent representation in the trained autoencoder both (1) retains only relevant features for the target application in input images, and (2) maps images that are similar (from the perspective of the target application) to nearby locations in latent space.

The loss function minimizes the difference between an original and a restored image based on a distance metric during autoencoder training. The most commonly used metric is a simple ℓ^2 distance between the autoencoder's input and output:

$$L(\theta) = \sum_{x \in S} ||x - D_\theta(E_\theta(x))||_2^2, \tag{1}$$

where S is a set of training inputs; θ is the encoder and decoder parameters, for which values are to be set via training; and x and $D_\theta(E_\theta(x))$ are an input in S and its output (i.e., the restored version of x), respectively. However, optimizing with Equation (1) is inadequate for our purposes because it tends to generate different representations for an image x and the rotated image $R(x)$, as shown in Figure 2, with the result that the two images end up in different clusters. Since any particular physically driven cloud pattern can occur in different orientations, we want an autoencoder that assigns cloud types to images consistently, regardless of orientation. Other ML techniques that combine dimensionality reduction with clustering algorithms have not addressed the issue of rotation–invariance within their training process. For example, while non-negative matrix factorization (NMF) [30] can approximate input data into a low-dimensional matrix—i.e., produce a dimensionally reduced representation similar to an autoencoder—that can be used for clustering, applications of NMF are not invariant to image orientation.

We have addressed this problem in prior work by defining a rotation-invariant loss function [18] that generates similar latent representations, agnostic to orientation, for similar morphological clouds (Figure 2b). This RI autoencoder, motivated by the shifted transform invariant autoencoder of Matsuo et al. [31], uses a loss function L that combines both a rotation-invariant loss, L_{inv}, to learn the rotation invariance needed to map different orientations of identical input images into a uniform orientation, and a restoration loss, L_{res}, to learn the spatial structure needed to restore structural patterns in inputs with high fidelity. The two loss terms are combined as follows, with values for the scalar weights λ_{inv} and λ_{res} chosen as described below:

$$L = \lambda_{\text{inv}} L_{\text{inv}} + \lambda_{\text{res}} L_{\text{res}}, \tag{2}$$

The rotation-invariant loss function L_{inv} computes, for each image in a minibatch, the difference between the restored original and the 72 images obtained by applying a set $\mathcal{R}$ of 72 scalar rotation operators, each of which rotates an input by a different number of degrees in the set $\{0, 5, ..., 355\}$:

$$L_{\text{inv}}(\theta) = \frac{1}{N} \sum_{x \in S} \sum_{R \in \mathcal{R}} ||D_\theta(E_\theta(x)) - D_\theta(E_\theta(R(x)))||_2^2. \tag{3}$$

Thus, minimizing Equation (3) yields values for θ that produce similar latent representations for an image, regardless of its orientation.

The restoration loss, $L_{\text{res}}(\theta)$, learns the spatial substructure in images by computing the sum of minimum differences over the minibatch:

$$L_{\text{res}}(\theta) = \sum_{x \in S} \min_{R \in \mathcal{R}} ||R(x) - D_\theta(E_\theta(x))||_2^2. \tag{4}$$

Thus, minimizing Equation (4) results in values for θ that preserve spatial structure in inputs.

Our RI autoencoder training protocol [18], which sweeps over $(\lambda_{\text{inv}}, \lambda_{\text{res}})$ values, identifies $(\lambda_{\text{inv}}, \lambda_{\text{res}}) = (32, 80)$ as the coefficients for the two loss terms that best balance the transform-invariant and restoration loss terms. We note that the specific values of the two coefficients, not just their relative values, matter. For example, the values $(\lambda_{\text{inv}}, \lambda_{\text{res}}) = (32, 80)$ give better results than $(\lambda_{\text{inv}}, \lambda_{\text{res}}) = (3.2, 8.0)$.

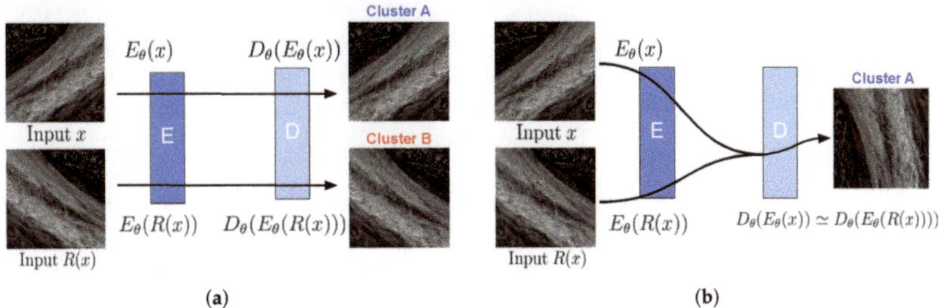

Figure 2. Illustration of the learning process when training (**a**) a conventional autoencoder with Equation (1) vs. (**b**) a rotation-invariant autoencoder with Equation (2). Because a conventional autoencoder reflects orientation in the latent representation, two input images that are identical in texture but different in orientation are assigned to different clusters, A and B. The rotation-invariant autoencoder produces a latent representation that is agnostic to orientation, allowing clustering to group both together.

The neural network architecture is the other factor needed to achieve rotation invariance: Following the heuristic approach of deep convolutional neural networks, we designed an encoder and decoder that stack five blocks of convolutions, each with three convolutional layers activated by leaky ReLU [32], and with batch normalization [33] applied at the final convolutional layer in each block before activation. We train our RI autoencoder on our one million training patches for 100 epochs by using stochastic gradient descent with a learning rate of 10^{-2} on 32 NVIDIA V100 GPUs in the Argonne National Laboratory ThetaGPU cluster.

Cluster Cloud Patterns. Once we have applied the trained autoencoder to a set of patches to obtain latent representations, we can then cluster those latent representations to identify the centroids that will define our cloud clusters. We use hierarchical agglomerative clustering (HAC) [16] for this purpose, and select Ward's method [34] for the linkage metric, so that HAC minimizes the variance of square distances as it merges clusters from bottom to top. We have shown in previous work [35] that HAC clustering results outperform those obtained with other common clustering algorithms.

Given N data points, a naive HAC approach requires $\mathcal{O}(N^2)$ memory to store the distance matrix used when calculating the linkage metric to construct the tree structure [36] —which would be impractical for the one million patches in OC-Patches$_\text{AE}$. Thus, we use a smaller set of patches, OC-Patches$_\text{HAC}$, comprising 74911 ocean-cloud patches from the year 2003 (the first year in which both Terra and Aqua satellites ran for the entire year concurrently) for the clustering phase. We apply our trained encoder to compute latent representations for each patch in OC-Patches$_\text{HAC}$ and then run HAC to group those latent representations into k^* clusters, in the process identifying k^* cluster centroids and assigning each patch in OC-Patches$_\text{HAC}$ a cluster label, $1..k^*$. The sequential scikit-learn [37] implementation of HAC that we use in this work takes around 10 hours to cluster the 74911 OC-Patches$_\text{HAC}$ patches on a single core. While we could use a parallelizable HAC algorithm [38–40] to increase the quantity of data clustered, this would not address the intrinsic limitation of our clustering process given the 801 terabytes of MODIS data.

3.3. Stage 3: Evaluate Clusters Generated by RICC

A challenge when employing unsupervised learning is to determine how to evaluate results. While a supervised classification problem involves a perfect ground truth against which to output can be compared, an unsupervised learning system produces outputs whose utility must be more creatively evaluated. Therefore, we defined in previous work a series of evaluation protocols to determine whether the cloud classes derived from a set of cloud images are meaningful and useful [18]. We seek cloud clusters that: (1) are *physically reasonable* (i.e., embody scientifically relevant distinctions); (2) capture information on *spatial distributions*, such as textures, rather than only mean properties; (3) are *separable* (i.e., are cohesive, and separated from other clusters, in latent space); (4) are *rotationally invariant* (i.e., insensitive to image orientation); and (5) are *stable* (i.e., produce similar or identical clusters when different subsets of the data are used). We summarize in Table 4 these criteria and the quantitative and qualitative tests that we have developed to validate them.

Table 4. Our five evaluation criteria protocol, as described in Kurihana et al. [18], and protocols for meeting them. In that work, we used the first four criteria to demonstrate that our quantitative and qualitative evaluation protocols can distinguish useful from non-useful autoencoders, even when common ML metrics such as ℓ^2 loss show insignificant differences. In the current work, we describe a protocol to ensure meeting the last criterion, stability.

Criterion	Test	Requirement
Physically reasonable	Cloud physics	Non-random distribution; median inter-cluster correlation < 0.6
	Spatial coherence	Spatially coherent clusters
Spatial distribution	Smoothing	Low adjusted mutual information (AMI) score
	Scrambling	Low AMI score
Separable	Separable clusters	No crowding structure
Rotationally invariant	Multi-cluster	AMI score closer to 1.0
	Significance of cluster stability	Ratio of Rand Index $G/R \geq 1.01$
Stable	Similarity of clusterings	Higher Adjusted Rand Index (ARI)
	Similarity of intra-cluster textures	Lower weighted average mean square distance
	Clusters capture seasonal cycle	Minimal seasonal texture difference

In our previous work [18], we showed that an analysis using RICC to separate cloud images into 12 clusters satisfies the first four of these criteria. In this work, we describe how we evaluate the last criterion, *stability*. Specifically, we evaluate the extent to which RICC clusters cloud textures and physical properties in a way that is stable against variations in the specific cloud patches considered, and that groups homogeneous textures within each cluster. We describe this process in Section 4 in the context of how we estimate the *optimal* number of clusters for this dataset when maximizing stability and similarity in clustering. For the remaining criteria, the clusters necessarily remain rotationally invariant, and we present in Section 5 results further validating that the algorithm, when applied to a global dataset, produces clusters that show physically reasonable distinctions, are spatially coherent, and involve distinct textures (i.e., learn spatial information).

3.4. Stage 4: Assign Cluster Labels to Patches

We have so far trained our RI autoencoder on the 1 million patches in OC-Patches$_{\text{AE}}$ and applied HAC to the 74,911 patches in OC-Patches$_{\text{HAC}}$ to obtain a set of k^* cluster centroids, $\mu = \{\mu_1, \ldots, \mu_{k^*}\}$, where k^* is the number of clusters defined in Section 3.3. We next want to assign a cluster label to each of the 198 million patches in OC-Patches. We do this by identifying for each patch x_i the cluster centroid μ_k with the smallest Euclidean distance to its latent representation, $z(x_i)$. We use Euclidean distance as our metric because our HAC algorithm uses Ward's method with Euclidean distance. That is, we calculate the cluster label assignment $c_{k,i}$ for the i-th patch as:

$$c_{k,i} = \underset{k=\{1,\ldots,k^*\}}{\arg\min} \, ||z(x_i) - \mu_k||_2. \tag{5}$$

This label prediction or *inference* process is easily parallelized. We use the Parsl parallel Python library [41], which enables scalable execution on many processors via simple

Python decorators, for this purpose. We observe an execution time of 533 seconds per day of MODIS imagery (~13,000 patches) on 256 cores of the Argonne Theta supercomputer.

4. Evaluating Cluster Stability

Cluster stability is an important property for a cloud classification algorithm [15]. A clustering method is said to be *stable* for a dataset, D, and a number of clusters, k, if it produces similar or identical clusters when applied to different subsets of D. As noted in Table 4, we define four tests to evaluate this criterion:

1. We measure *clustering similarity* by generating clusterings for different subsets of the same dataset, and calculating the average distance between those clusterings.
2. We measure *clustering similarity significance* by comparing each clustering similarity score to that obtained when our clustering method is applied to data from a uniform random distribution.
3. We measure *intra-cluster texture similarity* by calculating the average distance between latent representations in each cluster.
4. We measure *seasonal stability* by comparing intra-cluster texture similarity for patches from January and July.

We are concerned not only to determine whether our clustering method, RICC, generates clusters that are stable, but also to identify the optimal number of clusters, k^*, to use for AICCA. In determining that number, we must consider all four tests just listed: we want a high clustering similarity, a high significance (certainly greater than 1), a low intra-cluster similarity score, and low intra-seasonal texture differences.

For all of our stability tests, we work with $D = \{\texttt{OC-Patches}$ from 2003 to 2021, inclusive$\}$. $|D| \approx 180$ M. (We do not consider data from 2000–2002 because Terra and Aqua were not operating at the same time for an entire year-long observation during that period). We create a holdout subset H with number of patches $N_H = 14{,}000$, and create 30 random subsets S_i with $N_R = 56{,}000$ by sampling without replacement from $D \setminus H$. This procedure ensures that the different S_i are mutually exclusive and that there is no intersection between our holdout set H and the random subsets. The ratio $N_H : N_R$ of $20 : 80$ is standard practice. We then create our 30 test datasets as $H \cup S_i$ for $\forall i \in \{1, \ldots, 30\}$.

In the remainder of this section, we describe four stability tests, whose results are shown in Figures 3 and 4. These tests lead us to choose 42 as the optimal number of clusters. We also conduct additional evaluations of whether the result of using RICC with 42 clusters creates cloud classes that have reasonable texture and physical properties, when compared to similar exercises with suboptimal numbers of clusters.

4.1. Stability Test 1: Clustering Similarity

We measure clustering similarity by first generating clusterings for different subsets of the target dataset and then calculating the average pairwise distance between those clusterings. This approach is documented as Algorithm A1 in Appendix A.2. As described above, we work with sets $H \cup S_i$, $i \in 1..30$, to generate 30 different clustering assignments via a trained RICC. We compute the adjusted Rand index, ARI (Appendix A.1), as a measure of pairwise distance between pairs of clusterings. We average among the 30 clusterings generated by the models $\{\text{RICC}_k^i, i \in 1..30\}$ to determine the mean clustering similarity for that specific cluster number k, and then calculate the ARI for all $\binom{30}{2} = 435$ combinations of those 30 clusterings to determine the mean ARI score G. See Appendix A.2 for details.

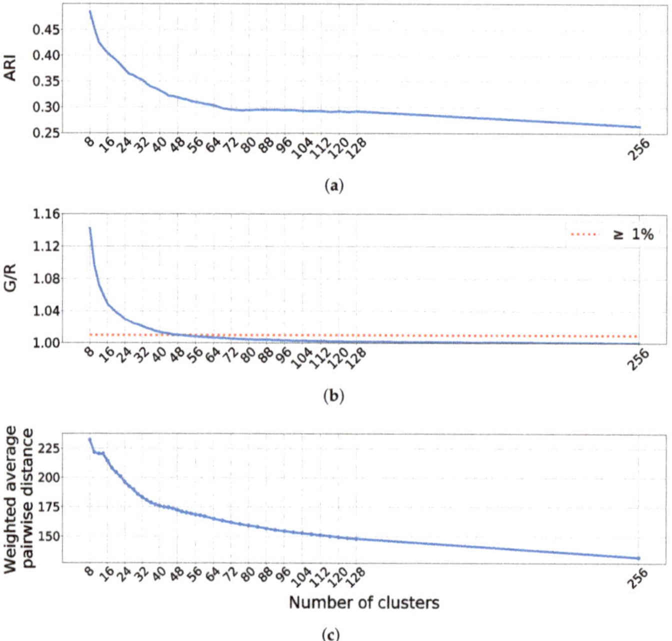

Figure 3. Plots for the first three stability criteria metrics of Table 4, each as a function of the number of clusters. (**a**) *Clustering similarity*: Adjusted Rand Index (ARI) as a measure of similarity of clusterings generated by RICC models trained on different subsets of patches. (**b**) *Clustering similarity significance*: The blue line represents the ratio of the mean Rand Index based on RICC applied to our holdout patches $\{x \mid x \in H\}$ (G) to the mean Rand Index from HAC applied to random uniform distributions (R). The red dashed line is G/R $\geq$ 1.01, indicating that the stability of cluster label assignments produced from RICC is $\geq$1% better than results of simply clustering random uniform data. (**c**) *Intra-cluster texture similarity*: The blue line shows the weighted average of the mean squared Euclidean distance between pairs of patches within each cluster. Lower values suggest more homogeneous textures and physical features within each cluster. The use of three similarity tests allows for achieving both stability and maximality criteria when grouping clusters.

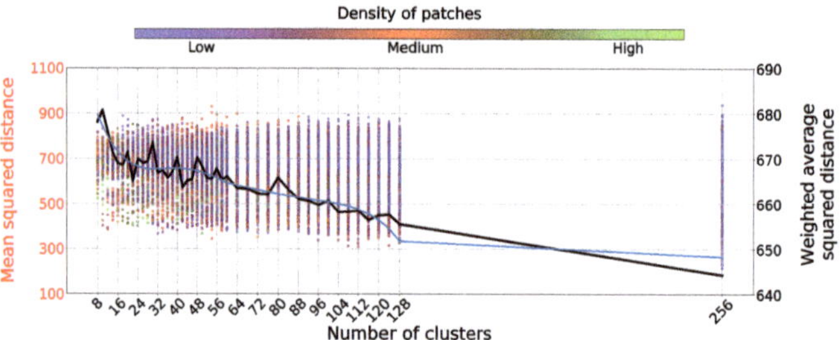

Figure 4. Seasonal stability test comparing the intra-seasonal variance of textures within each cluster as a function of the number of clusters. Each of $9 \cdot k$ colored dots for each value of k gives the average squared distance (left y-axis) between July and January patches as described in the text; the color indicates cluster density, a measure of cluster size. The black line shows the mean WASD (right y-axis) from nine trials as described in text. The blue line shows a smoothed WASD curve obtained by applying a Savitzky–Golay filter with a degree six polynomial. The minimum WASD value in $40 \leq k^* \leq 48$ occurs at $k = 42$, motivating our choice for AICCA.

The optimal number of clusters k^* should have $G > 0$, and a higher score indicates that patches are more stably grouped into the same clusters. Figure 3a shows that the mean ARI drops from 0.48 at eight clusters to 0.32 at 48 clusters, and then continues to decline to below 0.3 after 68 clusters. Although the ARI score of 0.32 with 48 clusters is far from the perfect score of 1, previous literature [42] on the relative association of ARI scores and supervised learning measures for multiclass datasets reports that an ARI of 0.29 corresponds to 63.13% in the classification correct percentage rate (COR) in the configuration of supervised learning, and that an ARI of 0.46 corresponds to 62.4% in COR. In addition, visual inspection suggests that the clusters produced by the RICC stably group similar cloud patterns.

4.2. Stability Test 2: Significance of Similarities

Having determined how cluster similarity scores vary with the number of clusters, we next turn to the question of whether these values are significant. Following Von Luxburg [43], we compare cluster similarity scores, as shown in Algorithm A2 in Appendix A.3, against those obtained when the same method is applied to data generated not by our trained autoencoder but from a random uniform distribution clustered with the same HAC method. We then compute the mean clustering similarity score G from our patches and R from the data from the random uniform distribution for each k for all 435 combinations, though here we use the Rand index (as described in Appendix A.1) rather than ARI, as we are not comparing scores across k. We can then compare how the ratio between those two values varies with number of clusters. A ratio > 1 indicates that cluster assignments are more stably grouped than would be expected by chance; a value of 1 indicates that there is no benefit to adding extra clusters.

We expect the ratio G/R to be more than 1 if RICC cluster assignments are more stable than than those obtained on the null reference distribution. We set a threshold of $G/R \geq 1.01$, meaning that the results obtained with RICC should be 1% or more better than those with the null distribution. Figure 3b shows the significance of the stability values G/R as a function of the number of clusters k. The significance curve drops to 1.01 at 50 clusters, indicating an optimal cluster number $k^* < 50$.

4.3. Stability Test 3: Intra-Cluster Texture Similarity

A stable clustering should group patches with similar textures within the same cluster. To determine whether a clustering has this property, we examine how the average distance between latent representations within each cluster changes when we apply RICC to create different numbers of clusters. The mean distance between pairs of latent representations in a cluster relates to their similarity of texture, as our RI autoencoder learns texture features and encodes those features in latent representations. Specifically, we calculate the mean squared Euclidean distance between the latent representations computed for patches in our holdout set H.

For a clustering with k clusters, let n_c be the number of elements in cluster c, and $y_1 .. y_{n_c}$ be the patches in that cluster. As cluster sizes can vary, we weight each cluster's mean distance by $w_c = n_c / \sum_{i=1}^{k} n_i$, to obtain a weighted average mean squared Euclidean distance:

$$d_k = \sum_{c=1}^{k} \left(w_c \sum_{i=1}^{m} \sum_{j>i}^{m} \frac{||z(y_i) - z(y_j)||_2^2}{\frac{m}{2}(m-1)} \right) \text{ where } m = \min(n_c, N_p), \tag{6}$$

where z represents the latent representations generated by our RI autoencoder, and N_p is the maximum number of patches to consider in the distance calculation—a limitation used to accelerate calculations. We set $N_p = 200$ for our tests. Note that, when the total number of clusters is large, some individual clusters may have a size less than this limit.

We calculate Equation (6) for k from 8 to 256 for each of our 30 clusterings of test subsets $\{\text{RICC}_k^1(H), \dots, \text{RICC}_k^{30}(H)\}$, and then compute the mean value across clusterings. The resultant weighted average distance decreases monotonically with the cluster number k:

see Figure 4c, as does the metric G/R from test 2, but the trends have opposite implications: lower values are worse in test 2 but better in test 3. A lower distance value indicates that cloud texture and physical properties are more homogeneous within a given cluster, meaning the resultant AICCA dataset provides a more consistent cloud diagnostic. The implication is that the optimal number of clusters k^* will be approximately the largest number that satisfies our criterion in test 2.

In Figure 4c, the distance metric sharply decreases from 8 to 36 clusters, but the slope then flattens and values are almost unchanged between 40–48 clusters. That is, the pairwise similarity of latent representations drastically increases between 8 and 36 clusters but becomes less different among the range between 40–48 clusters. Selection of a k value from within this range would not change the result significantly. Since test 2 provides an upper bound of $k^* < 50$, the results of test 3 suggest that the optimal number of clusters lies in $40 \leq k^* \leq 48$.

To summarize: We observe that, as G/R decreases, ARI also declines, and that our G/R threshold requires $k^* < 50$. We observe that a cluster number in the range $40 \leq k \leq 48$ satisfies all four stability criteria. We have validated that these choices also satisfy criteria 1–4 in Table 4.

4.4. Stability Test 4: Seasonal Variation of Textures within Clusters

The results of the three tests above indicate that choices in the range $40 \leq k^* \leq 48$ will yield clusters that not only are stably assigned but also group similar cloud texture patterns. Our final test investigates whether clusters produced via RICC show similar patterns regardless of season: we compare intra-cluster texture similarity between OC-Patches from January and July. If differences are small, the number of clusters used is sufficient to accommodate the large seasonal changes in cloud morphology.

We use RICC with the autoencoder trained on OC-Patches$_{AE}$ and cluster centroids based on OC-Patches$_{HAC}$, for different numbers of clusters k, as before. For each k, we then apply the trained RICC$_k$ model to the patches in OC-Patches$_{HAC}$ to assign a label $c \in \{1, .., k\}$ to each patch, and for each c, extract the latent representations for m_c^s randomly selected July patches and m_c^w randomly selected January patches with that label (with m_c^s and m_c^w being at most 100 in these analyses, but less if a particular cluster has fewer January or July patches, respectively), compute an intra-cluster texture similarity score for each set of July and January patches, and (as in Section 4.3) weight each cluster mean by the actual m_c^s or m_c^w so that we can consider texture similarities from many clusters without results being dominated by trivial clusters that we observe to group fewer similar patches due to undersampling. We then sum the scores to obtain the overall weighted averaged squared distance (WASD) for k clusters. In summary:

$$\text{WASD}_k = \sum_{c=1}^{k} \left(w_c \sum_{i=1}^{m_c^s} \sum_{j=1}^{m_c^w} \frac{||z(y_i^s) - z(y_j^w)||_2^2}{m_c^s \cdot m_c^w} \right) \tag{7}$$

where w_c and z are as defined in Section 4.3 and $y^s = \{y_1^s \,..\, y_{m_c^s}^s\}$ and $y^w = \{y_1^w \,..\, y_{m_c^w}^w\}$ are the January and July patches in cluster c, respectively.

We expand the analysis to account for two additional potential sources of bias. Because the specific days used in OC-Patches$_{HAC}$ may affect our results, we assemble two additional versions of OC-Patches$_{HAC}$, selecting two days without replacement from each season in 2003, as before. The resulting OC-Patches$_{HAC-2}$ and OC-Patches$_{HAC-3}$ have 77,235 and 76,143 patches, respectively. Similarly, to account for any effect of the random selection of the m^s summer and m^w winter patches, we repeat the analysis of Equation (7) three times for each of OC-Patches$_{HAC}$, OC-Patches$_{HAC-2}$, and OC-Patches$_{HAC-3}$. In this way, we obtain a total of $9 \cdot k$ mean squared distance values and nine WASD values for each k in the range 8 to 256. These are shown as the dots in Figure 4. The WASD curve (black) decreases with increasing cluster number k, implying as expected that higher cluster numbers allow for better capturing of seasonal changes. Because a smoothed version of the WASD curve (blue) has a minimum of $k = 42$ over the range $40 \leq k \leq 48$, we choose 42 clusters as the optimum number and use this value in the inference step of Section 3.4. Given that the WMO cloud classes define approximately 28 subcategories, the 42 AICCA clusters should not overwhelm users who use AICCA to investigate cloud transitions.

4.5. Sanity Check: Comparison of RICCs with Different Number of Clusters to ISCCP Classes

As a final step, to confirm the utility of the choice of 42 classes, we consider whether and how RICC clusters associate with the nine ISCCP classes. We compare and contrast the frequencies of co-occurrence of (a) RICC clusters and (b) ISCCP classes, and evaluate how this relationship varies with cluster number used, considering not only the selected $k = 42$ but also $k = 10, 64$, and 256.

Recall that each of the nine ISCCP classes is defined by a distinct range of cloud optical thickness (COT) and cloud top pressure (CTP) values [3]: high, medium, and low clouds, and thin, medium, and thick clouds. To compare RICC clusters with ISCCP classes, we calculate the relative frequency of occurrence (RFO) of RICC clusters across the same two-dimensional COT–CTP space, a standard approach to evaluating unsupervised learning algorithms [44–46]. For this evaluation, we use the cluster assignments obtained with RICC when trained on $OC\text{-}Patches_{AE}$ and $OC\text{-}Patches_{HAC}$ to produce the AICCA dataset, as described in Section 3. We take the Terra satellite ocean-cloud patches for January and July 2003, and for each cluster, use the mean and standard deviation of the COT and CTP values for its patches to define a rectangular region for that cluster within two-dimensional COT-CTP space that extends for one standard deviation on either side of the mean. We then calculate the number of clusters that are associated with each of the nine ISCCP classes by counting the number of clusters that overlap with that region of COT-CTP space and dividing this number by the total number of cluster-class overlaps for all clusters and classes. Note that the latter number will typically be greater than the number of clusters because a single cluster can extend over multiple ISCPP classes.

This analysis shows a similar proportionality between RICC unsupervised learning clusters and ISCCP observation-based classes. Table 5 compares the resulting proportions of RICC clusters (for each value of cluster number k) with the simple mapping of all patches to ISCCP classes based on their COT and CTP values (top line). In all cases, the Stratocumulus (Sc) class is the largest single category, and medium-thickness clouds (Sc, As, Cs) predominate at each altitude level.

Stratocumulus (Sc) account for approximately 30% of RICC cluster overlaps, while the proportion of cloud observations in this category is over 50%. Similarly, for all k values, relatively few RICC clusters are assigned to high clouds, as expected since these make up only ~15% of total cloud occurrences. The thin and medium ISCCP classes (Cu, Sc, Ac, As), which account for 78.4% of cloud occurrence in the MODIS dataset, are represented by a similar proportion of RICC cluster overlaps: 74.45%, 70.44%, and 71.40% for $k = 42, 64$, and 256 clusters, respectively. There is no physical reason that cluster overlaps and cloud occurrence frequencies need be exactly the same: if, for example, all low medium-thickness clouds were identical in texture, we would expect that they would be assigned to a single cluster. However, the similarity of proportions suggests that AICCA captures physically meaningful distinctions among cloud types.

4.6. Discussion of Stability Protocol Results

We have used the stability protocol described in this section to determine the number of clusters that both achieves a stable grouping of patches and maximizes the richness of the information contained in our clusters. Recall that Von Luxburg's normalized stability protocol [43] simply minimizes an instability metric to determine the number of clusters that maximize stability. In contrast, we combine four tests—adjusted cluster similarity, normalized stability, weighted intra-cluster distance, and seasonal texture differences—to address the stability criterion. We used these tests to evaluate whether the cloud clusters produced by our unsupervised learning approach can provide meaningful insights for climate science applications.

This use of multiple similarity tests is essential to achieving our goal of both stability and maximality when grouping clusters. The *clustering similarity* test gives a mean score of scaled values calculated by ARI as a measurement of the degree of stability in OC-Patches. While this value is easy to understand when the resulting mean ARI is close to 1 (i.e., OC-Patches are always clustered into the same cluster group), ARI when applied to real world data could result in a value that is close to neither 0 nor 1 [42].

Table 5. ISCCP: Relative frequencies of occurrence based on mean COT and CTP values for OC-Patches from January and July, 2003. AICCA: Relative frequencies of occurrence of RICC clusters over each of the nine ISCCP cloud classes [3], as determined by counting the number of clusters that overlap (as determined by the mean, plus or minus one standard deviation, of COT and CTP values for patches within each cluster) with each class, divided by the total number of cluster-class overlaps. We allow double counts if a cluster overlaps more than one ISCCP class. Results are given for k=10, 42, 64, and 256 clusters, and for just January patches, just July patches, and both January and July patches. The AICCA values that are closest to the frequencies from MODIS column are in boldface. Recall that MODIS values are based on frequencies of *patches* over COT-CTP space, while the AICCA values are based on frequencies of *clusters* over COT-CTP space. Note that frequencies in each line add to 100, modulo rounding. We observe that the AICCA cluster frequencies are roughly proportional to the ISCCP category frequencies, although they consistently underestimate the Sc class (by 20%) and overestimate Cu and As classes.

	Height		Low			Medium			High		
	Thickness		Thin	Med	Thick	Thin	Med	Thick	Thin	Med	Thick
Dataset	Month	k	Cu	Sc	St	Ac	As	Ns	Ci	Cs	Dc
ISCCP	Jan & July 2003		5.29	53.94	2.93	3.65	15.50	2.12	3.39	10.52	2.60
AICCA$_k$	Jan 2003	10	11.42	25.71	8.57	**5.71**	**22.85**	8.57	2.85	8.57	5.71
		42	12.50	29.16	4.16	10.00	25.00	5.00	2.50	8.33	3.33
		64	10.38	**34.41**	**3.24**	7.79	26.62	**3.24**	0.64	**9.74**	3.89
		256	**9.06**	32.90	4.45	8.90	25.27	5.08	**3.65**	8.58	**2.06**
	July 2003	10	13.33	23.33	3.33	**6.66**	**16.66**	3.33	6.66	16.66	10.00
		42	10.30	30.92	4.12	7.21	20.61	4.12	**3.09**	**13.40**	6.18
		64	10.20	30.61	1.36	10.20	19.04	2.72	5.44	14.28	6.12
		256	**9.31**	**32.16**	**2.46**	8.78	21.61	**1.93**	5.97	13.53	**4.21**
	Jan & July 2003	10	12.50	25.00	6.25	3.12	**18.75**	3.12	9.37	15.62	6.25
		42	10.67	**32.03**	**2.91**	7.76	24.27	**2.91**	**2.91**	**11.65**	4.85
		64	8.80	29.55	3.77	8.17	23.89	5.03	5.03	11.94	**3.77**
		256	**8.42**	31.57	3.63	7.93	23.47	3.96	5.45	11.73	3.80

The *significance of similarities* test enables us to find the number of clusters after which there is reduced merit, from the perspective of stability against the null reference distribution, in adding more clusters. Normalized stability thus provides statistical support for eliminating certain cluster numbers, especially when the first test produces an ARI value that is close to neither 0 nor 1.

We introduce the *similarity of intra-cluster textures* test because common approaches to estimating an optimal number of clusters, such as the elbow method [47], silhouette method [48], and gap statistics [49], seek to determine the *minimum* number of clusters needed to characterize a dataset, which is not our goal. In our application, achieving a minimum number of clusters might result in the merging of sub-clusters with unique textures and slightly different physical properties. By minimizing the intra-cluster difference shown in Figure 4c until the slope of the curve of distance becomes small, the third test causes the lower bound on the optimal number of clusters to increase to $40 \leq k^*$, avoiding oversimplifications.

Finally, the *seasonal stability* test provides a further validation of our choice of k^*. A too-small number of clusters is likely to result in dissimilar July or January patches being mapped to the same cluster. We see in Figure 4 a local minimum in weighted average intracluster seasonal difference.

A disadvantage of our stability protocol is that, unlike other heuristic approaches [43,49], it does not always determine a unique optimal number of clusters. Indeed, our stability protocol in Section 4 concludes that $40 \leq k^* \leq 48$. Determining a single optimal number in the range sandwiched by the results of the four tests ultimately requires a subjective choice, based for example on the structure of cloud clusters in OC-Patches$_{HAC}$. In this study, we chose 42 as the number in the range $40 \leq k^* \leq 48$ that minimizes the seasonal variation of textures within clusters: see Section 4.4—although we note that a different selection of OC-Patches in OC-Patches$_{HAC}$ could motivate a different value.

5. Results

Having determined in Section 4 an optimal number of clusters, k^*, we then validate the scientific utility of AICCA$_{42}$ by evaluating the relationship between cloud class labels and their physical properties and spatial patterns. We have previously verified that the cloud clusters produced by RICC are physically reasonable using a limited subset of the MODIS data [18]. This section provides a similar analysis on a far more complete dataset of 589500 Terra ocean-cloud patches for January 2003 and July 2003. The goal is to confirm that AICCA$_{42}$ diagnoses meaningful physical properties for use in climate science applications.

5.1. Seasonal Variability of Cloud Cluster Regimes

Because the Earth is not symmetric, its clouds show strong seasonal variability not only in any given location but in the global mean. In this section, we show that the physical properties of AICCA$_{42}$ clusters are reasonable and remain stable even if the dataset is restricted to a single month. This analysis builds on those in Sections 4.4 and 4.5. In Section 4.4, we used intra-cluster seasonal differences as a criterion for choosing an optimal k of 42. In Section 4.5, we showed that RICC distributed those clusters in the COT-CTP space that defines established ISCCP classifications roughly in accordance with actual frequencies of cloud occurrence. We now plot the cluster distribution in COT-CTP space, and show that it is indeed reasonably constant across seasons (Figure 5). Note that, in assigning cluster labels, we sort the clusters first on CTP and then on the global occurrence of the clusters within each 50 hPa pressure bin.

As expected based on prior results, Figure 5 shows that most AICCA$_{42}$ clusters fall in the low cloud range (680–1100 hPa cloud top pressure) with low to medium optical thickness (2–20): Compare to Table 5. These results are broadly consistent with those of Jin et al. [50], who performed a simple clustering analysis with the joint histogram of optical thickness and cloud top pressure, though they obtained relatively more clusters associated with high clouds (four of their 11 clusters, vs. five of 42 in this work). The distribution of clusters is largely unchanged even when only January or July data are used in clustering. For example, the cumulus (Cu: left bottom) and stratocumulus (Sc: center bottom) regimes comprise 30 clusters in the full-year analysis, 30 in July only, and 32 in January only.

Using 42 clusters clearly allows RICC to capture richer cloud information than in the limited set of nine ISCCP cloud classes. In our previous work [18], we found that 12 clusters were insufficient to achieve a clear separation between high and low clouds. In this work, the clusters from our cloud fields can distinguish the full range of physical properties here (from high to low CTP and thick to thin COT), though thin clouds are included only because our cloud clusters defined by means and error bars (i.e., standard deviation of the cloud parameter) cover more than one ISCCP class. The choice of a cluster number of 42 produces a reasonable trade-off.

5.2. Comparing AICCA$_{42}$ and ISCCP Classifications

We now investigate further how AICCA$_{42}$ distributes clusters in COT-CTP space, and compare to observed occurrence frequencies. A limitation of the ISCCP cloud classification scheme is that the stratocumulus clouds whose behavior is of the greatest concern to climate scientists, and which comprise 54% of the MODIS dataset (Table 5), are lumped into a single ISCCP class (Figure 6a–c). A major motivation for AICCA$_{42}$ is to provide greater interpretive detail for understanding these low, marine clouds.

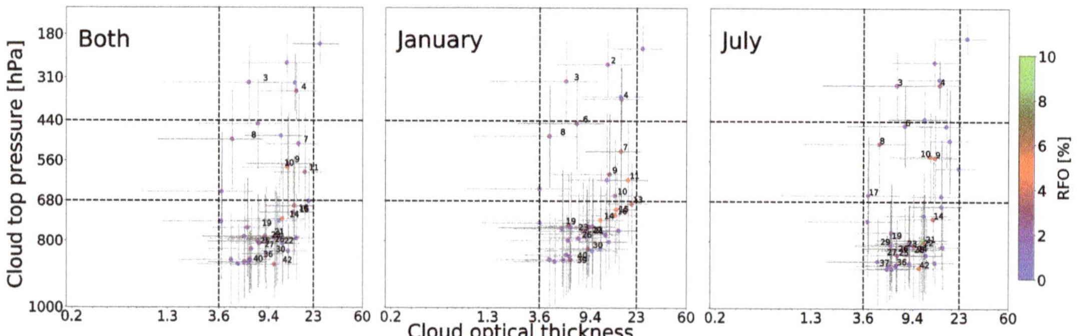

Figure 5. Distributions of cluster properties for AICCA$_{42}$ in COT–CTP space, where COT is cloud optical thickness (dimensionless) and CTP cloud top pressure (hPa). We show January and July 2003 (left), January only (center), and July only (right). Dots indicate mean values for each cluster and error bars the standard deviation of cluster properties. Data point colors indicate the relative frequency of occurrence (RFO) of each individual cluster in the dataset. Note that, in assigning cluster labels, we sort the clusters first on CTP and then on the global occurrence of the clusters within each 50 hPa pressure bin. Thus, small cluster numbers (e.g., #1) represent high-altitude cloud, and within a similar CTP range (e.g., 500 hPa–550 hPa), smaller numbers represent the more dominant patterns within the bin. For clarity, we show only the 21 clusters with the highest RFOs. For comparison, dashed lines divide the COT-CTP space into the nine regions corresponding to ISCCP cloud classes. AICCA$_{42}$ captures a greater variety of cloud types than do the ISCCP categories, with most of the clusters at low altitude (high CTP). January and July panels are similar, indicating that AICCA$_{42}$ adequately captures seasonal variation in cloud properties.

As shown in previous sections, AICCA$_{42}$ does provide a richer sampling of the stratocumulus (Sc) regime. AICCA$_{42}$ allocates 71% of cluster centers to the stratocumulus regime (Figure 5; 30 of 42 classes), or 32% of their relative occurrence frequency inclusive of overlaps (Table 5; see Section 4.5 for description of methodology). While Table 5 provided only mean values for each ISCCP class, Figure 6d shows the full distributions. As we would hope, AICCA$_{42}$ partitions cloud information more finely at low cloud altitudes and moderate cloud thickness (Sc), while still sampling every part of COT-CPT space.

5.3. Separation of Ice and Liquid Phases

We showed in previous work that RICC-generated clusters can differentiate between clouds that are dominated by ice vs. liquid phase. (See Figure 10 in Kurihana et al. [18]). We extend this analysis here and demonstrate that the same discrimination occurs in the larger AICCA$_{42}$ dataset. Figure 7 shows for each cloud class the average percentage of cloud pixels that are identified as an ice phase in the MOD06 cloud properties. As expected, cloud classes centered at high altitude (low CTP) are predominantly ice, those at middle altitudes are mixed, and those at low altitude are predominantly liquid. The lowest classes have <3% ice labels, and note that MOD06 cloud properties themselves have some error rate. The gradient in ice content across mid-level clouds, the region of transition from liquid to ice, also matches physical expectations. Note that while our ice phase ratio metric predominantly captures mixed-phase clouds, in which ice and liquid coexist in a single meteorological event (for our purposes, a patch), it is also affected by cases where a cluster contains a mix of pure-ice and pure-liquid clouds.

In summary, the AICCA$_{42}$ classes are sufficiently homogeneous to provide meaningful interpretation. These results support the physical reasonableness of the AICCA dataset.

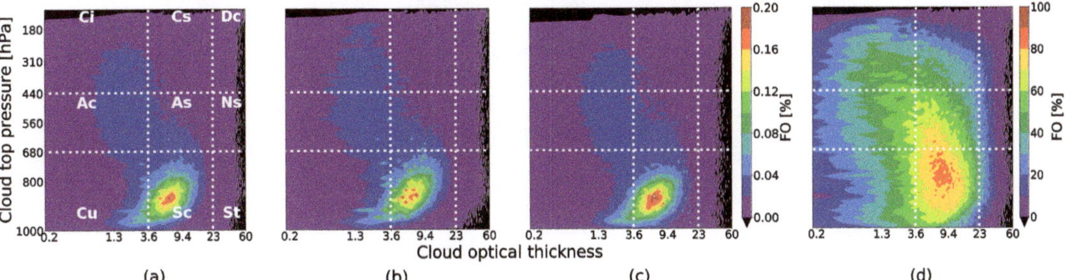

Figure 6. Heatmaps of the relative frequency of occurrence in COT-CTP space for (**a**) observed patches from both January and July, (**b**) values only from January, (**c**) values only from July, and (**d**) cluster counts inclusive of overlap from AICCA$_{42}$. Distributions are smoothed; resolution is 0.5 for COT and 10 hPa for CTP. Panels for observed frequencies (**a**–**c**) and cluster density (**d**) are expected to have different values. For example, in (**a**), a heatmap value of 0.1% indicates 5895 patches fall in a given histogram bin. In (**d**), a heatmap value of 71% indicates that 30 of 42 clusters overlap with that histogram bin over the range of one standard deviation. The data used here are those used throughout Section 5: all ocean-cloud patches from January and July 2003 from the Terra instrument. White dashed lines show the boundaries of the nine ISCCP cloud classes [3,51]: Cirrus (Ci), Cirrostratus (Cs), Deep convection (Dc), Altocumulus (Ac), Altostratus (As), Nimbostratus (Ns), Cumulus (Cu), Stratocumulus (Sc), and Stratus (St). AICCA$_{42}$ clusters cover all nine ISCCP classes, with the largest representation in the Stratocumulus (Sc) category where occurrence also peaks.

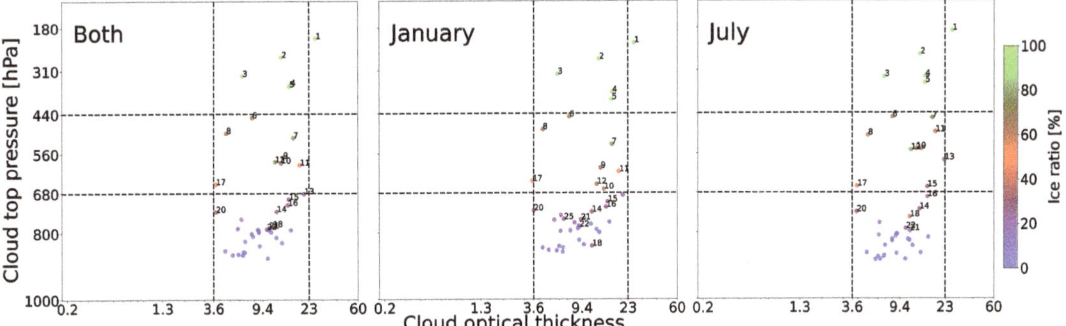

Figure 7. Test of whether AICCA$_{42}$ captures expected variations in cloud microphysics, i.e., the ice and liquid fractions for individual cloud classes. The figure is constructed in the same way as Figure 5, but with each color marker now showing the cluster's mean *ice phase ratio*, defined as the mean within-cluster percentage of cloud pixels denoted as ice phase. We omit all pixels labeled as "undetermined" in MOD06; many of these are internally mixed phase but the proportions cannot be determined. AICCA$_{42}$ cloud classes are sufficiently restricted that they capture the expected microphysics, with higher ice fractions in higher-altitude clouds.

5.4. Case Studies: Spatial Distribution of Cloud Textures and Associated Cluster Labels

To provide a visual example of the power of AICCA$_{42}$ classes in interpreting cloud processes, we examine two case studies involving swaths of MODIS imagery, both dominated by marine stratocumulus, off the west coast of South America: see Figure 8a. Note that the swath labeled B is from January and that labeled C from July.

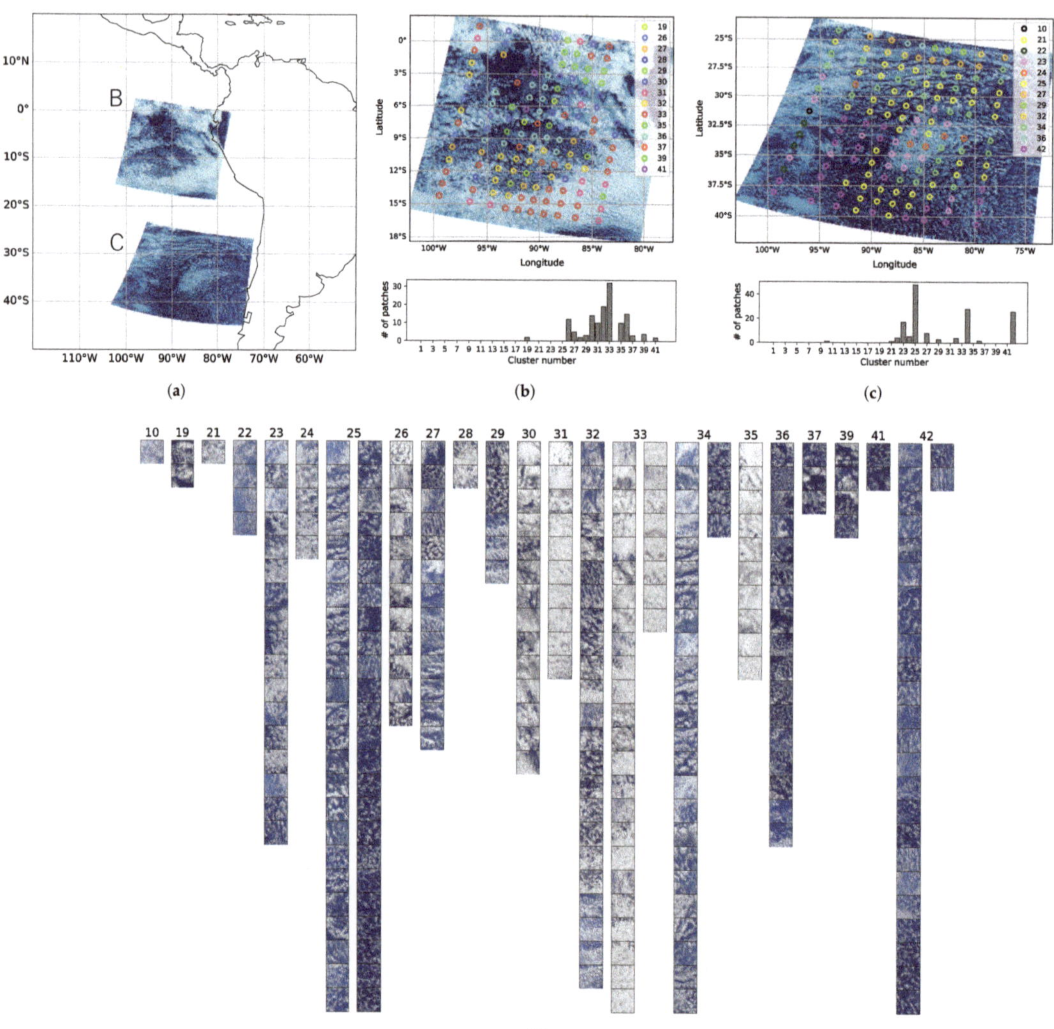

Figure 8. (**a**) Geographical location of two example MODIS swaths (B and C) off the west coast of South America, both from the Terra instrument but at different times; (**b**) Swath B, 133 ocean-cloud patches between 18° S to 3° N, 76° S to 104° S, observed on 15 January 2003, with each patch represented by a dot with color indicating its associated class label in the range 1..42; (**c**) Swath C, 147 ocean-cloud patches between 44° S to 23° S, 72° S to 103° S, observed on 20 July 2003, similarly labeled. Note that not all clusters appear in each swath. Histograms in (**b**,**c**) show the distribution of cloud class labels; note there is little overlap; (**d**) MODIS true color images [52] for all ocean-cloud patches labeled in (**b**,**c**), grouped by cluster number. Note the visual similarity of cloud textures within each cluster. AICCA$_{42}$ produces spatially coherent cluster assignments, groups visually similar textures, provides rich detail by subdividing stratocumulus clouds into multiple classes, and identifies subtle spatial and/or temporal differences.

The two example swaths show the richness and diversity of stratocumulus patterns. The more equatorial swath B (Figure 8b) shows regions of both open- and closed-cell stratocumulus clouds, and sharp transition regions. The mid-latitude summertime swath C (Figure 8c) is dominated by open-cell stratocumulus clouds, with broad transitional regions and only small patches of classic closed-cell.

$AICCA_{42}$ cluster labels capture important aspects of these distributions. As usual, we label only patches with >30% cloud pixels; each such patch is marked with a dot in Figure 8b,c, with the color denoting the cluster label. The cloud classes assigned are geographically contiguous and reflect clear visual distinctions in cloud texture (Figure 8d). They also capture important and subtle distinctions. Each swath contains 12–14 unique classes, but only four are shared between both. That is, cloud classes of otherwise similar visual appearance are strongly differentiated in space and/or time. Open-cell stratocumulus in swath B is assigned to classes #32 and #36, but that in swath C largely to #25, #34, and #42. Similarly, closed-cell stratocumulus in swath B is assigned to classes #30, #31, #33, and #35, none of which are present in swath C. Instead, the smaller areas of closed-cell stratocumulus in swath C are labeled as class #24. These results suggest that real-world stratocumulus cloud textures involve subtle but important spatial and/or temporal distinctions and that $AICCA_{42}$ is capturing those distinctions.

5.5. Use Case: Geographic Distribution of Cluster Label Occurrence

In this last study, we examine the geographic distribution of $AICCA_{42}$ cluster labels. Using the same dataset as in the other part of this section, we show in Figure 9 mean incidences for each of the 42 cloud types in the dataset used throughout Section 5, gridded on a 1° global grid We see strong geographic distinctions among cluster labels, with some occurring only in the tropics and others only at high latitudes. Some show even finer geographic restrictions. For example, cloud classes #1–#3 are localized primarily in the West Pacific warm pool, all likely associated with tropical deep convection, though ranging in altitude (232–367 hPa CTP) and thickness (24–6 COT). (Classes are numbered in order of their mean altitude; see Section 5.1 for details). By contrast, the stratocumulus cloud labels discussed for Figure 8 show different distributions. Those most clearly associated with classic closed-cell stratocumulus—#30, #33, and #35—are as expected primarily localized to small areas on the west coasts of continents. The most predominant open-cell Sc cloud labels in Figure 8—#25, #32, and #36—are more widely distributed but with strong latitudinal dependence. The six clusters just described are all low in altitude (mean CTP of 803–901 hPa) and moderate in thickness (mean COT of 8.4–13.6 thickness for the closed-cell classes and 5.7–7.1 for the open-cell). All would therefore be labeled as Sc in the ISCCP classification; $AICCA_{42}$ reveals their striking differences. Note that, because our example dataset includes both January and July 2003, these graphs include both summer and wintertime occurrences. When displayed as an animation of monthly means, the geographic distinctions become even sharper, with patterns migrating seasonally with the sun's position.

To highlight the *texture* distinctions in the Sc cloud classes just discussed, we show in Figure 10 the true color images [52] corresponding to the 20 patches closest to the OC-Patches$_{HAC}$ centroid for each of the six clusters. Patches shown for each cluster are visually similar, and the different clusters have distinct differences in not only cloud pixel density but also spatial arrangement, even within the broad open cell (top row, #25, #32, and #36) and closed cell (#30, #33, and #35) categories. These distinctions show that $AICCA_{42}$ is separating stratocumulus clouds by texture as well as by mean properties across the patch.

The strong localization of some cloud classes near the poles raises concern that they may be affected by the presence of sea ice. We have restricted analysis to ocean clouds to avoid the complications of surface effects—the ocean provides a dark and homogeneous background—but parts of the high-latitudes ocean are covered in wintertime ice. Because two of the MODIS bands used in our cloud clustering system, bands 6 (1.6 µm) and 7 (2.12 µm), are also used by the MODIS snow and ice detection algorithm [53], the resulting AICCA dataset can inadvertently include some surface information in the latent representation. To check for contamination, we use a MODIS cloud product that describes the presence of a snow and ice background for each pixel (MOD06). Only one cloud class may experience significant interference: #12, which forms in local winter. (Sea ice makes up 16/31% of its labeled pixels in January/July). The other polar cloud classes appear in local summer. Sea ice effects therefore do not appear to drive the labeling of geographically distinct cloud classes that appear in polar oceans.

These results suggest that $AICCA_{42}$ identifies real and important differences between cloud types and can help climate scientists understand the drivers of distinct cloud patterns and regimes.

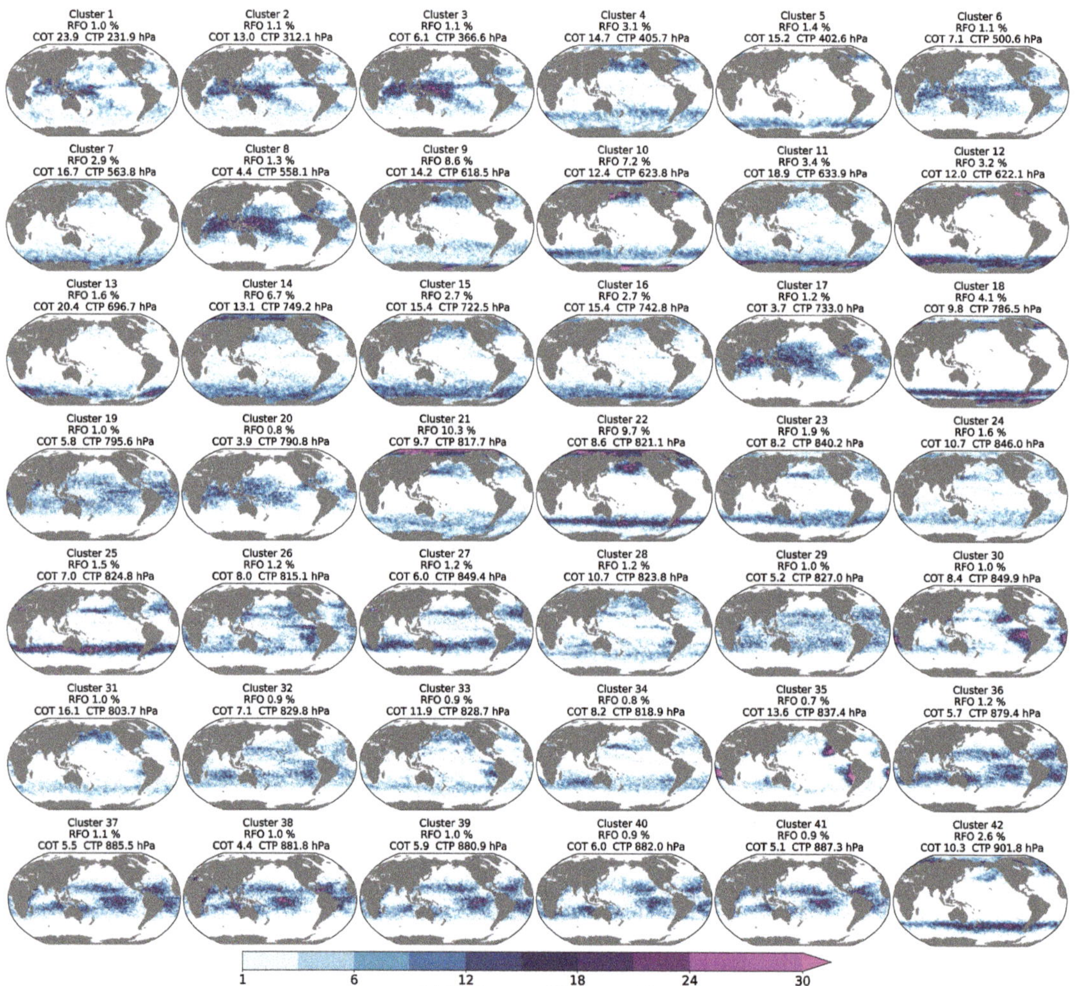

Figure 9. An example application of AICCA. We plot the relative frequency of occurrence (RFO) for each of the 42 $AICCA_{42}$ clusters, using all data from January and July, 2003. Land is in grey, and areas where RFO < 1.0% are in white. Surtitles show global mean RFO, cloud optical thickness (COT), and cloud top pressure (CTP) for the given cluster. Clusters show striking geographic distinctions, and those with roughly similar spatial patterns have different mean physical properties, suggesting meaningful physical distinctions. The 99 percentile of RFO values (RFO $\geq$ 1 %) of #30 is 29.85 % and the value of #35 is 36.58 %.

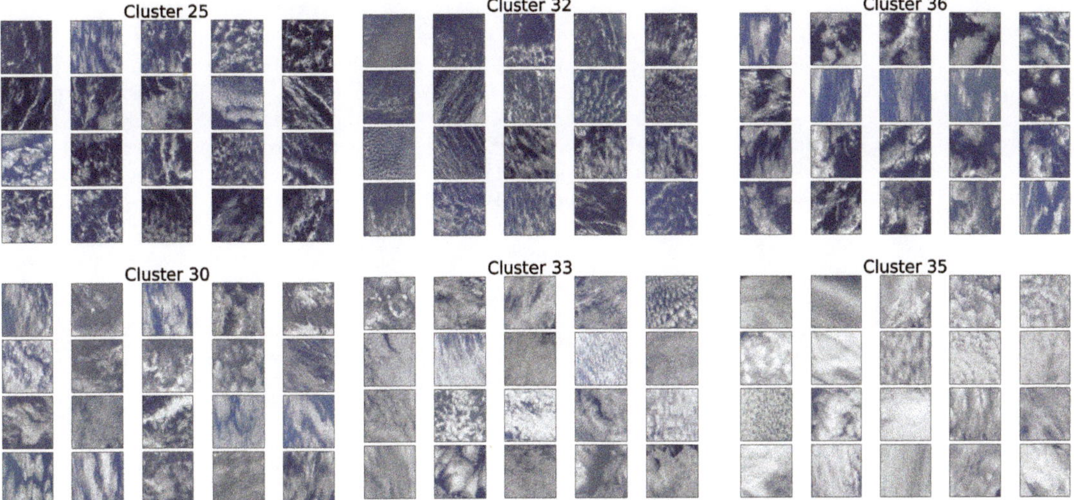

Figure 10. Selected MODIS true color images [52] for the six clusters that dominate open-cell (upper row: #25, #32, and #36) and closed-cell (lower row: #30, #33, and #35) stratocumulus clouds in Figure 9. Surtitles show the cluster numbers. We show the 20 patches closest to OC-Patches$_{HAC}$ centroids. Note how AICCA discriminates between textures (e.g., compare the fine-scale detail of #32 to the more coarsely aggregated #36) even for patches of similar mean cloud properties.

6. Conclusions

We have introduced an AI-driven cloud classification atlas, AICCA$_{42}$ that provides the first global-scale unsupervised classification of clouds in MODIS satellite imagery. AICCA$_{42}$ provides a compact form of the information available in multi-spectral satellite images, reducing 801 TB of MODIS products to 54.2 gigabytes of cloud labels and, for diagnostic purposes, four cloud properties from MOD06 (cloud optical thickness, cloud top pressure, cloud phase, and cloud effective radius). We have described the complete workflow used to generate the dataset, the five criteria used to assess its success (physically reasonable, spatial distributions, separable, rationally invariant, and stable), and the novel protocol developed to determine the optimal number of clusters that meets the stability requirement.

The new stability protocol is needed because our goal differs from the norm in clustering studies, which generally seek to determine the *minimum* number of clusters needed to characterize a dataset. Instead, we seek to maximize the richness of information captured by determining the *maximum* number of clusters that remain stable to changes in the training set. The protocol of four tests suggests an optimal cluster number of $k^* = 42$, and our seasonal stability sanity check confirms that this number is sufficient to capture the full seasonal diversity of global cloud textures. The resulting atlas of cloud classes greatly enhances the richness of information provided over the traditional 9-class ISCCP scheme, especially for climate-critical cloud types: for example, 30 of the AICCA$_{42}$ classes are devoted to stratocumulus, whose behavior is a key uncertainty in climate projections [54].

Preliminary analysis of the AICCA$_{42}$ atlas suggests its power for science. Its cloud classes meaningfully group physical properties such as altitude or optical thickness, and also capture distinct textures and patterns. Cloud classes show strikingly different geographical distributions, with distributions evolving seasonally. Some classes can be matched to known cloud processes: deep convection in the West Pacific warm pool, for example, or marine stratocumulus decks that form off the west coast of continents. In other cases, cloud classes capture distinctions not previously appreciated, and can lead to new lines of scientific inquiry. We conclude that (1) our methodology has explanatory power, in that it captures regionally unique cloud classes, and (2) 42 clusters is a useful number for a global analysis.

The AICCA approach also opens up possibilities in other areas. For example, increasing computing power means the spatial scale of climate simulations has shrunk to the point where their output can resolve complex cloud textures [55]. Unsupervised cloud classification can help in assessing whether models capture those textures correctly. More broadly, advances in remote sensing instrumentation mean that many fields have seen large increases in data volume. We have shown here that AI-based methods using a convolutional autoencoder can effectively identify novel patterns in spatial data. Unsupervised learning offers the possibility of unlocking large satellite datasets and making them tractable for analysis.

Author Contributions: Conceptualization, E.J.M. and I.T.F.; methodology, T.K., E.J.M. and I.T.F.; software, T.K.; validation, T.K., E.J.M. and I.T.F.; formal analysis, T.K.; investigation, T.K.; resources, I.T.F.; data curation, T.K.; writing—original draft preparation, T.K.; writing—review and editing, E.J.M. and I.T.F.; visualization, T.K.; supervision, E.J.M. and I.T.F.; project administration, E.J.M. and I.T.F.; funding acquisition, E.J.M. and I.T.F. All authors have read and agreed to the published version of the manuscript.

Funding: This work was supported in part by the AI for Science program of the Center for Data and Computing at the University of Chicago; by the Center for Robust Decision-Making on Climate and Energy Policy (RDCEP), under NSF Grant No. SES-1463644; and by the U.S. Department of Energy, Office of Science, Advanced Scientific Computing Research, under Contract DE-AC02-06CH11357.

Institutional Review Board Statement: Not applicable.

Data Availability Statement: The AICCA data are available upon request until 1 May 2023, on which date all data will be made available without restriction at: https://github.com/RDCEP/clouds#download-aicca-dataset (accessed on 26 August 2022). The rotation-invariant autoencoder can be found at: https://acdc.alcf.anl.gov/dlhub/?q=climate (accessed on 16 September 2022). Our codebases and Jupyter notebooks are available in Github: https://github.com/RDCEP/clouds (accessed on 2 November 2022).

Acknowledgments: The authors thank Rebecca Willett and Davis Gilton for contributing to the RICC design; Scott Sinno for providing MODIS products from the NASA Center for Climate Simulation; NASA for access to MODIS data; Zhuozhao Li, Sandro Fiore, and Donatello Elia for collaboration and feedback on the AICCA workflow; and the Argonne Leadership Computing Facility and the University of Chicago's Research Computing Center for access to computing resources.

Conflicts of Interest: The authors declare no conflict of interest.

Appendix A

Appendix A.1. Rand Index and Adjusted Version for Chance

We describe the Rand index used in Section 4. Let $U = \{U_1, \ldots, U_r\}$ and $V = \{V_1, \ldots, V_c\}$ be two clustering partitions of a set of N objects $O = \{o_1, \ldots, o_{N_p}\}$, such that $\bigcup_{i=1}^{r} U_i = \bigcup_{j=1}^{c} V_j = O$, and $U_i \cup U_{i'} = \varnothing$ as well as $V_j \cup V_{j'} = \varnothing$ for $1 \le i \le r$ and $1 \le j \le c$. We count how many of the $\binom{N}{2}$ possible pairings of elements in O are in the same or different clusters in U and V:

- P_{11}: number of element pairs that are in the *same* clusters in both U and V;
- P_{10}: number of element pairs that are in *different* clusters in U, but in the *same* cluster in V;
- P_{01}: number of element pairs that are in the *same* cluster in U, but in *different* clusters in V; and
- P_{00}: number of element pairs that are in *different* clusters in both U and V.

The Rand index then computes the fraction of correct cluster assignments:

$$\mathrm{RandI}(U, V) = \frac{P_{11} + P_{00}}{P_{11} + P_{10} + P_{01} + P_{00}} = \frac{P_{11} + P_{00}}{\binom{N}{2}} \tag{A1}$$

It has value 1 if all pairs of labels are grouped correctly and 0 if none are correct. The metric is independent of the absolute values of the labels: that is, it allows for permutations.

To illustrate how the Rand index works, consider the two clusterings: $A = \{d_1\}, \{d_2, d_3\}$ and $B = \{d_1, d_2\}, \{d_3\}$ of the dataset $D = \{d_1, d_2, d_3\}$. Here, $N = 3$, and there are $\binom{3}{2} = 3$ possible pairings of the three dataset elements: $(d_1, d_2), (d_1, d_3), (d_2, d_3)$. Thus: $P_{11} = 0$, as no pair is in the same cluster in both A and B; $P_{10} = 1$, as d_1 and d_2 are in different clusters in A but the same cluster in B; $P_{01} = 1$, as d_2 and d_3 are in different clusters in A but the same cluster in B; and $P_{00} = 1$, as d_1 and d_3 are in different clusters in both A and B. Hence, the Rand index by Equation (A1) of A and B is $(0 + 1)/3 = 0.33$.

A difficulty with the Rand index is that its value tends to increase with the number of clusters, hindering comparisons across different numbers of clusters. In order to permit comparisons of Rand index values across different numbers of clusters, the **adjusted Rand index** (ARI) [56] corrects for co-occurrences due to chance:

$$\text{ARI}(U, V) = \frac{\binom{N}{2}(P_{11} + P_{00}) - [(P_{11} + P_{10})(P_{11} + P_{01}) + (P_{01} + P_{00})(P_{10} + P_{00})]}{\binom{N}{2}^2 - [(P_{11} + P_{10})(P_{11} + P_{01}) + (P_{01} + P_{00})(P_{10} + P_{00})]}, \quad \text{(A2)}$$

where the P_{xy} are as defined above.

Appendix A.2. Clustering Similarity Test

We present as Algorithm A1 our implementation of the *clustering similarity test*. As described in Section 4.1, we use as the input dataset D all ocean-cloud patches from 2003–2021, inclusive. We define a holdout set, H, for evaluation (line 1), and use as our "perturbed versions" N subsets selected without replacement from $D \setminus H$ (line 3). Then, for each number of clusters, k, in the range $8 \leq k \leq k_{\max}$, we: train RICC on each subset (line 8); apply the trained RICC to generate a clustering for the holdout set (line 6); use the adjusted Rand index, ARI, to evaluate pairwise distances between those clusterings (line 10); and average among the 30 clusterings generated by the RICC models $\{\text{RICC}_k^i, i \in 1..30\}$ to determine the mean clustering similarity for that specific cluster number k. Finally, we calculate the ARI for all $\binom{30}{2} = 435$ combinations of those 30 clusterings and determine the mean ARI score $G_8..G_{k_{max}}$ (line 12).

Algorithm A1 Pseudocode for the clustering similarity test described in Section 4.1.

Input: D: { OC-Patches for 2003–2021, inclusive }
Output: $G_8, \ldots, G_{k_{\max}}$: Clustering similarity scores for cluster counts from 8 to k_{max}.

1:   $H := \{x \mid x \in D\}$ where $|H| = N_H$ ▷ Select holdout set to be used for evaluation
2:   **for** i from 1 to N **do**
3:      Select a subset $S_i := \left\{x \mid x \in D \setminus H \setminus \bigcup_{j=1}^{i-1} S_j\right\}$ with $|S_i| = N_R$
4:      **for** k from 8 to $k_{\max}$ **do**
5:        $\text{RICC}_k^i \leftarrow$ Train RICC with k clusters on $S_i \cup H$
6:        $C_k^i \leftarrow \text{RICC}_k^i(H)$ ▷ Determine cluster assignments in H with RICC_k^i
7:      **end for**
8:   **end for**
9:   **for** k from 8 to $k_{\max}$ **do**
10:      $G_k = \dfrac{1}{\binom{N}{2}} \sum_{(i,j) \in \binom{N}{2}} \text{ARI}\left(C_k^i, C_k^j\right)$ ▷ Mean similarities for RICC clusters
11:   **end for**
12:   Return clustering similarity scores $\{G_8, \ldots, G_{k_{max}}\}$

Appendix A.3. Stability Significance Test

Algorithm A2 implements the stability significance test described in Section 4.2. For each k in the range $8..k_{\max}$, we first compute clusterings (line 9) as in the clustering similarity test of Appendix A.2 and then compute the mean Rand index score (see Appendix A.1)

$G_8..G_{k_{max}}$ (line 15). To produce random label assignments, we first prepare 30 datasets that are sampled from random uniform distributions $\mathcal{U} \in [-2\sigma, 2\sigma]$ (line 6). We then apply HAC to the random data to generate random labels (line 11), from which we also calculate the Rand Index for 435 combinations, giving the mean scores $R_8..R_{k_{max}}$ (line 16). Finally, we compare how the ratio $\frac{G_k}{R_k}$ varies with number of clusters, k (line 19).

Algorithm A2 Pseudocode for the stability significance test described in Section 4.2.

Input: D: { OC-Patches for 2003–2021, inclusive }, trained rotation-invariant autoencoder AE

Output: $\{\frac{G_8}{R_8}, \ldots, \frac{G_{k_{max}}}{R_{k_{max}}}\}$: cluster similarity significance scores

1:   $H := \{x \mid x \in D\}$ where $|H| = N_H$ $\triangleright$ Select holdout set to be used for evaluation

2:   $z = \{AE(x) : x \in H\}$ $\triangleright$ Use trained autoencoder to compute latent representations

3:   $\sigma = \sqrt{\frac{1}{N_H} \sum_{j=1}^{N_H} \left(z_j - \bar{z}\right)^2}$ $\triangleright$ Calculate standard deviation σ for latent representations

4:   **for** i from 1 to N **do**

5:       Select a subset $S_i := \left\{x \mid x \in D \setminus H \setminus \bigcup_{j=1}^{i-1} S_j\right\}$ with $|S_i| = N_R$

6:       Sample $U_i := \left\{u \mid u \in \mathcal{U}[-2\sigma, 2\sigma]\right\}$ with $|U_i| = N_H$, $\mathcal{U}$ a random uniform distribution.

7:       **for** k from 8 to k_{max} **do**

8:          $\text{RICC}_k^i \leftarrow$ Train RICC on $S_i \cup H$

9:          $\text{RICC}_k^i(H) \leftarrow$ Determine cluster assignments in H

10:         $\text{HAC}_k^i \leftarrow$ Train HAC on U_i

11:         $\text{HAC}_k^i(U_i) \leftarrow$ Determine cluster assignments in U_i

12:       **end for**

13: **end for**

14: **for** k from 8 to k_{max} **do** $\triangleright$ Calculate averages of cluster similarities

15:       $G_k = \frac{1}{\binom{N}{2}} \sum_{(i,j)\in\binom{N}{2}} \left[\text{RandI}\left(\text{RICC}_k^i(H), \text{RICC}_k^j(H)\right)\right]$ $\triangleright$ Mean similarities for RICC clusters

16:       $R_k = \frac{1}{\binom{N}{2}} \sum_{(i,j)\in\binom{N}{2}} \left[\text{RandI}\left(\text{HAC}_k^i(U_i), \text{HAC}_k^j(U_j)\right)\right]$ $\triangleright$ Mean similarities for random clusters

17:       Calculate $\frac{G_k}{R_k}$, ratio of stability between RICC and random samples

18: **end for**

19: Return cluster similarities significance scores, $\{\frac{G_8}{R_8}, \ldots, \frac{G_{k_{max}}}{R_{k_{max}}}\}$

References

1. Rossow, W.B.; Schiffer, R.A. ISCCP cloud data products. *Bull. Am. Meteorol. Soc.* **1991**, *71*, 2–20. [CrossRef]
2. Rossow, W.B.; Walker, A.W.; Garder, L.C. Comparison of ISCCP and other cloud amounts. *J. Clim.* **1993**, *6*, 2394–2418. [CrossRef]
3. Rossow, W.B.; Schiffer, R.A. Advances in understanding clouds from ISCCP. *Bull. Am. Meteorol. Soc.* **1999**, *80*, 2261–2288. [CrossRef]
4. World Meteorological Organization. International Cloud Atlas. Available online: https://cloudatlas.wmo.int/ (accessed on 1 November 2022).
5. Wood, R. Stratocumulus clouds. *Mon. Weather Rev.* **2012**, *140*, 2373–2423. [CrossRef]
6. Zhang, J.; Liu, P.; Zhang, F.; Song, Q. CloudNet: Ground-based cloud classification with deep convolutional neural network. *Geophys. Res. Lett.* **2018**, *45*, 8665–8672. [CrossRef]
7. Rasp, S.; Schulz, H.; Bony, S.; Stevens, B. Combining crowd-sourcing and deep learning to understand meso-scale organization of shallow convection. *arXiv* **2019**, arXiv:1906.01906. [CrossRef].
8. Zantedeschi, V.; Falasca, F.; Douglas, A.; Strange, R.; Kusner, M.; Watson-Parris, D. Cumulo: A Dataset for Learning Cloud Classes. NeurIPS Workshop on Tackling Climate Change with Machine Learning. 2019. Available online: https://www.climatechange.ai/papers/neurips2019/11 (accessed on 1 November 2022).
9. Yuan, T.; Song, H.; Wood, R.; Mohrmann, J.; Meyer, K.; Oreopoulos, L.; Platnick, S. Applying deep learning to NASA MODIS data to create a community record of marine low-cloud mesoscale morphology. *Atmos. Meas. Tech.* **2020**, *13*, 6989–6997. [CrossRef]
10. Marais, W.J.; Holz, R.E.; Reid, J.S.; Willett, R.M. Leveraging spatial textures, through machine learning, to identify aerosols and distinct cloud types from multispectral observations. *Atmos. Meas. Tech.* **2020**, *13*, 5459–5480. [CrossRef]
11. Stevens, B.; Bony, S.; Brogniez, H.; Hentgen, L.; Hohenegger, C.; Kiemle, C.; L'Ecuyer, T.S.; Naumann, A.K.; Schulz, H.; Siebesma, P.A.; et al. Sugar, gravel, fish and flowers: Mesoscale cloud patterns in the trade winds. *Q. J. R. Meteorol. Soc.* **2020**, *146*, 141–152. [CrossRef]
12. Visa, A.; Iivarinen, J.; Valkealahti, K.; Simula, O. Neural network based cloud classifier. In *Industrial Applications of Neural Networks*; World Scientific: Singapore, 1998; pp. 303–309. [CrossRef]

13. Tian, B.; Shaikh, M.K.; Azimi-Sadjadi, M.R.; Haar, T.H.; Reinke, D. A study of cloud classification with neural networks using spectral and textural features. *IEEE Trans. Neural Netw.* **1999**, *10*, 138–151. [CrossRef]

14. Denby, L. Discovering the importance of mesoscale cloud organization through unsupervised classification. *Geophys. Res. Lett.* **2020**, *47*, e2019GL085190. [CrossRef]

15. Kurihana, T.; Foster, I.T.; Willett, R.; Jenkins, S.; Koenig, K.; Werman, R.; Barros Lourenco, R.; Neo, C.; Moyer, E.J. Cloud classification with unsupervised deep learning. In Proceedings of the 9th International Workshop on Climate Informatics, Paris, France, 2–4 October 2019. [CrossRef]

16. Johnson, S.C. Hierarchical clustering schemes. *Psychometrika* **1967**, *32*, 241–254. [CrossRef] [PubMed]

17. Hinton, G.E.; Krizhevsky, A.; Wang, S. Transforming auto-encoders. In *International Conference on Artificial Neural Networks*; Springer: Berlin/Heidelberg, Germany, 2011; pp. 44–51. [CrossRef]

18. Kurihana, T.; Moyer, E.J.; Willett, R.; Gilton, D.; Foster, I.T. Data-driven cloud clustering via a rotationally invariant autoencoder. *IEEE Trans. Geosci. Remote Sens.* **2021**, *60*, 4103325. [CrossRef]

19. Adams, D. *The Hitchhikers Guide to the Galaxy*; Random House: New York, NY, USA, 1979.

20. MODIS Characterization Support Team. *MODIS/Aqua 1km Calibrated Radiances Product*; Goddard Space Flight Center: Greenbelt, MD, USA, 2017. [CrossRef]

21. MODIS Characterization Support Team. *MODIS/Terra 1km Calibrated Radiances Product*; Goddard Space Flight Center: Greenbelt, MD, USA, 2017. [CrossRef]

22. Rakwatin, P.; Takeuchi, W.; Yasuoka, Y. Stripe noise reduction in MODIS data by combining histogram matching with facet filter. *IEEE Trans. Geosci. Remote Sens.* **2007**, *45*, 1844–1856. [CrossRef]

23. Rew, R.; Davis, G. NetCDF: An interface for scientific data access. *IEEE Comput. Graph. Appl.* **1990**, *10*, 76–82. [CrossRef]

24. Kluyver, T.; Ragan-Kelley, B.; Pérez, F.; Granger, B.; Bussonnier, M.; Frederic, J.; Kelley, K.; Hamrick, J.; Grout, J.; Corlay, S.; et al. Jupyter Notebooks—A publishing format for reproducible computational workflows. In *Positioning and Power in Academic Publishing: Players, Agents and Agendas*; Loizides, F., Schmidt, B., Eds.; IOS Press: Amsterdam, The Netherlands, 2016; pp. 87–90.

25. Kurihana, T. Rotation-Invariant Cloud Clustering Code. 2022. Available online: https://github.com/RDCEP/clouds (accessed on 1 November 2022).

26. Chard, R.; Li, Z.; Chard, K.; Ward, L.; Babuji, Y.; Woodard, A.; Tuecke, S.; Blaiszik, B.; Franklin, M.J.; Foster, I.T. DLHub: Model and data serving for science. In Proceedings of the IEEE International Parallel and Distributed Processing Symposium, Rio de Janeiro, Brazil, 20–24 May 2019; pp. 283–292. [CrossRef]

27. Chard, K.; Tuecke, S.; Foster, I.T. Efficient and secure transfer, synchronization, and sharing of big data. *IEEE Cloud Comput.* **2014**, *1*, 46–55. [CrossRef]

28. Chard, R.; Babuji, Y.; Li, Z.; Skluzacek, T.; Woodard, A.; Blaiszik, B.; Foster, I.T.; Chard, K. FuncX: A federated function serving fabric for science. In Proceedings of the 29th International Symposium on High-performance Parallel and Distributed Computing, Stockholm, Sweden, 23–26 June 2020; pp. 65–76. [CrossRef]

29. Hinton, G.E.; Richard, S.Z. Autoencoders, minimum description length and Helmholtz free energy. In *Advances in Neural Information Processing Systems 6*; Morgan-Kaufmann: Burlington, MA, USA, 1994; pp. 3–10.

30. Lee, D.D.; Seung, H.S. Learning the parts of objects by non-negative matrix factorization. *Nature* **1999**, *401*, 788–791. [CrossRef]

31. Matsuo, T.; Fukuhara, H.; Shimada, N. Transform invariant auto-encoder. In Proceedings of the IEEE/RSJ International Conference on Intelligent Robots and Systems, Vancouver, BC, Canada, 24–28 September 2017; pp. 2359–2364. [CrossRef]

32. Nair, V.; Hinton, G.E. Rectified linear units improve restricted Boltzmann machines. *Int. Conf. Mach. Learn.* **2010**. Available online: https://icml.cc/Conferences/2010/papers/432.pdf (accessed on 1 November 2022).

33. Ioffe, S.; Szegedy, C. Batch normalization: Accelerating deep network training by reducing internal covariate shift. *Int. Conf. Mach. Learn.* **2015**, *37*, 448–456. [CrossRef]

34. Ward, J.H., Jr. Hierarchical grouping to optimize an objective function. *J. Am. Stat. Assoc.* **1963**, *58*, 236–244. [CrossRef]

35. Jenkins, S.; Moyer, E.J.; Foster, I.T.; Kurihana, T.; Willett, R.; Maire, M.; Koenig, K.; Werman, R. Developing unsupervised learning models for cloud classification. *AGU Fall Meet. Abstr.* **2019**, A51U-2673.

36. Moertini, V.S.; Suarjana, G.W.; Venica, L.; Karya, G. Big data reduction technique using parallel hierarchical agglomerative clustering. *IAENG Int. J. Comput. Sci.* **2018**, *45*, 1.

37. Varoquaux, G.; Buitinck, L.; Louppe, G.; Grisel, O.; Pedregosa, F.; Mueller, A. Scikit-learn: Machine learning without learning the machinery. *Getmobile Mob. Comput. Commun.* **2015**, *19*, 29–33. [CrossRef]

38. Jin, C.; Liu, R.; Chen, Z.; Hendrix, W.; Agrawal, A.; Choudhary, A. A scalable hierarchical clustering algorithm using Spark. In Proceedings of the IEEE First International Conference on Big Data Computing Service and Applications, Redwood City, CA, USA, 30 March 2015–2 April 2015; pp. 418–426. [CrossRef]

39. Sumengen, B.; Rajagopalan, A.; Citovsky, G.; Simcha, D.; Bachem, O.; Mitra, P.; Blasiak, S.; Liang, M.; Kumar, S. Scaling hierarchical agglomerative clustering to billion-sized datasets. *arXiv* **2021**. [CrossRef]

40. Monath, N.; Dubey, K.A.; Guruganesh, G.; Zaheer, M.; Ahmed, A.; McCallum, A.; Mergen, G.; Najork, M.; Terzihan, M.; Tjanaka, B.; et al. Scalable hierarchical agglomerative clustering. In Proceedings of the 27th ACM SIGKDD Conference on Knowledge Discovery and Data Mining, Singapore, 14–18 August 2021; pp. 1245–1255. [CrossRef]

41. Babuji, Y.; Woodard, A.; Li, Z.; Katz, D.S.; Clifford, B.; Kumar, R.; Lacinski, L.; Chard, R.; Wozniak, J.M.; Foster, I.T.; et al. Parsl: Pervasive parallel programming in Python. In Proceedings of the 28th International Symposium on High-Performance Parallel and Distributed Computing, Phoenix, AZ, USA, 22–29 June 2019; pp. 25–36. [CrossRef]

42. Santos, J.M.; Embrechts, M. On the use of the adjusted Rand index as a metric for evaluating supervised classification. In *International Conference on Artificial Neural Networks*; Springer: Berlin/Heidelberg, Germany, 2009; pp. 175–184. [CrossRef]

43. Von Luxburg, U. *Clustering Stability: An Overview*; Now Publishers Inc.: Norwell, MA, USA, 2010. [CrossRef]

44. Tselioudis, G.; Rossow, W.; Zhang, Y.; Konsta, D. Global weather states and their properties from passive and active satellite cloud retrievals. *J. Clim.* **2013**, *26*, 7734–7746. [CrossRef]

45. McDonald, A.; Parsons, S. A comparison of cloud classification methodologies: Differences between cloud and dynamical regimes. *J. Geophys. Res. Atmos.* **2018**, *123*, 11–173. [CrossRef]

46. Schuddeboom, A.; McDonald, A.J.; Morgenstern, O.; Harvey, M.; Parsons, S. Regional regime-based evaluation of present-day general circulation model cloud simulations using self-organizing maps. *J. Geophys. Res. Atmos.* **2018**, *123*, 4259–4272. [CrossRef]

47. Bholowalia, P.; Kumar, A. EBK-means: A clustering technique based on elbow method and k-means in WSN. *Int. J. Comput. Appl.* **2014**, *105*, 17–24.

48. Rousseeuw, P.J. Silhouettes: A graphical aid to the interpretation and validation of cluster analysis. *J. Comput. Appl. Math.* **1987**, *20*, 53–65. [CrossRef]

49. Tibshirani, R.; Walther, G.; Hastie, T. Estimating the number of clusters in a data set via the gap statistic. *J. R. Stat. Soc. Ser. Stat. Methodol.* **2001**, *63*, 411–423. [CrossRef]

50. Jin, D.; Oreopoulos, L.; Lee, D. Simplified ISCCP cloud regimes for evaluating cloudiness in CMIP5 models. *Clim. Dyn.* **2017**, *48*, 113–130. [CrossRef]

51. ISCCP Definition of Cloud Types. Available online: https://isccp.giss.nasa.gov/cloudtypes.html (accessed on 1 May 2022).

52. Gumley, L.; Descloitres, J.; Schmaltz, J. *Creating Reprojected True Color MODIS Images: A Tutorial*; University of Wisconsin—Madison: Madison, WI, USA, 2003; Volume 19.

53. Riggs, G.A.; Hall, D.K.; Román, M.O. *MODIS Snow Products Collection 6 User Guide*; National Snow and Ice Data Center: Boulder, CO, USA, 2015; Volume 66. Available online: https://modis-snow-ice.gsfc.nasa.gov/uploads/C6_MODIS_Snow_User_Guide.pdf (accessed on 1 November 2022).

54. Schneider, T.; Kaul, C.M.; Pressel, K.G. Possible climate transitions from breakup of stratocumulus decks under greenhouse warming. *Nat. Geosci.* **2019**, *12*, 163–167. [CrossRef]

55. Norman, M.R.; Bader, D.A.; Eldred, C.; Hannah, W.M.; Hillman, B.R.; Jones, C.R.; Lee, J.M.; Leung, L.; Lyngaas, I.; Pressel, K.G.; et al. Unprecedented cloud resolution in a GPU-enabled full-physics atmospheric climate simulation on OLCF's Summit supercomputer. *Int. J. High Perform. Comput. Appl.* **2022**, *36*, 93–105. [CrossRef]

56. Hubert, L.; Arabie, P. Comparing partitions. *J. Classif.* **1985**, *2*, 193–218. [CrossRef]

remote sensing

Article

Applying Deep Learning in the Prediction of Chlorophyll-a in the East China Sea

Haobin Cen [1], Jiahan Jiang [1], Guoqing Han [1], Xiayan Lin [1,*], Yu Liu [1,2], Xiaoyan Jia [1], Qiyan Ji [1] and Bo Li [1,3]

[1] Marine Science and Technology College, Zhejiang Ocean University, Zhoushan 316300, China
[2] Southern Marine Science and Engineering Guangdong Laboratory (Zhuhai), Zhuhai 519000, China
[3] Science Foundation of Donghai Laboratory, Zhoushan 316021, China
* Correspondence: linxiayan@zjou.edu.cn

Abstract: The ocean chlorophyll-a (Chl-a) concentration is an important variable in the marine environment, the abnormal distribution of which is closely related to the hazards of red tides. Thus, the accurate prediction of its concentration in the East China Sea (ECS) is greatly important for preventing water eutrophication and protecting the coastal ecological environment. Processed by two different pre-processing methods, 10-year (2011–2020) satellite-observed chlorophyll-a data and logarithmic data were used as the long short-term memory (LSTM) neural network training datasets in this study. The 2021 data were used for comparison to prediction results. The past 15 days' data were used to predict the concentration of chlorophyll-a for the five following days. Results showed that the predictions obtained by both pre-processing methods could simulate the seasonal distribution of the Chl-a concentration in the ECS effectively. Moreover, the prediction performance of the model driven by the original values was better in the medium- and low-concentration regions. However, in the high-concentration region, the prediction of extreme concentrations by the two data-driven LSTM models showed underestimation, considering that the prediction performance of the model driven by the original values was better. Results of sensitivity experiments showed that the prediction accuracy of the model decreased considerably when the backward prediction time step increased. In this study, the neural network was driven only by chlorophyll-a, whose concentration in the ECS was forecasted, and the effect of other relevant marine elements on Chl-a was not considered, which is the current weakness of this study.

Keywords: LSTM; chlorophyll-a; East China Sea

Citation: Cen, H.; Jiang, J.; Han, G.; Lin, X.; Liu, Y.; Jia, X.; Ji, Q.; Li, B. Applying Deep Learning in the Prediction of Chlorophyll-a in the East China Sea. *Remote Sens.* 2022, 14, 5461. https://doi.org/10.3390/rs14215461

Academic Editors: Ana B. Ruescas, Veronica Nieves and Raphaëlle Sauzède

Received: 15 September 2022
Accepted: 28 October 2022
Published: 30 October 2022

Publisher's Note: MDPI stays neutral with regard to jurisdictional claims in published maps and institutional affiliations.

1. Introduction

In marine ecosystems, marine phytoplankton chlorophyll-a (Chl-a) can effectively reflect the biomass of marine primary producers and the photosynthetic carbon sequestration capacity of marine primary productivity [1–5], which are fundamental to marine ecosystems. The prediction of the marine chlorophyll-a concentration and the analysis of its spatial and temporal changes are not only useful for the study of marine primary productivity, but also important for the study of carbon cycling in the ocean–atmosphere system [6,7], red tide hazard monitoring [8–10], environmental monitoring [11], ocean currents (such as upwelling and coastal currents) [12,13], as well as fishery management and the estimation of aquaculture production [14].

The chlorophyll-a concentration is influenced by many factors, such as climatic factors, namely light, temperature, precipitation, and wind speed [15,16], and geographical factors [17,18]. In addition, in the early period, a relative paucity of data relating to the Chl-a concentration was observed, leading to a high level of uncertainty in its prediction. The prediction methods could be broadly categorized into two methods. The first is the statistical method, which was first proposed by Vollenweider [19], who used statistical models to predict the issue of eutrophication. Kiyofuji et al. [20] developed a statistical spatiotemporal model to predict the distribution of chlorophyll-a in the Sea of Japan on the basis of SeaWiFS data. Although the model was able to predict its distribution effectively during summer and early autumn, this traditional statistical method could only solve the average concentration of a particular element and could not simulate the effect of relevant factors on chlorophyll-a. The second method is based on ecological dynamics, the properties of water bodies, and establishing a theoretical analysis model to predict the concentration of chlorophyll-a [21–23]. Using data collected monthly, Liu et al. [24] used multivariate statistical methods to simulate the effects of multiple chemical variables on chlorophyll-a in Lake Qilu. This method considers the interactions between elements in nature and includes many parameters, thereby causing difficulty in accurate modeling or parameterization due to the diversity of water quality variables in the ocean.

The research on remote sensing monitoring of chlorophyll-a and remote sensing inversion has become increasingly sophisticated with the development of satellite remote sensing technology. Moreover, a large amount of water quality data can be obtained as the access to information becomes more diverse. A new trend in recent years has been the use of machine learning methods for water quality variable prediction. Machine learning methods can capture the characteristics of the input data to explore the potential relationships between variables, and narrow the difference between predictions and observations by updating the parameters in the model. The most widely used machine learning methods at present include artificial neural networks (ANN, [25–29]), support vector machine (SVM, [30–32]), decision tree (DT), random forest (RF, [32–35]), and regression, etc. Deep learning (DL) is a special type of machine learning [36]. Zhang et al. [37] proposed a new prediction approach for algal blooms on the basis of deep learning to represent and predict highly dynamic and complex phenomena. Most current studies use independent deep learning models for chlorophyll-a concentration prediction. Several deep learning models, such as the recurrent neural network (RNN) and its variant, the long short-term memory neural network (LSTM), are commonly used in time-series forecasting. Both approaches have good performance in dealing with time-series information problems. Compared with the traditional RNN, LSTM does not have the problem of gradient disappearance in the process of training long-term sequences. Therefore, the LSTM model can effectively predict the chlorophyll-a concentration [38–40]. Yossof et al. [41] used an LSTM model and a convolutional neural network (CNN) model to predict harmful algal blooms on the western coast of Sabah. The results show that the LSTM model outperforms the CNN model in terms of prediction accuracy. Barzegar et al. [42] first built a coupled CNN–LSTM model to predict water quality variables in Small Prespa Lake, Greece, and the results showed that the hybrid CNN–LSTM model was better than the independent model in predicting the chlorophyll-a concentration.

The Eastern China Sea (ECS) area is under the influence of the East Asian monsoon; the chlorophyll-a concentration in the ECS has evident seasonal variation characteristics and is influenced by land runoff, mainly from the Yangtze River [43–45]. The distribution of it in the East China Sea is also influenced by the Kuroshio, with high-temperature and high-salt seawater [46,47]. The Eastern China Sea area has a long coastline, of which the Zhejiang coast is one of the famous upwelling areas in China, and it has important fishing grounds, such as the Zhoushan and Yushan fishing grounds [48]. With the rapid development of coastal cities in recent years, the frequency of red tides in the ECS has increased substantially [14,49], not only polluting the marine environment of this region but also severely damaging the fishery resources, leading to huge economic losses [50]. Therefore, accurate prediction of the chlorophyll-a concentration in this area is important for the prevention of eutrophication and the protection of the offshore ecosystem.

Machine learning methods have been applied to research on forecasting ocean elements, such as storm surges [51], harmful algal blooms (HAB), and sea surface temperature (SST) in the ECS. Xu et al. [52] used the SVM model to predict the occurrence of red tides in Haizhou Bay in the ECS. Xiao et al. [53,54] used a combined LSTM–AdaBoost model and a convolutional LSTM (ConvLSTM) model to predict the SST field in the ECS, respectively.

The results showed that the LSTM–AdaBoost and ConvLSTM models have good promise in accurately predicting the short- and medium-term SST fields.

At present, no research has used machine learning to predict the chlorophyll-a concentration in the East China Sea area. Thus, this study first uses the LSTM neural network to predict the concentration in this region. The specific objectives of this research are (1) comparing the effects of different processing methods for chlorophyll-a data on the forecast result; and (2) evaluating the prediction results of the LSTM neural network in the ECS on the basis of the previous step by using the optimal processing method for chlorophyll-a data.

The rest of this paper is organized as follows, Section 2 describes the satellite data and LSTM neural network used in this study, Section 3 presents the experimental results and detailed discussion, and Section 4 draws the conclusions obtained from this study.

2. Materials and Methods

2.1. Materials

This study uses the ocean color data product (OCEANCOLOUR_GLO_BGC_L4_MY_009 _104) provided by the Copernicus Marine Environment Monitoring Service (CMEMS, http://www.copernicus.eu/ (accessed on 11 July 2022)). This product integrates data from SeaWiFS, MODIS-Aqua, MODIS-Terra, MERIS, VIIRS-SNPP, OLCI-S3A&S3B, and other satellites. The time resolution is 1 day, the spatial resolution is 4 km × 4 km, and the time span is from September 1997 to the present. The spatial range of chlorophyll-a data used in our study is 22°N–33°N, 120°E–131°E, and the time range is from 2011 to 2021, of which the data from 2011 to 2020 are used as the training dataset, and the chlorophyll-a data from 2021 are used as the test dataset.

2.2. Methods

2.2.1. LSTM Neural Network

LSTM was proposed by Hochreiter and Schmidhuber in 1997 [55] as a variant neural network of RNN for long-time-series training. It can effectively solve the gradient disappearance problem, which easily occurs in the training process of the traditional RNN. The internal network structure of the LSTM unit is more complex than that of the traditional RNN. The information in the current unit is processed by the input gate, forgetting gate, and output gate, and then the historical unit information is selected to be either "forgotten" or "remembered".

The two most important states in the LSTM cell structure are the cell state $c(t)$ and the hidden state $h(t)$. The cell state transmits information through different gates, thereby enhancing the dependency among long-time-series information; the cell structure is shown in Figure 1.

First, the function of the forget gate f_t is to select which information needs to be discarded in the current state of the cell. f_t is calculated as follows:

$$f_t = \sigma\left(W_{fh}h_{t-1} + W_{fx}x_t + b_f\right) \tag{1}$$

where $\sigma(\cdot)$ represents the sigmoid activation function, W_{fh} and W_{fx} represent the corresponding weight parameters, x_t represents the input at moment t, h_{t-1} represents the hidden state of the cell at moment $t-1$, and b_f is the bias term.

Second, the function of the input gate i_t is to remember the candidate cell state selectively, thereby updating the cell state at the current moment, and a new cell state $\tilde{C}_t$ is generated by the following calculation formula:

$$i_t = \sigma(W_{ih}h_{t-1} + W_{ix}x_t + b_i) \tag{2}$$

$$\tilde{C}_t = \tanh(W_{ch}h_{t-1} + W_{cx}x_t + b_c) \tag{3}$$

$$C_t = f_t \times C_{t-1} + i_t \times \tilde{C}_t \tag{4}$$

where $\sigma(\cdot)$ represents the sigmoid activation function; W_{ih}, W_{ix}, W_{ch}, and W_{cx} represent the corresponding weight parameters, x_t represents the input at moment t, h_{t-1} represents the hidden state of the cell at moment $t-1$, $\widetilde{C}_t$ represents the candidate state at the current moment, and b_i and b_c are bias terms.

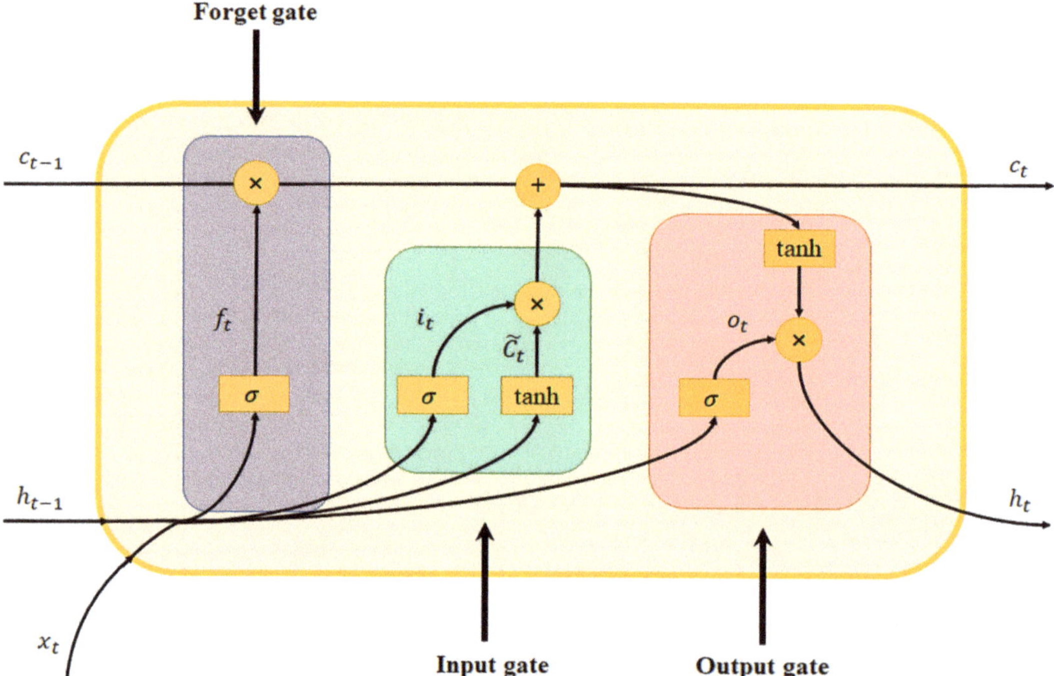

Figure 1. Structure of the LSTM unit.

Then, the output gate o_t is used to determine the output component of the cell state through the sigmoid function, whereas the cell state is processed through tanh and multiplied with the output gate o_t to obtain the new hidden state h_t; the calculation formula is as follows:

$$o_t = \sigma(W_{oh}h_{t-1} + W_{ox}x_t + b_o) \tag{5}$$

$$h_t = o_t \times \tanh(c_t) \tag{6}$$

where $\sigma(\cdot)$ represents the sigmoid activation function, W_{oh} and W_{ox} represent the corresponding weight parameters, x_t represents the input at moment t, and b_o represents the bias term.

2.2.2. Architecture of the LSTM Model for Chl-a Forecasts

In this study, a regional chlorophyll-a concentration prediction model is established on the basis of the LSTM neural network, including an input layer, three LSTM layers, a dropout layer, and a dense layer, as shown in Figure 2. Dropout is a method to control the complexity of the model. In each training batch, a certain number of hidden nodes are set to 0 to reduce the interaction between hidden nodes, thereby preventing the model from overfitting [56,57]. During training, we use tanh as the activation function to generate the output of hidden neurons. Adam optimization is a stochastic gradient descent method based on the adaptive estimation of first- and second-order moments; compared with other stochastic optimization algorithms, the Adam algorithm has more advantages in practical applications [58]. Therefore, in this study, we adopt the Adam optimization algorithm to minimize the error between predicted and observed values.

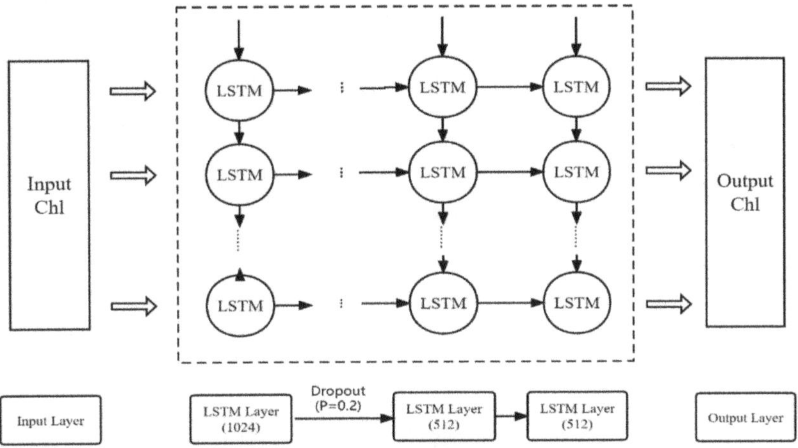

Figure 2. Architecture of the LSTM model for Chl-a forecasts.

2.2.3. Data Pre-Processing

To investigate the effect of different input data on the prediction results of the LSTM model, one group used the original data as input to the model, and the other group used the logarithmic data as input to the LSTM model. Both groups used the data of the previous 15 days to predict the value of the next 5 days. Data from 2011 to 2020 were used to generate the corresponding training and validation datasets, where the ratio of the data volume of the training dataset to the validation dataset was 4:1. Data from 2021 were used as a test dataset to make predictions for chlorophyll-a, which was excluded from the model training to ensure relative independence between the training and test datasets. To explore the influence of input length on the prediction results of the LSTM model, under the condition that hyperparameters, such as the number of hidden layers, the neurons, and the learning rate, do not change, the prediction length was controlled to 1 day, and the input length was set to 7, 10, and 15 days, respectively. Similarly, to explore the influence of the prediction length on the prediction results of the model when the other hyperparameters remain unchanged, the input length was controlled to 15 days, and the prediction length was set to 1, 3, and 5 days, respectively. The training dataset used in the training model needed to be standardized. In the process of standardizing the data, we used the MinmaxScaler function imported from the sklearn library to scale the data of the training dataset to $(-1, 1)$ to obtain the standardized training dataset.

2.2.4. Evaluation Functions

To compare the performance of the different methods further, the following indicators were used in this study to evaluate model performance: root mean square error ($RMSE$), standard deviation (STD), coefficient of determination (R^2), and absolute error (AE). The formulas are shown below.

$$RMSE = \sqrt{\frac{1}{n} \sum_{t=1}^{n} (Y_t - y_t)^2} \tag{7}$$

$$S = \sqrt{\frac{\sum_{t=1}^{n} (x_t - \overline{x})^2}{n-1}} \tag{8}$$

$$R^2 = 1 - \frac{\sum_{t=1}^{n}(Y_t - y_t)}{\sqrt{\sum_{t=1}^{n}(Y_t - \overline{Y})^2}} \tag{9}$$

$$AE = |Y_t - y_t| \tag{10}$$

where Y represents the satellite-observed value, $\overline{Y}$ represents the average of the satellite-observed values, y represents the model-predicted value, and y represents the average of the model-predicted chlorophyll-a values. x represents the value from satellite observations or model forecasts, and $\overline{x}$ represents the average of the values from satellite observations or model forecasts. Small $RMSE$ and AE values indicate the high forecast accuracy of the model. The closer the value of S to the STD of the observed values, the better the prediction performance of the model. The closer the value of R^2 to 1, the higher the fitness between the predicted and observed values.

3. Results

The chlorophyll-a concentration in the East China Sea varies widely from nearshore to offshore due to the influence of surface runoff. It also has substantial seasonal variations due to environmental factors, such as monsoons and ocean currents. Therefore, this study selected four points, marked as L1 (32.1°N, 122.2°E), L2 (28.0°N, 123.4°E), L3 (30.8°N, 124.9°E), and L4 (23.8°N, 126.9°E), as shown in Figure 3, to analyze the chlorophyll-a concentration predicted by the LSTM model. L1 was selected because the annual mean of the chlorophyll-a concentration at this location is higher, as well as the standard deviation of the concentration. L2 and L3 were selected because these points are located in the median area of the annual mean chlorophyll-a concentration; the coefficient of chlorophyll-a variation is higher at L2. L4 was selected because the concentration in location L4 is lower in the distant sea area.

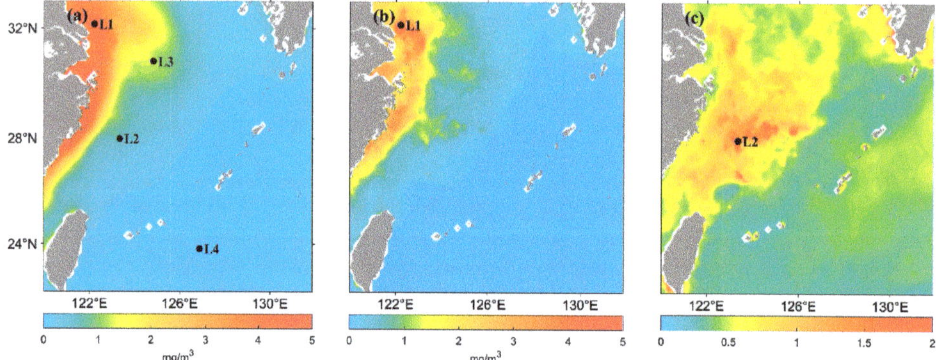

Figure 3. (**a**) Spatial distribution of annual mean Chl-a concentration from 2021 satellite observations in ECS. (**b**) Spatial distribution of the standard deviation of the Chl-a concentration from 2021 satellite-observed data in ECS. (**c**) Spatial distribution of the coefficient of variation in the Chl-a concentration from 2021 satellite-observed data in ECS. L1, L2, L3, L4 are the four different points selected.

3.1. LSTM Prediction Results under Different Data Pre-Processing Methods

First, this study discussed the effect of chlorophyll-a data obtained from different data pre-processing methods on the prediction performance of the LSTM model. Original and logarithmic data of the past 15 days were used as the inputs to the neural network to predict the chlorophyll-a concentration for the following one day.

Figure 4 shows the variation in concentration predicted by the LSTM model at different locations and the real concentration observed by a satellite over time. The red line indicates the data from satellite observations, the blue line is the concentration predicted when using the original data as input to the neural network, and the green line is the concentration predicted when using the logarithmic data as input to the neural network. Figure 4a–d show that both data processing methods can accurately predict the variations in the chlorophyll-a concentration. In terms of the prediction of the extremum, when using logarithmic data as input to the neural network, the predicted extremum of the concentration is smaller than the satellite observations; when using original data as the input to the neural network, the extremum of the concentration is better predicted in the regions with medium and low concentrations (Figure 4b–d). In addition, both LSTM models can better predict the concentration of chlorophyll-a at times when its value changes gently. In the region with a higher concentration (Figure 4a), the predicted values of both models severely underestimate the extremum of it in the two time periods when the concentrations reach their peak (Figure 4a). Figure 4b shows that a similar underestimation occurs around April 1 and during the chlorophyll-a peak at the end of October, when logarithmic data are used as input to the neural network. According to Figure 4d, the predicted values obtained when using logarithmic data as input to the model are underestimated most of the time.

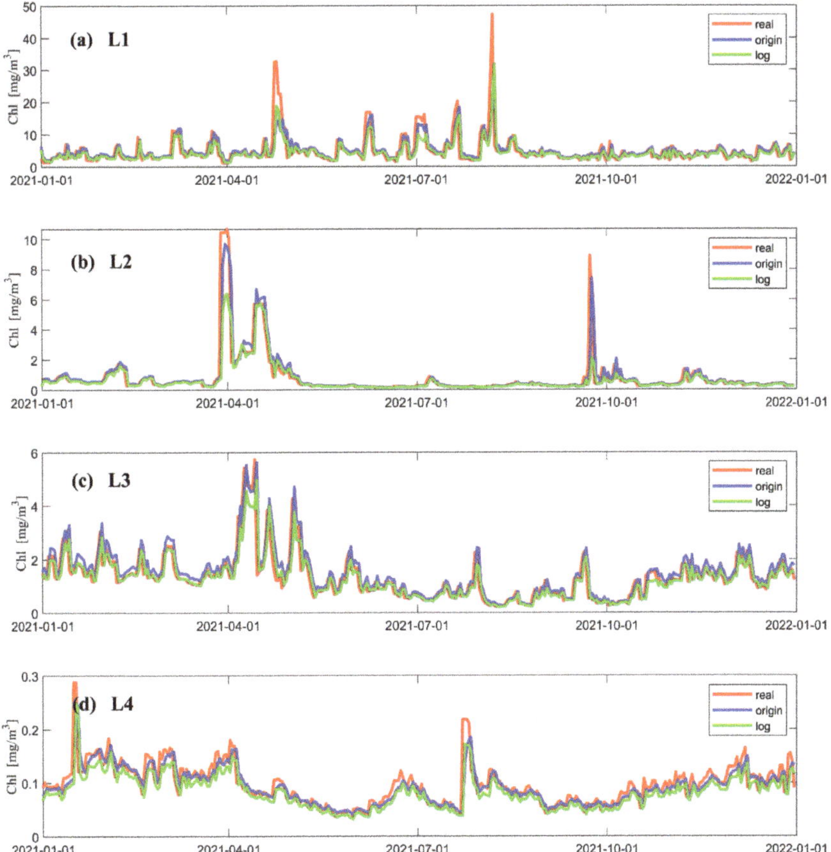

Figure 4. Chl-a data from satellite observations at different points and the results predicted by LSTM models using two different data processing methods, where the red solid line shows the data from satellite observations, the blue solid line shows the result data predicted using the original data as the input to the neural network, and the green solid line shows the results predicted using the logarithmic data as the input to the neural network. (**a**–**d**) show comparison of prediction results with satellite observations for each of the four points in 2021. The vertical coordinate is the Chl-a concentration, and the horizontal coordinate is the number of days.

According to Figure 5, it can be seen that, most of the time, the error between the observed and predicted value of the two LSTM models is small. However, the neural network does not predict the concentrations well when transient and drastic changes in concentrations occur. Moreover, the time points at which the errors are larger are mostly concentrated at times when the concentrations undergo dramatic changes. It can be seen from Figure 5d that when using logarithmic data as input data in the LSTM model, the predicted values of chlorophyll-a are underestimated most of the time.

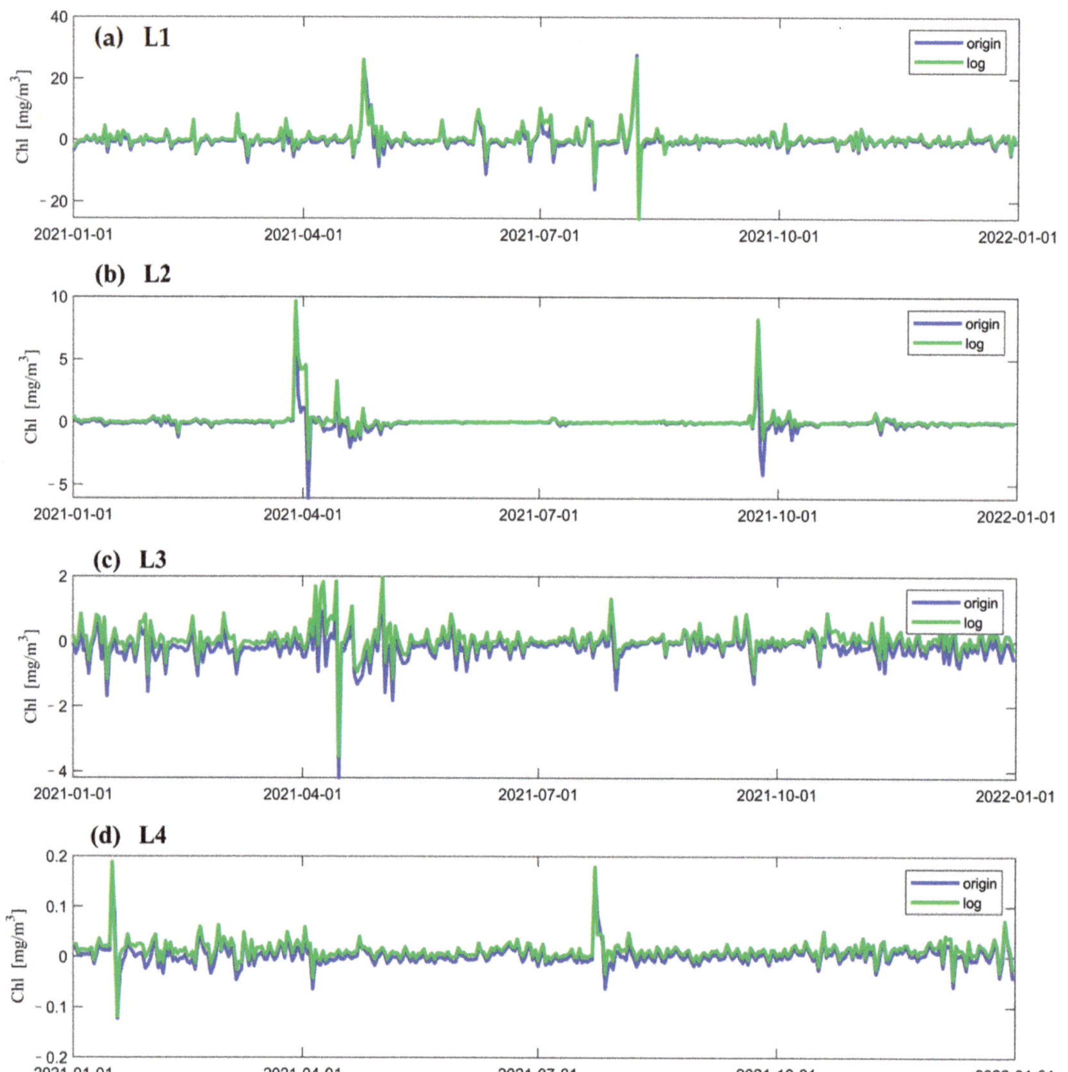

Figure 5. The error between Chl-a data from satellite observations at different points and the results predicted by LSTM models using two different data processing methods; the error is obtained by subtracting the observed value from the LSTM forecast value. (**a**–**d**) show the errors at the four points L1, L2, L3, and L4 in 2021. The vertical coordinate is the Chl-a concentration and the horizontal coordinate is the number of days.

The chlorophyll-a distribution in the East China Sea area has substantial seasonal variation. Figure 6 shows the seasonal distribution of the values from satellite observations and the predicted values of the two LSTM models using the two different data processing methods. The predicted results of both neural networks can accurately simulate the seasonal variation, but the predicted values are lower than the observed values when using logarithmic data as the input in the high-value nearshore region. The seasonal distribution of the predicted values has better accuracy on the nearshore and offshore when the original data are used as input. Figure 7 shows that when using the original data as the input, the AE between the predicted values and the observed values is small. The inaccuracies are mainly concentrated in the high- and medium-concentration regions; they are mostly less than 0.5 mg/m^3. When using the logarithmic data as input, the AE is relatively large, especially in the high-value nearshore region and on the offshore, where the concentration is low; the error between predicted and observed values is small.

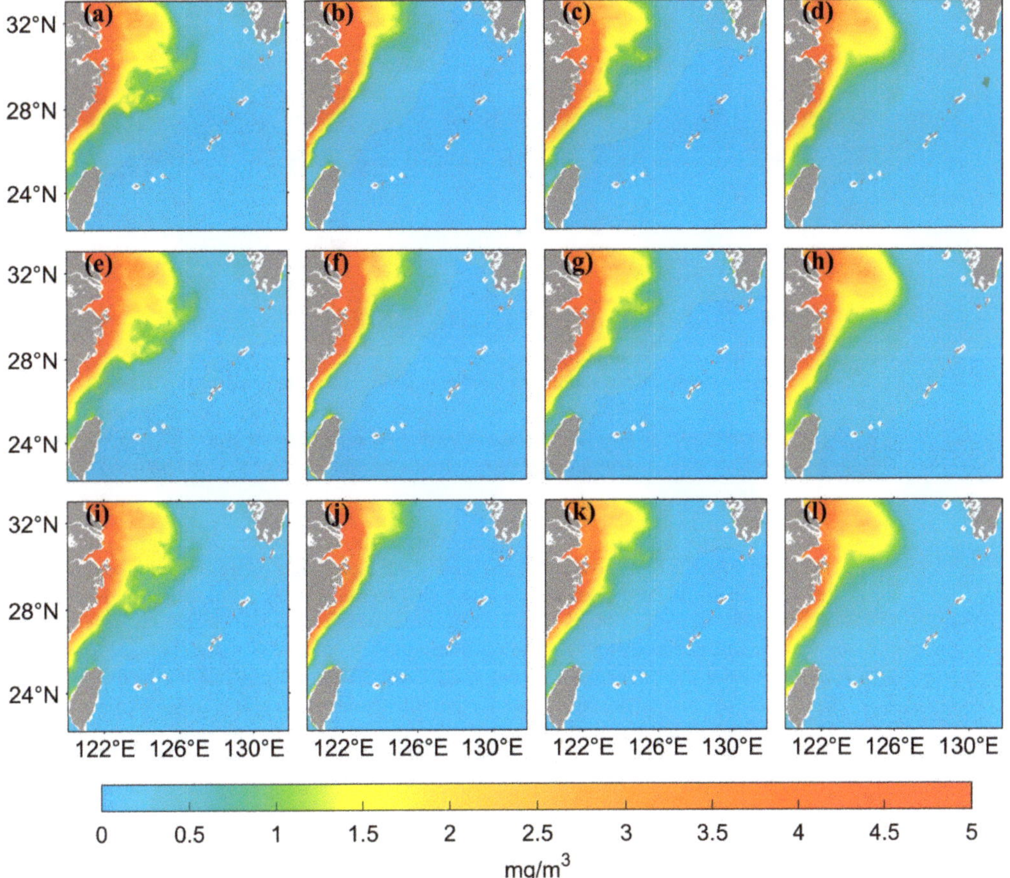

Figure 6. Spatial distribution of satellite-observed values and predicted values from LSTM models using two different data processing methods for four seasons. March to May represents Spring, June to August represents Summer, September to November represents Autumn, and December to February represents Winter. (**a–d**) represent the distribution of the data from satellite observations, in the order of spring, summer, autumn, and winter. (**e–h**) represent the distribution of the values predicted by the LSTM model using the original values as input, and (**i–l**) represent the distribution of the values predicted by the LSTM model using the logarithmic values as input.

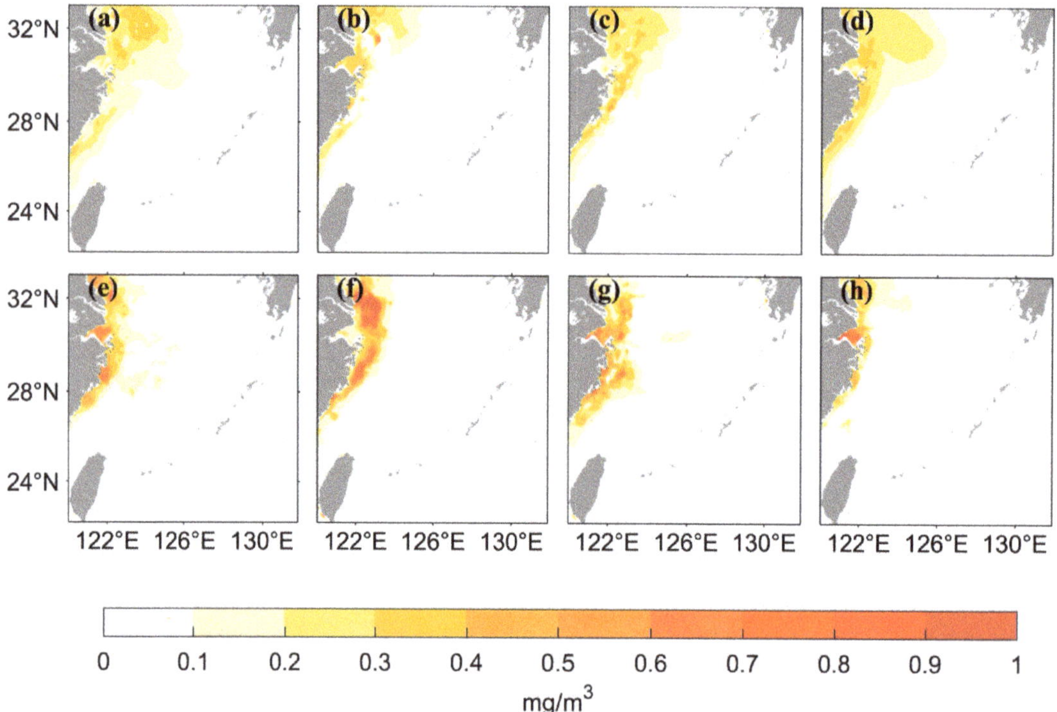

Figure 7. Spatial distribution of AE between Chl-a values predicted by two different LSTM models and Chl-a values observed by satellite for four seasons. (**a–d**) represent the distribution of AE obtained by LSTM model using the original values as input, in the order of spring, summer, autumn, and winter. (**e–h**) represent the distribution of AE obtained by LSTM model using the logarithmic values as input.

Figure 8 shows that the spatial distribution of $RMSE$ for concentrations predicted by the two different LSTM models has high agreement overall. Figure 8c shows that the $RMSE$ of the prediction results using the original data as input is larger in most high and median regions of the nearshore compared with that using logarithmic data as input. In this study, the $RMSE$s for the prediction results of the two models were divided by the values of their average, and the spatial distribution of this result was drawn, as shown in Figure 9. The predicted results using the original data are larger in most areas of the median region compared with those using logarithmic data as input, whereas the opposite is true in most areas of the low-value region in the distant ocean.

Figure 10 shows that, in terms of STD, the forecast results at three locations (i.e., high-value area (L1), medium-value area with a small coefficient of variation (L2), and low-value area (L4)) have better prediction performance when using the original data as input to the neural network, whereas the correlation coefficients between the observed values and the predicted results of the two different models at the four positions do not differ considerably. Although the correlation coefficient values between the observed values and the predictions using the two different models are close in L1, their correlation coefficient values are low (only 0.64); combined with R^2 in Table 1, the LSTM model is prone to errors when predicting the concentration of chlorophyll-a in high-value areas. In L3, although the difference in the $RMSE$ and correlation coefficients of the observed values and the predictions of the two models is relatively small, the prediction performance using the logarithmic data as input is better in terms of R^2 in the medium area with a large coefficient of variation. The predictions of the two different LSTM models in the four different locations, except L3, indicate that the neural network using the original data as input has better prediction performance for the three other points based on R^2. Therefore, we use the original data as the input of the neural network for further work in the subsequent sections.

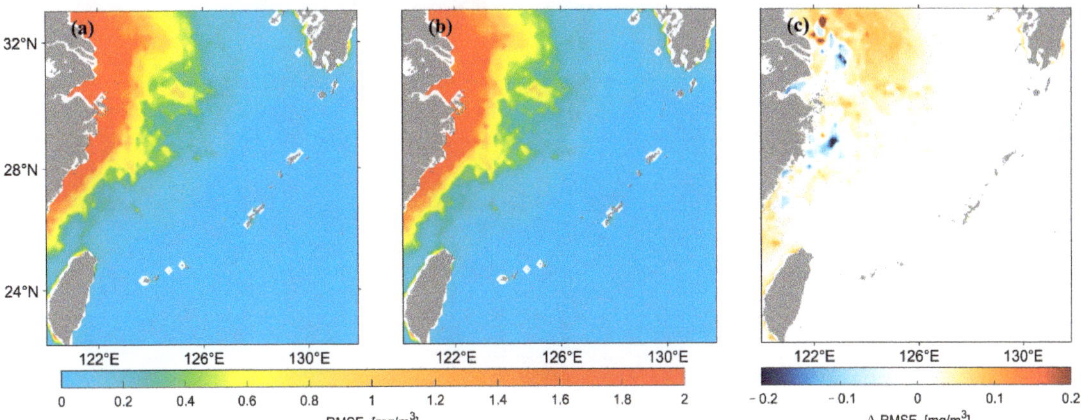

Figure 8. Spatial distribution of *RMSE* of Chl-a predicted by the two LSTM models. (**a**) is the result predicted using the original data as input, (**b**) is the result predicted using the logarithmic data as input, (**c**) represents (**a**,**b**).

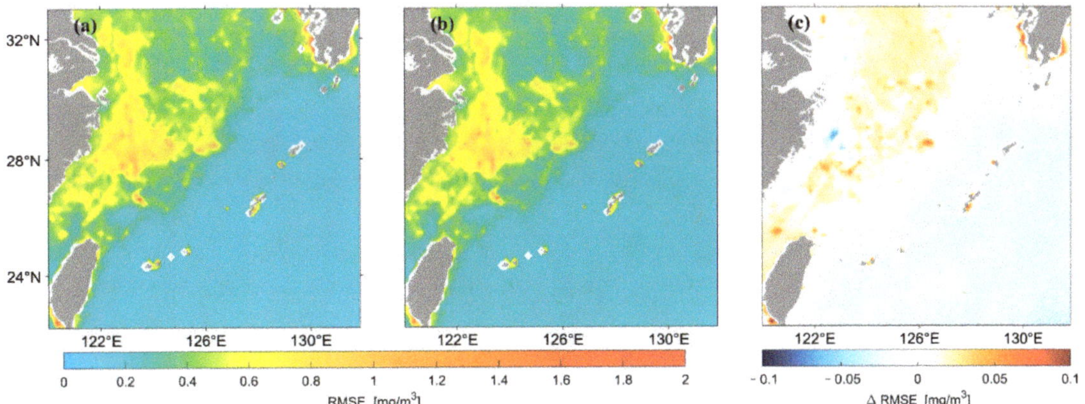

Figure 9. Spatial distribution of *RMSE* divided by the average of Chl-a predicted by the two LSTM models. (**a**) is the result predicted using the original data as input, (**b**) is the result predicted using the logarithmic data as input, (**c**) represents (**a**,**b**).

3.2. LSTM Prediction Results with Different Input and Output Lengths

Table 2 shows that the neural network has the best prediction performance when forecasting 1 day at four different locations, after which the prediction performance of the neural network decreases as the number of forecast days increases.

According to Table 3, in terms of *RMSE* and *STD*, the prediction performance of the neural network in the region with a high concentration (L1) and that with a medium concentration and low coefficient of variation (L2) was optimal when the input length was 15 days. In the medium-concentration area with a large coefficient of variation (L3) and the low-concentration area (L4), the *RMSE* and *STD* were close when the input length was 15 and 7 days. In addition, the prediction performance of the neural network with the input length of 7 days was slightly better than that with the input length of 15 days. In terms of correlation coefficient, the predicted results in the region of high concentration (L1) and that in the region of medium concentration with a small coefficient of variation correlated best with observations when the input length was 15 days. Moreover, in the region of medium concentration with a large coefficient of variation (L3) and in the region of low concentration (L4), the predictions correlated best with observations when the input length was 7 days. In terms of R^2, the high-concentration area (L1) and medium-concentration area (L2, L3) had the optimal prediction performance when the input length was 15 days; by contrast, in the low-concentration area (L4), the model had the best prediction performance when the input length was 7 days.

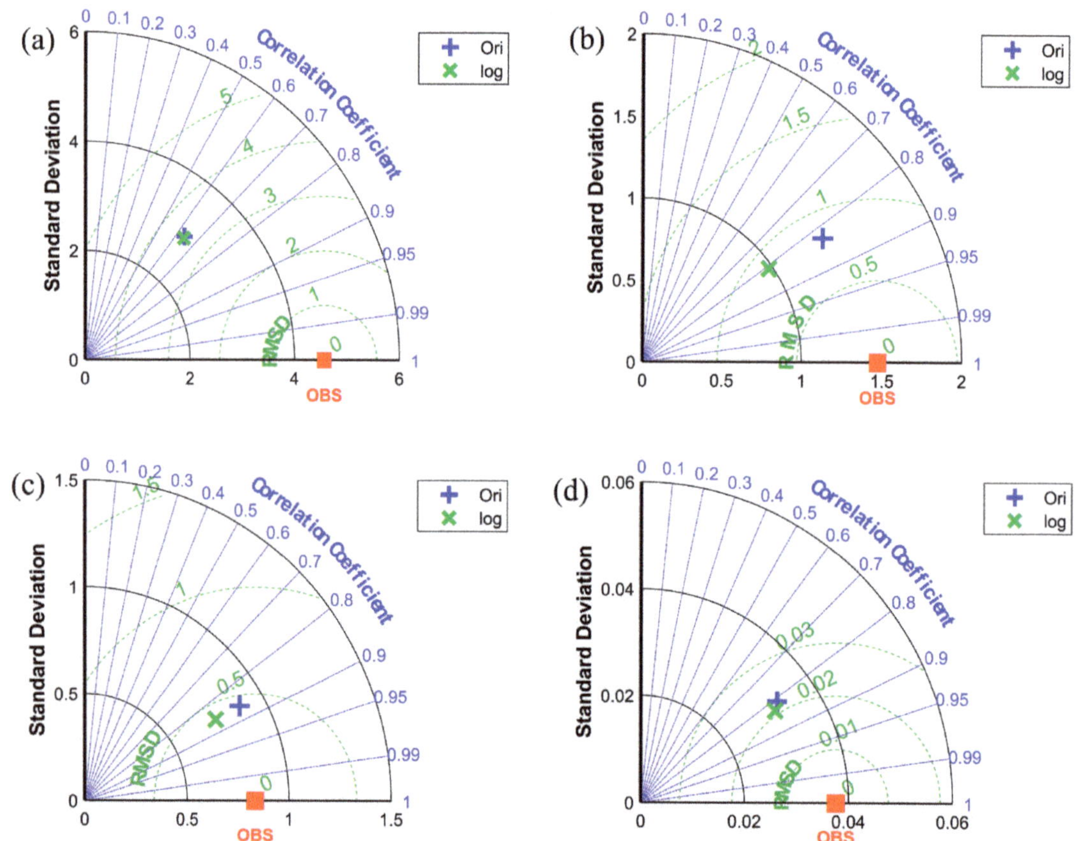

Figure 10. The Taylor diagram of different positions. (**a**–**d**) represent the results of L1, L2, L3, and L4.

Table 1. Values of R^2 for the four points. ori indicates results using the original data as input, log indicates results using logarithmic data as input, and obs indicates the observed value.

	L1	L2	L3	L4
R^2	ori: 0.4133	ori: 0.6806	ori: 0.6732	ori: 0.6482
	log: 0.3936	log: 0.66273	log: 0.7337	log: 0.5681

Table 2. Values of $RMSE$, STD, COR, and R^2 at the four points for different output lengths. Here, 1d indicates the forecast results for one day backward, 3d indicates the forecast results for three days backward, and 5d indicates the forecast results for five days backward.

	L1	L2	L3	L4
RMSE	1d: 3.5012 3d: 4.6445 5d: 4.6475	1d: 0.8304 3d: 1.2453 5d: 1.4237	1d: 0.4779 3d: 0.6867 5d: 0.7653	1d: 0.0223 3d: 0.0303 5d: 0.0320
STD	1d: 2.9369 3d: 1.6825 5d: 1.2218	1d: 1.3581 3d: 1.0272 5d: 0.8789	1d: 0.8796 3d: 0.7922 5d: 0.7349	1d: 0.0324 3d: 0.0312 5d: 0.0300
COR	1d: 0.6431 3d: 0.1495 5d: 0.1051	1d: 0.8306 3d: 0.5510 5d: 0.3504	1d: 0.8368 3d: 0.6755 5d: 0.5661	1d: 0.8116 3d: 0.6366 5d: 0.5841
R^2	1d: 0.4133 3d: −0.0330 5d: −0.0356	1d: 0.6806 3d: 0.2820 5d: 0.0619	1d: 0.6732 3d: 0.3252 5d: 0.1599	1d: 0.6482 3d: 0.3553 5d: 0.2802

Table 3. Values of $RMSE$, STD, COR, and R^2 at four points for different input lengths. Here, 15d indicates the results using data of the first 15 days to predict, 10d indicates the results using data of the first 10 days, and 7d indicates the results using data of the first seven days.

	L1	L2	L3	L4
RMSE	15d: 3.5012 10d: 3.5972 7d: 3.5265	15d: 0.8304 10d: 0.8702 7d: 0.8530	15d: 0.4799 10d: 0.4845 7d: 0.4796	15d: 0.0223 10d: 0.0292 7d: 0.0217
STD	15d: 2.9369 10d: 3.2444 7d: 3.0379	15d: 1.3581 10d: 1.3997 7d: 1.3791	15d: 0.8796 10d: 0.9039 7d: 0.8783	15d: 0.0324 10d: 0.0263 7d: 0.0369
COR	15d: 0.6431 10d: 0.6235 7d: 0.6372	15d: 0.8306 10d: 0.8169 7d: 0.8228	15d: 0.8368 10d: 0.8591 7d: 0.8613	15d: 0.8116 10d: 0.8313 7d: 0.8316
R^2	15d: 0.4133 10d: 0.3807 7d: 0.4048	15d: 0.6806 10d: 0.6492 7d: 0.6629	15d: 0.6732 10d: 0.6640 7d: 0.6708	15d: 0.6482 10d: 0.3955 7d: 0.6663

4. Conclusions

The difference between nearshore and offshore chlorophyll-a concentrations can be large, with high and low values of concentration often varying by several orders of magnitude; thus, most of the relevant studies initially processed the concentration values logarithmically. To explore whether different input data affect the prediction performance of the LSTM neural network, this study uses two different data pre-processing methods, using the data of the previous 15 days as input to the neural network and intelligently estimating the concentration of the next 5 days. In the nearshore with a high concentration, the predicted results of the neural network that is driven by original data are closer to the actual satellite observational values, and the predicted results of the neural network that is driven by logarithmic data are smaller than the observed values. The error is mainly in the nearshore with high and median concentrations; the AE between the concentrations predicted by original data and the observed values was small, i.e., less than 0.5 mg/m^3, in most areas. By contrast, the AE between the results predicted using the logarithmical data and the observed values was larger, especially in some high-concentration regions of the nearshore areas, where the AE was as high as 1 mg/m^3. Analysis of the $RMSE$ and R^2 of the prediction results from different LSTM models indicated that the prediction performance of the model driven by the original data was improved in the region with a high concentration, the region with a medium concentration and a large coefficient of variation, and the region with a low concentration. Moreover, the prediction performance of the LSTM model driven by logarithmic data was improved in the region with a medium concentration and low coefficient of variation. With all the factors considered, the prediction results are improved when the original data are used as input to the LSTM model.

In addition, different inputs and forecast lengths affect the prediction performance of the LSTM model. As the forecast length increases, the prediction accuracy of the neural network decreases remarkably. The prediction accuracy starts to decrease by the third day of forecasting downwards, and the best prediction accuracy is achieved at the forecast length of 1 day. Increasing the input length can increase the prediction performance of the neural network to a certain extent, and the optimal result is obtained when the input length is 15 days in the high- and medium-concentration regions. Furthermore, the optimal result is obtained at a 7-day input length in the low-concentration region.

5. Discussion

Previous studies on the prediction of chlorophyll-a concentrations used several methods, such as statistical models and ANNs. In this study, we established an intelligent forecast model for chlorophyll-a in the East China Sea on the basis of the LSTM algorithm and discussed its forecast performance. It is a novel prediction method and has achieved good results. However, this study only used chlorophyll-a as the input to drive the neural network, whereas, in the real ocean, many factors, such as temperature, precipitation, and wind speed, may affect the concentration. Therefore, in future studies, we will attempt to consider multiple variables to drive the LSTM neural network to improve the prediction performance of the model for the prediction of the chlorophyll-a concentration in the ECS further.

Author Contributions: Conceptualization, H.C., Q.J. and G.H.; methodology, H.C.; software, H.C. and X.J.; validation, H.C., G.H. and X.L.; formal analysis, H.C. and G.H.; investigation, H.C. and G.H.; resources, X.L.; data curation, Y.L. and J.J.; writing—original draft preparation, H.C.; writing—review and editing, G.H. and X.L.; visualization, H.C.; supervision, X.L.; project administration, X.L.; funding acquisition, X.L., Y.L. and B.L.; reply to review comments, X.J., Q.J. and B.L. All authors have read and agreed to the published version of the manuscript.

Funding: This study was supported by the projects of the Southern Marine Science and Engineering Guangdong Laboratory (Zhuhai) (SML2020SP007, 311020004), the National Natural Science Foundation of China (41806030, 41806004), the Basic Scientific Research Business Expenses of Zhejiang Provincial Universities (2020J00007), the Science Foundation of Donghai Laboratory (DH-2022KF0208), and the Open Foundation from Marine Sciences in the First-Class Subjects of Zhejiang (OFMS006).

Data Availability Statement: Publicly available datasets were analyzed in this study. Chlorophyll-a data from satellite observations can be found at https://resources.marine.copernicus.eu/products (accessed on 11 July 2022).

Conflicts of Interest: The authors declare no conflict of interest.

References

1. Iriarte, J.; González, H.; Liu, K.; Rivas, C.; Valenzuela, C. Spatial and Temporal Variability of Chlorophyll and Primary Productivity in Surface Waters of Southern Chile (41.5–43 S). *Estuar. Coast. Shelf Sci.* **2007**, *74*, 471–480. [CrossRef]
2. Lee, Y.J.; Matrai, P.A.; Friedrichs, M.A.; Saba, V.S.; Antoine, D.; Ardyna, M.; Asanuma, I.; Babin, M.; Bélanger, S.; Benoît-Gagné, M.; et al. An Assessment of Phytoplankton Primary Productivity in the Arctic Ocean from Satellite Ocean Color/in Situ Chlorophyll-a Based Models. *J. Geophys. Res. Ocean.* **2015**, *120*, 6508–6541. [CrossRef] [PubMed]
3. Arrigo, K.R.; Matrai, P.A.; Van Dijken, G.L. Primary Productivity in the Arctic Ocean: Impacts of Complex Optical Properties and Subsurface Chlorophyll Maxima on Large-Scale Estimates. *J. Geophys. Res. Ocean.* **2011**, *116*, C11022. [CrossRef]
4. Ardyna, M.; Gosselin, M.; Michel, C.; Poulin, M.; Tremblay, J.-É. Environmental Forcing of Phytoplankton Community Structure and Function in the Canadian High Arctic: Contrasting Oligotrophic and Eutrophic Regions. *Mar. Ecol. Prog. Ser.* **2011**, *442*, 37–57. [CrossRef]

5. Ardyna, M.; Babin, M.; Gosselin, M.; Devred, E.; Bélanger, S.; Matsuoka, A.; Tremblay, J.-É. Parameterization of Vertical Chlorophyll a in the Arctic Ocean: Impact of the Subsurface Chlorophyll Maximum on Regional, Seasonal, and Annual Primary Production Estimates. *Biogeosciences* **2013**, *10*, 4383–4404. [CrossRef]
6. Sharada, M.; Yajnik, K. Seasonal Variation of Chlorophyll and Primary Productivity in Central Arabian Sea: A Macrocalibrated Upper Ocean Ecosystem Model. *Proc. Indian Acad. Sci.-Earth Planet. Sci.* **1997**, *106*, 33–42. [CrossRef]
7. Thomalla, S.; Fauchereau, N.; Swart, S.; Monteiro, P. Regional Scale Characteristics of the Seasonal Cycle of Chlorophyll in the Southern Ocean. *Biogeosciences* **2011**, *8*, 2849–2866. [CrossRef]
8. Hao, Y.; Tang, D.; Yu, L.; Xing, Q. Nutrient and Chlorophyll a Anomaly in Red-Tide Periods of 2003–2008 in Sishili Bay, China. *Chin. J. Oceanol. Limnol.* **2011**, *29*, 664–673. [CrossRef]
9. Ishizaka, J.; Kitaura, Y.; Touke, Y.; Sasaki, H.; Tanaka, A.; Murakami, H.; Suzuki, T.; Matsuoka, K.; Nakata, H. Satellite Detection of Red Tide in Ariake Sound, 1998–2001. *J. Oceanogr.* **2006**, *62*, 37–45. [CrossRef]
10. Zhang, C.; Zeng, Y.; Zhang, X.; Pan, W.; Lin, J. Ocean Chlorophyll a Derived from Satellite Data with Its Application to Red Tide Monitoring. *J. Appl. Meteorol. Sci.* **2007**, *18*, 821–831.
11. Papenfus, M.; Schaeffer, B.; Pollard, A.I.; Loftin, K. Exploring the Potential Value of Satellite Remote Sensing to Monitor Chlorophyll-a for US Lakes and Reservoirs. *Environ. Monit. Assess.* **2020**, *192*, 1–22. [CrossRef] [PubMed]
12. D'Croz, L.; O'Dea, A. Variability in Upwelling along the Pacific Shelf of Panama and Implications for the Distribution of Nutrients and Chlorophyll. *Estuar. Coast. Shelf Sci.* **2007**, *73*, 325–340. [CrossRef]
13. Grodsky, S.A.; Carton, J.A.; McClain, C.R. Variability of Upwelling and Chlorophyll in the Equatorial Atlantic. *Geophys. Res. Lett.* **2008**, *35*, L03610. [CrossRef]
14. Zhao, D.; Zhao, L.; Zhang, F.; Zhang, X. Temporal Occurrence and Spatial Distribution of Red Tide Events in China's Coastal Waters. *Hum. Ecol. Risk Assess.* **2004**, *10*, 945–957. [CrossRef]
15. Chen, M.; Li, J.; Dai, X.; Sun, Y.; Chen, F. Effect of Phosphorus and Temperature on Chlorophyll a Contents and Cell Sizes of Scenedesmus Obliquus and Microcystis Aeruginosa. *Limnology* **2011**, *12*, 187–192. [CrossRef]
16. Wu, Q.; Xia, X.; Li, X.; Mou, X. Impacts of Meteorological Variations on Urban Lake Water Quality: A Sensitivity Analysis for 12 Urban Lakes with Different Trophic States. *Aquat. Sci.* **2014**, *76*, 339–351. [CrossRef]
17. Carneiro, F.M.; Nabout, J.C.; Vieira, L.C.; Roland, F.; Bini, L.M. Determinants of Chlorophyll-a Concentration in Tropical Reservoirs. *Hydrobiologia* **2014**, *740*, 89–99. [CrossRef]
18. de Oliveira Marcionilio, S.M.L.; Machado, K.B.; Carneiro, F.M.; Ferreira, M.E.; Carvalho, P.; Vieira, L.C.G.; de Moraes Huszar, V.L.; Nabout, J.C. Environmental Factors Affecting Chlorophyll-a Concentration in Tropical Floodplain Lakes, Central Brazil. *Environ. Monit. Assess.* **2016**, *188*, 1–9. [CrossRef]
19. Vollenweider, R.A. Input-Output Models. *Schweiz. Z. Hydrol.* **1975**, *37*, 53–84. [CrossRef]
20. Kiyofuji, H.; Hokimoto, T.; Saitoh, S.-I. Predicting the Spatiotemporal Chlorophyll-a Distribution in the Sea of Japan Based on SeaWiFS Ocean Color Satellite Data. *IEEE Geosci. Remote Sens. Lett.* **2006**, *3*, 212–216. [CrossRef]
21. Jørgensen, S.E.; Mejer, H.; Friis, M. Examination of a Lake Model. *Ecol. Model.* **1978**, *4*, 253–278. [CrossRef]
22. Wu, Z.; Wang, X.; Chen, Y.; Cai, Y.; Deng, J. Assessing River Water Quality Using Water Quality Index in Lake Taihu Basin, China. *Sci. Total Environ.* **2018**, *612*, 914–922. [CrossRef] [PubMed]
23. Wang, L.; Wu, Y.; Xu, J.; Zhang, H.; Wang, X.; Yu, J.; Sun, Q.; Zhao, Z. Nonlinear Dynamic Numerical Analysis and Prediction of Complex System Based on Bivariate Cycling Time Stochastic Differential Equation. *Alex. Eng. J.* **2020**, *59*, 2065–2082. [CrossRef]
24. Liu, Y.; Guo, H.; Yang, P. Exploring the Influence of Lake Water Chemistry on Chlorophyll a: A Multivariate Statistical Model Analysis. *Ecol. Model.* **2010**, *221*, 681–688. [CrossRef]
25. Kim, M.E.; Shon, T.S.; Shin, H.S. Forecasting Algal Bloom (Chl-a) on the Basis of Coupled Wavelet Transform and Artificial Neural Networks at a Large Lake. *Desalination Water Treat.* **2013**, *51*, 4118–4128. [CrossRef]
26. Wang, H.; Yan, X.; Chen, H.; Chen, C.; Guo, M. Chlorophyll-a Predicting Model Based on Dynamic Neural Network. *Appl. Artif. Intell.* **2015**, *29*, 962–978. [CrossRef]
27. Wei, B.; Sugiura, N.; Maekawa, T. Use of Artificial Neural Network in the Prediction of Algal Blooms. *Water Res.* **2001**, *35*, 2022–2028. [CrossRef]
28. Lee, J.H.; Huang, Y.; Dickman, M.; Jayawardena, A.W. Neural Network Modelling of Coastal Algal Blooms. *Ecol. Model.* **2003**, *159*, 179–201. [CrossRef]
29. Tian, W.; Liao, Z.; Zhang, J. An Optimization of Artificial Neural Network Model for Predicting Chlorophyll Dynamics. *Ecol. Model.* **2017**, *364*, 42–52. [CrossRef]
30. Jimeno-Sáez, P.; Senent-Aparicio, J.; Cecilia, J.M.; Pérez-Sánchez, J. Using Machine-Learning Algorithms for Eutrophication Modeling: Case Study of Mar Menor Lagoon (Spain). *Int. J. Environ. Res. Public Health* **2020**, *17*, 1189. [CrossRef]
31. Liao, Z.; Zang, N.; Wang, X.; Li, C.; Liu, Q. Machine Learning-Based Prediction of Chlorophyll-a Variations in Receiving Reservoir of World's Largest Water Transfer Project—A Case Study in the Miyun Reservoir, North China. *Water* **2021**, *13*, 2406. [CrossRef]
32. Li, X.; Sha, J.; Wang, Z.-L. Application of Feature Selection and Regression Models for Chlorophyll-a Prediction in a Shallow Lake. *Environ. Sci. Pollut. Res.* **2018**, *25*, 19488–19498. [CrossRef] [PubMed]
33. Jia, W.; Cheng, J.; Hu, H. A Cluster-Stacking-Based Approach to Forecasting Seasonal Chlorophyll-a Concentration in Coastal Waters. *IEEE Access* **2020**, *8*, 99934–99947. [CrossRef]

34. Kim, K.-M.; Ahn, J.-H. Machine Learning Predictions of Chlorophyll-a in the Han River Basin, Korea. *J. Environ. Manag.* **2022**, *318*, 115636. [CrossRef] [PubMed]

35. Yajima, H.; Derot, J. Application of the Random Forest Model for Chlorophyll-a Forecasts in Fresh and Brackish Water Bodies in Japan, Using Multivariate Long-Term Databases. *J. Hydroinformatics* **2018**, *20*, 206–220. [CrossRef]

36. Guo, Q.; Jin, S.; Li, M.; Yang, Q.; Xu, K.; Ju, Y.; Zhang, J.; Xuan, J.; Liu, J.; Su, Y.; et al. Application of Deep Learning in Ecological Resource Research: Theories, Methods, and Challenges. *Sci. China Earth Sci.* **2020**, *63*, 1457–1474. [CrossRef]

37. Zhang, F.; Wang, Y.; Cao, M.; Sun, X.; Du, Z.; Liu, R.; Ye, X. Deep-Learning-Based Approach for Prediction of Algal Blooms. *Sustainability* **2016**, *8*, 1060. [CrossRef]

38. Rostam, N.A.P.; Malim, N.H.A.H.; Abdullah, R.; Ahmad, A.L.; Ooi, B.S.; Chan, D.J.C. A Complete Proposed Framework for Coastal Water Quality Monitoring System With Algae Predictive Model. *IEEE Access* **2021**, *9*, 108249–108265. [CrossRef]

39. Cho, H.; Park, H. Merged-LSTM and Multistep Prediction of Daily Chlorophyll-a Concentration for Algal Bloom Forecast. In Proceedings of the IOP Conference Series: Earth and Environmental Science, Kaohsiung, Taiwan, 1–4 July 2019; Volume 351, p. 012020. Available online: https://iopscience.iop.org/article/10.1088/1755-1315/351/1/012020/meta (accessed on 14 September 2022).

40. Zheng, L.; Wang, H.; Liu, C.; Zhang, S.; Ding, A.; Xie, E.; Li, J.; Wang, S. Prediction of Harmful Algal Blooms in Large Water Bodies Using the Combined EFDC and LSTM Models. *J. Environ. Manag.* **2021**, *295*, 113060. [CrossRef]

41. Yussof, F.N.; Maan, N.; Md Reba, M.N. LSTM Networks to Improve the Prediction of Harmful Algal Blooms in the West Coast of Sabah. *Int. J. Environ. Res. Public Health* **2021**, *18*, 7650. [CrossRef]

42. Barzegar, R.; Aalami, M.T.; Adamowski, J. Short-Term Water Quality Variable Prediction Using a Hybrid CNN–LSTM Deep Learning Model. *Stoch. Environ. Res. Risk Assess.* **2020**, *34*, 415–433. [CrossRef]

43. Gong, G.-C.; Wen, Y.-H.; Wang, B.-W.; Liu, G.-J. Seasonal Variation of Chlorophyll a Concentration, Primary Production and Environmental Conditions in the Subtropical East China Sea. *Deep. Sea Res. Part II Top. Stud. Oceanogr.* **2003**, *50*, 1219–1236. [CrossRef]

44. Ji, C.; Zhang, Y.; Cheng, Q.; Tsou, J.; Jiang, T.; San Liang, X. Evaluating the Impact of Sea Surface Temperature (SST) on Spatial Distribution of Chlorophyll-a Concentration in the East China Sea. *Int. J. Appl. Earth Obs. Geoinf.* **2018**, *68*, 252–261. [CrossRef]

45. Chen, C.-C.; Shiah, F.-K.; Chiang, K.-P.; Gong, G.-C.; Kemp, W.M. Effects of the Changjiang (Yangtze) River Discharge on Planktonic Community Respiration in the East China Sea. *J. Geophys. Res. Ocean.* **2009**, *114*, C03005. [CrossRef]

46. Hsueh, Y. The Kuroshio in the East China Sea. *J. Mar. Syst.* **2000**, *24*, 131–139. [CrossRef]

47. Guo, X.; Zhu, X.-H.; Wu, Q.-S.; Huang, D. The Kuroshio Nutrient Stream and Its Temporal Variation in the East China Sea. *J. Geophys. Res. Ocean.* **2012**, *117*, C01026. [CrossRef]

48. Lou, X.; Shi, A.; Xiao, Q.; Zhang, H. Satellite Observation of the Zhejiang Coastal Upwelling in the East China Sea during 2007–2009. In Proceedings of the Remote Sensing of the Ocean, Sea Ice, Coastal Waters, and Large Water Regions 2011; Volume 8175, pp. 454–460. Available online: https://www.spiedigitallibrary.org/conference-proceedings-of-spie/8175/81751M/Satellite-observation-of-the-Zhejiang-Coastal-upwelling-in-the-East/10.1117/12.898140.short (accessed on 28 October 2022).

49. Lou, X.; Hu, C. Diurnal Changes of a Harmful Algal Bloom in the East China Sea: Observations from GOCI. *Remote Sens. Environ.* **2014**, *140*, 562–572. [CrossRef]

50. Peng, D.; Yang, Q.; Yang, H.-J.; Liu, H.; Zhu, Y.; Mu, Y. Analysis on the Relationship between Fisheries Economic Growth and Marine Environmental Pollution in China's Coastal Regions. *Sci. Total Environ.* **2020**, *713*, 136641. [CrossRef]

51. Chen, K.; Kuang, C.; Wang, L.; Chen, K.; Han, X.; Fan, J. Storm Surge Prediction Based on Long Short-Term Memory Neural Network in the East China Sea. *Appl. Sci.* **2021**, *12*, 181. [CrossRef]

52. Xu, Y.; Cheng, C.; Zhang, Y.; Zhang, D. Identification of Algal Blooms Based on Support Vector Machine Classification in Haizhou Bay, East China Sea. *Environ. Earth Sci.* **2014**, *71*, 475–482. [CrossRef]

53. Xiao, C.; Chen, N.; Hu, C.; Wang, K.; Xu, Z.; Cai, Y.; Xu, L.; Chen, Z.; Gong, J. A Spatiotemporal Deep Learning Model for Sea Surface Temperature Field Prediction Using Time-Series Satellite Data. *Environ. Model. Softw.* **2019**, *120*, 104502. [CrossRef]

54. Xiao, C.; Chen, N.; Hu, C.; Wang, K.; Gong, J.; Chen, Z. Short and Mid-Term Sea Surface Temperature Prediction Using Time-Series Satellite Data and LSTM-AdaBoost Combination Approach. *Remote Sens. Environ.* **2019**, *233*, 111358. [CrossRef]

55. Hochreiter, S.; Schmidhuber, J. Long Short-Term Memory. *Neural Comput.* **1997**, *9*, 1735–1780. [CrossRef] [PubMed]

56. Srivastava, N.; Hinton, G.; Krizhevsky, A.; Sutskever, I.; Salakhutdinov, R. Dropout: A Simple Way to Prevent Neural Networks from Overfitting. *J. Mach. Learn. Res.* **2014**, *15*, 1929–1958.

57. LeCun, Y.; Bengio, Y.; Hinton, G. Deep Learning. *Nature* **2015**, *521*, 436–444. [CrossRef]

58. Kingma, D.P.; Ba, J. Adam: A Method for Stochastic Optimization. *arXiv* **2014**, arXiv:1412.6980.

Article

Vertically Resolved Global Ocean Light Models Using Machine Learning

Pannimpullath Remanan Renosh *, Jie Zhang, Raphaëlle Sauzède and Hervé Claustre

Laboratoire d'Océanographie de Villefranche, Institut de la Mer de Villefranche, Sorbonne Université, CNRS INSU, 06230 Villefranche-sur-Mer, France; raphaelle.sauzede@imev-mer.fr (R.S.); herve.claustre@imev-mer.fr (H.C.)
* Correspondence: renosh.pr@imev-mer.fr

Abstract: The vertical distribution of light and its spectral composition are critical factors influencing numerous physical, chemical, and biological processes within the oceanic water column. In this study, we present vertically resolved models of downwelling irradiance (ED) at three different wavelengths and photosynthetically available radiation (PAR) on a global scale. These models rely on the SOCA (Satellite Ocean Color merged with Argo data to infer bio-optical properties to depth) methodology, which is based on an artificial neural network (ANN). The new light models are trained with light profiles (ED/PAR) acquired from BioGeoChemical-Argo (BGC-Argo) floats. The model inputs consist of surface ocean color radiometry data (i.e., R_{rs}, PAR, and $k_d(490)$) derived by satellite and extracted from the GlobColour database, temperature and salinity profiles originating from BGC-Argo, as well as temporal components (day of the year and local time in cyclic transformation). The model outputs correspond to ED profiles at the three wavelengths of the BGC-Argo measurements (i.e., 380, 412, and 490 nm) and PAR profiles. We assessed the retrieval of light profiles by these light models using three different datasets: BGC-Argo profiles that were not used for the training (i.e., 20% of the initial database); data from four independent BGC-Argo floats that were used neither for the training nor for the 20% validation dataset; and the SeaBASS database (in situ data collected from various oceanic cruises). The light models show satisfactory predictions when thus compared with real measurements. From the 20% validation database, the light models retrieve light variables with high accuracies (root mean squared error (RMSE)) of 76.42 μmol quanta m^{-2} s^{-1} for PAR and 0.04, 0.08, and 0.09 W m^{-2} nm^{-1} for ED380, ED412, and ED490, respectively. This corresponds to a median absolute percent error (MAPE) that ranges from 37% for ED490 and PAR to 39% for ED380 and ED412. The estimated accuracy metrics across these three validation datasets are consistent and demonstrate the robustness and suitability of these light models for diverse global ocean applications.

Keywords: BGC-Argo; ED380; ED412; ED490; global ocean; light models; neural network; PAR

Citation: Renosh, P.R.; Zhang, J.; Sauzède, R.; Claustre, H. Vertically Resolved Global Ocean Light Models Using Machine Learning. *Remote Sens.* **2023**, *15*, 5663. https://doi.org/10.3390/rs15245663

Academic Editors: Chung-Ru Ho and Mark Bourassa

Received: 22 September 2023
Revised: 30 November 2023
Accepted: 3 December 2023
Published: 7 December 2023

1. Introduction

Incoming solar radiation, 40% of which originates from the visible part of the spectrum, stands as the main source of energy for the entire Earth system. In the ocean, this radiation propagates and attenuates from the surface to the depths. The characterization of this propagation critically depends on accurate estimation of the downwelling irradiance, ED (W m^{-2}), over various depths. This estimation serves as the core for understanding numerous surface and sub-surface oceanic processes, as well as for the quantification of key oceanic variables.

More specifically, knowledge of ED at different depths is crucial for the quantification of various photo-dependent processes, such as oceanic phytoplankton photosynthesis [1,2], which relies on photosynthetically available radiation (PAR) as an indication of the integration of irradiance over the visible domain (400–700 nm). Additionally, knowledge of ED is essential for determining the heating rate of the upper ocean [3,4], involving the entire spectrum from UV to infrared, and also for the photo-production or destruction of organic molecules [5], often driven by the energetic UV part of the spectrum.

The derivative of the ED with respect to depth, known as the diffuse attenuation coefficient, k_d (m^{-1}), is a reliable parameter that can be related to specific optically significant substances, such as chlorophyll-a concentration (Chla), the proxy for phytoplankton biomass [6,7] or colored dissolved organic matter (CDOM) [8], the proxy for dissolved organic carbon (DOC) [9].

For the computation of remote sensing reflectance (R_{rs}), in situ measurements of ED and upwelling irradiance (LU), which measure the radiant flux per unit area per unit solid angle (W m^{-2}sr^{-1}), are essential. R_{rs}, linked to the concentration of optically significant substances and accessible from satellite observations, is an apparent optical property (AOP) of fundamental importance in ocean-color-related science. Notably, ocean color products, including R_{rs}, as well as ocean surface heat flux, are labeled as essential oceanic variables within the framework of the Global Ocean Observation System (GOOS) program.

Most of the irradiance (multi- or hyper-spectral, PAR) profiles acquired so far essentially result from the deployment of irradiance profilers from ships. These measurements (and the subsequent derivation of k_d), along with the concurrent measurements of key biogeochemical variables (e.g., Chla) [10–12], have contributed to the establishment of reference databases. These databases have become the key for assessing the bio-optical and trophic status of oceanic environments [12,13] as well as supporting validation activities for satellite ocean color radiometric products [14].

The implementation of the BioGeoChemical(BGC)-Argo program, of which irradiance is one of the six core variables, has opened up a revolutionary way to acquire numerous irradiance profiles and develop internally consistent databases [15,16]. In particular, long time series are now available in highly remote oceanic areas as well as for the severe conditions encountered in high-latitude environments in winter. Apart from radiometric quantities, BGC-Argo also allows measurement of the profiles of bio-optical variables such as Chla and particle backscattering (b_{bp}, a proxy for the particulate organic carbon (POC)). As a consequence, BGC-Argo alleviates the seasonal and regional limits and biases observed in former bio-optical databases established through ship-based observation alone, thus filling observational gaps.

To clearly distinguish the bio-optical and biogeochemical characteristics of the upper water column, a precise determination of light parameters, particularly k_d, is essential. A variety of models, including numerical, analytical, and empirical approaches, are currently used to derive the vertical propagation of irradiance within the water column. Some of these models [17–21] primarily rely on the use of inherent optical properties (IOPs) and AOPs to derive subsurface light fields. Others [22] combine a clear-sky irradiance model [23] and a spectral bio-optical relationship linking Chla to $k_d(\lambda)$ [11], which is applied to vertical Chla profiles to propagate surface irradiance into the water column beneath. These models have been widely used for a variety of applications aiming to understand and quantify bio-optical or biogeochemical processes at a regional or global scale, particularly benefiting from ocean color radiometry measured by satellites. However, these models remain complex, and, more importantly, their inputs are not readily available for immediate use.

The unique, readily and openly accessible bio-optical database based on BGC-Argo measurements (e.g., [24]) has proven to be a pivotal starting point for refining bio-optical studies (e.g., [25]), as well as for the development of novel approaches. Among these, ref. [26] reports the development of a neural network method aimed at predicting the vertical distribution of b_{bp} for any geolocation in the open ocean. This neural network, named SOCA (Satellite Ocean Color merged with Argo data to infer bio-optical properties to depth), was trained and validated using the BGC-Argo database of temperature, salinity, and b_{bp} profiles. The SOCA method for b_{bp} estimation at depth requires satellite ocean color data combined with vertical profiles of temperature and salinity as inputs. The original method of [26] has been further refined (e.g., by including satellite altimetry data as additional predictors) and adapted for the estimation of both b_{bp} and Chla. Currently, this refined approach is presented as a standard three-dimensional gridded product delivered by the European Copernicus Marine Service [27]. SOCA-derived profiles of biogeochemical quantities, along with their uncertainties, offer a basis for valuable tools for overcoming existing observational gaps. These new products can potentially support a wide range of scientific activities, including ocean modeling.

The SOCA models have served as a proof of concept by successfully deriving, first, the b_{bp}, and then another bio-optical property measured from BGC-Argo floats (i.e., Chla). This achievement has boosted confidence in the methodology's effectiveness and its adaptability to various properties measured by the BGC-Argo floats. Building on this foundation established by the SOCA models, our study aims to introduce a similar approach, specifically tailored for retrieving vertically resolved light fields in the ocean. Referred to as SOCA-light, this model has been developed to estimate irradiance profiles at any geolocation in the open ocean (bathymetric depth greater than 1500 m). It relies on a unique database of PAR and ED profiles acquired by BGC-Argo floats over the last decade. This manuscript presents the development of SOCA-light, its validation, and explores its potential applications. This model represents a significant advancement in bio-optical studies, opening a new pathway for oceanographic research.

The manuscript is organized as follows: Section 2 introduces the data and methods used for the development and validation of the light models. The following section examines the performance of these models across several datasets, including BGC-Argo datasets as well as historical ones used to establish and validate numerous models. In this section, we additionally assess the capability of the light model to predict bio-optical products from the irradiance profile. In Section 4, the final section, we address the drawbacks, benefits, and future prospects of SOCA-light models.

2. Materials and Methods

2.1. Data

2.1.1. BGC-Argo Data

BGC-Argo floats [16] equipped with multi-spectral ocean color radiometers (Satlantic OCR-504, Satlantic Inc., Halifax, NS, Canada) measuring ED at 3 different wavelengths, i.e., 380, 412 and 490 nm, W m^{-2} nm^{-1}, and PAR, µmol quanta m^{-2} s^{-1}, were used for the present study. From among the synthetic BGC-Argo individual profiles available at the Coriolis Global Data Assembly Center (GDAC) [28], only radiometric measurements qualified in delayed-mode (DM) using the quality control and calibration procedures proposed by [29] were kept for the model development. These procedures identify and correct radiometric profiles for any sensor drift or temperature dependence. The correction relies on the acquisition of at least one night profile per year (for the assessment of sensor temperature dependence) and daily dark measurements when the float drifts at the 1000 dbar parking depth (for the assessment of sensor drift). Concurrently with radiometric profiles, DM-qualified profiles of pressure (P), temperature (T), and salinity (S), were also used for the present study. The P, T, and S profiles with a number of qualified measurements less than 5 in the upper 50 m and less than 15 in the upper 250 m were discarded from the present analysis. The geographical locations of all profiles (P, T, S, and PAR) used for the development and validation of the SOCA-light model for PAR are shown in Figure 1.

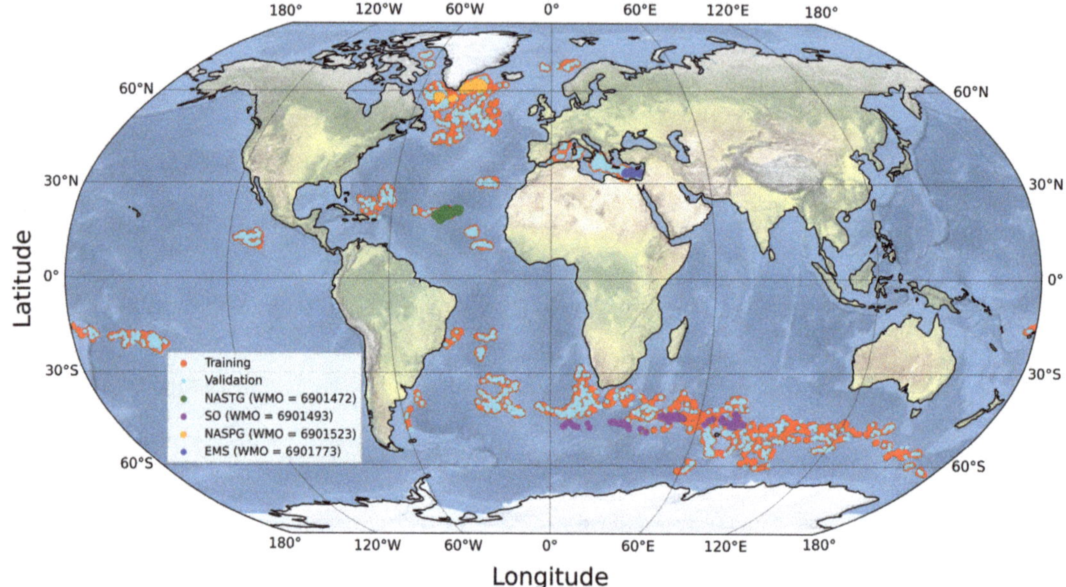

Figure 1. Geographical distribution of BGC-Argo profiles used for the development and validation of the SOCA-light model for photosynthetically available radiation (PAR) profiles. The details of the geographical distributions of profiles for other light variables (ED) are provided in Figures S1–S3 in Supplementary Information.

2.1.2. Satellite Ocean Color Data

For the neural network development and validation, and the extraction of monthly climatological light fields, we used satellite-based level-3 (L3) ocean color products of fully normalized remote sensing reflectance (R_{rs}), PAR, and $k_d(490)$ from GlobColour products. While (R_{rs} and $k_d(490)$ data were available from the Copernicus-GlobColour product, PAR (not similarly available) was directly downloaded from the GlobColour website (http://hermes.acri.fr, accessed on 17 February 2023). These global L3 products [30], which have a spatial resolution of 4 km, correspond to daily composites obtained from merged L3 Ocean Color outputs from different sensors, which ensures data continuity, improves spatial and temporal coverage, and reduces data noise [31]. The $k_d(490)$ product of GlobColour was computed from the corresponding merged Chla ($CHL\text{-}OC5$) product [32], using the following empirical equation [33].

$$k_d(490) = 0.0166 + 0.077298 \times CHL\text{-}OC5^{0.67155} \tag{1}$$

2.1.3. SeaBASS Data

The SeaWiFS Bio-Optical Archive and Storage System (SeaBASS) [34,35] is a high-quality in situ database of optical measurements, essential for satellite-data product validation and algorithm development. These data have been collected since 1998 using a variety of instrument packages (profilers, buoys, and hand-held instruments) from different manufacturers and operated on a variety of platforms, including ships and moorings. For our study, we specifically extracted profiles of ED at 380 nm (ED380), 412 nm (ED412), 490 nm (ED490), and PAR from the SeaBASS database. These profiles were collocated with ocean color and hydrological data from the ARMOR3D product (see below for details) (Figure 2). These extracted profiles were used to provide an independent assessment of the SOCA-light models developed in this study.

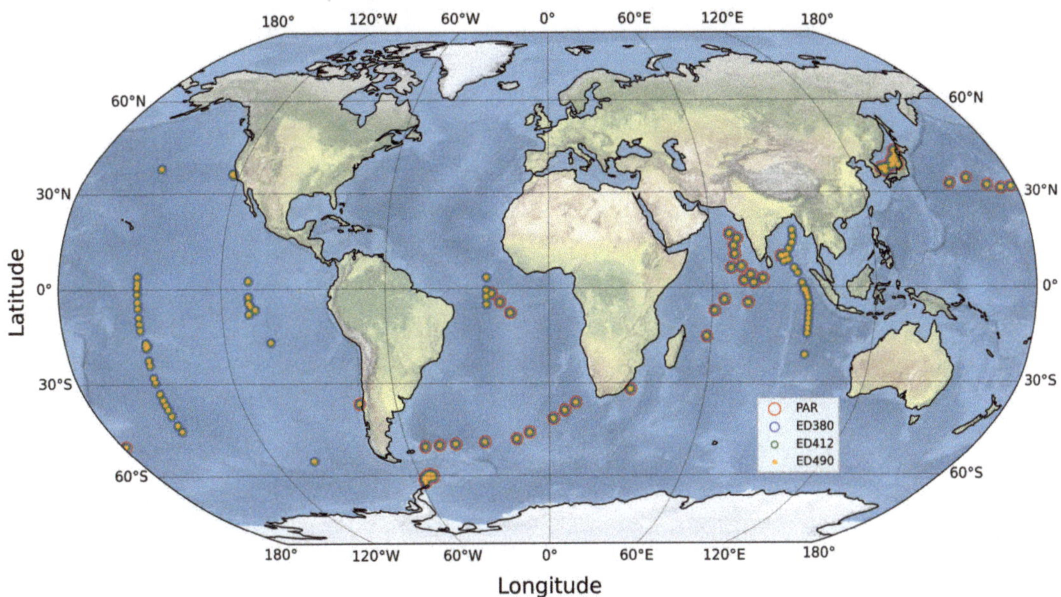

Figure 2. Geographical distribution of independent light-variable profiles (PAR, ED380, ED412, and ED490) available for validation from the SeaBASS database. Red circles represent locations of PAR profiles, blue circles correspond to ED380 profiles, green circles to ED412 profiles, and orange circles to ED490 profiles.

2.1.4. ARMOR3D Data

In this study, we used the ARMOR3D product [36,37], which provides temperature and salinity profiles at a resolution of $0.25° \times 0.25°$, encompassing 50 vertical levels within the upper 5500 m water column. This product additionally includes mixed layer depth (MLD) information. This ARMOR3D product [38] is available from the Copernicus Marine Service and was used in this study for (1) validation purposes; as temperature and salinity profiles were not available in the SeaBASS database, we used the ARMOR3D product collocated with light profiles, and (2) producing three-dimensional (3D) monthly climatological light fields; ARMOR3D monthly climatological temperature and salinity fields were used as inputs of the SOCA-light model.

2.1.5. Selection of the Database

BGC-Argo profiles, together with satellite products measuring ocean color, made up the initial database for neural network training and validation. The ocean color matchup was built by selecting the nearest available measurement both in time (within ±5 days) and space (within a 5×5 pixel area) relative to the float location and sampling time. Based on the monthly distribution of light profile acquisitions (Figure 3A), it appears that this database does not present any temporal bias in terms of the number of profiles per month globally. However, a seasonal geographical bias exists as fewer profiles exist for the northern and southern hemispheres during their respective winter months. This is due to the reduced number of matchups available because of increased cloud coverage during the winter. The present study uses all profiles sampled between 8 and 18 local hours. On an hourly basis, 97% of the profiles were sampled between 10 and 13 local hours (Figure 3). From this initial database, separate databases were created for each of the four models (i.e., PAR, ED380, ED412, and ED490).

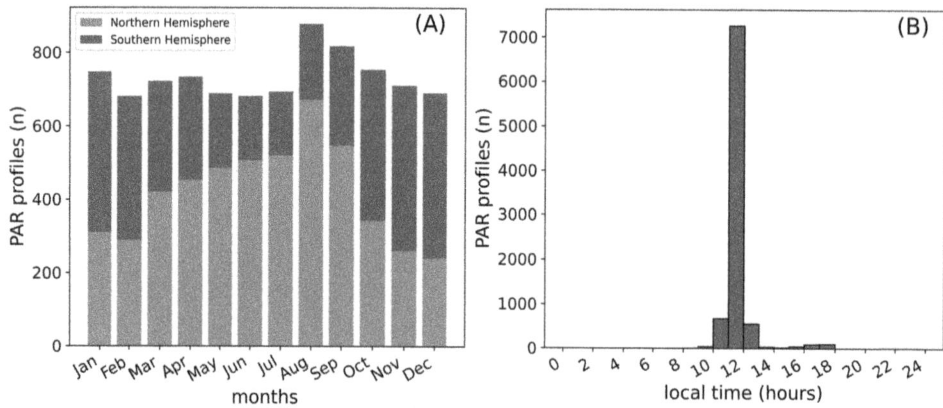

Figure 3. The temporal distribution (monthly (**A**) and hourly (**B**)) of PAR profiles used for this study.

The databases thus constituted were used for SOCA-light development, with 80% of profiles being used for model training and the remaining 20% for model validation. These training and validation databases were randomly selected. In parallel, the data from four floats with World Meteorological Organization (WMO) numbers 6901472, 6901493, 6901523, and 6901773 were kept aside for independent validation; i.e., these floats were not part of the training and validation processes. These four floats acquired multi-year measurements in four different oceanic areas considered to cover a large range of hydrological and bio-optical conditions typically representative of open ocean waters.

2.2. Methods

2.2.1. General Features of SOCA Models

Sauzède et al., (2016) [26] developed a machine-learning-based approach to extend surface bio-optical properties, such as the particulate backscattering coefficient (b_{bp}), to depth. This method, known as SOCA, relies on combining satellite ocean color observations with vertical physical information of the water column to infer the vertical distribution of the bio-optical variable b_{bp}. To train the SOCA neural network, concurrent profiles of BGC-Argo hydrological properties are matched with satellite ocean color data as inputs, while the corresponding BGC-Argo b_{bp} profiles are used as targeted outputs. This original SOCA method has been further refined by, for instance, including R_{rs} instead of satellite b_{bp} and Chla, using satellite altimetry products as additional predictors (to account for possible mesoscale influence) and adapting the method for the estimation of both b_{bp} and Chla. In this way, ocean color and hydrological products with different temporal scales (weekly fields and monthly climatologies) are used as inputs to these SOCA models, and the derived outputs are delivered as operational standard products by the Copernicus Marine Service [27].

2.2.2. The SOCA-Light Models

For this study, we developed a SOCA-type model based on a neural network, and more specifically, a multilayer perceptron (MLP). The MLP is a robust modeling tool used for supervised learning, employing multiple inputs and a known output value to train the model [39–41]. As a feedforward neural network, information flows unidirectionally from the MLP's input layer to its output layer, passing through one or more intermediate layers, also called hidden layers. Each layer is constructed from neurons, which are fundamental transfer functions that generate outputs when inputs are applied. Each connection between neurons has its own weight. The backpropagation algorithm then adjusts the weights of the neurons in each layer to minimize the loss function using a first-order gradient-based optimizer.

The SOCA-light models are largely derived from the generic SOCA methodology described in [26,27]. They consist of four models capable of predicting the vertical profiles of PAR, ED380, ED412, and ED490 at a given geolocation, using as inputs the data from matchups with satellite ocean color products and the vertical profiles of T and S. For SOCA-light, we have slightly modified the input variables used for other SOCA models (i.e., for Chla and b_{bp}) through the selection of key variables that depict the vertical propagation of light in the water column (i.e., first optical depth (Z_{pd})). In this way, while other SOCA models (Chla and b_{bp}) have used sea-level anomaly (SLA) as input to infer mesoscale processes that may impact the vertical distribution of phytoplankton biomass, in SOCA-light models we have removed SLA from the key variables. The four SOCA-light neural networks were trained using a database of concurrent profiles of temperature, salinity, and light variables (ED380, ED412, ED490, and PAR) collected by BGC-Argo floats and collocated with satellite-derived products. A schematic representation of all the SOCA-light models is shown in Figure 4.

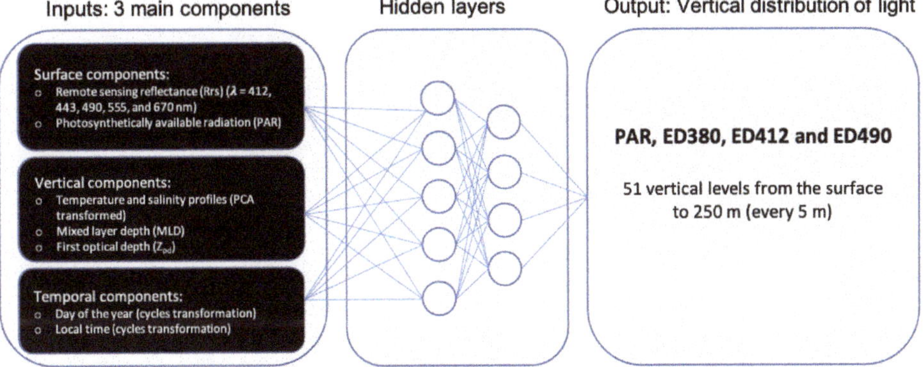

Figure 4. Schematic representation of the SOCA-light multilayer perceptron.

There are three main input components used for this model:

- **Surface components:** These encompass satellite-based surface estimates of R_{rs} at five different wavelengths (i.e., 412, 443, 490, 555, and 670 nm) and PAR.
- **Vertical components:** These rely on the first principal component analysis of salinity and temperature profiles. The principal components were selected on the basis of cumulative explained variance values less than or equal to 0.998. For temperature, this criterion is satisfied by five principal components, and for salinity, by four principal components. The mixed layer depth (MLD) was derived from density calculated from pressure, temperature and salinity profiles with a density differential threshold criterion of 0.03 kg m^{-3} with reference to the density at 10 m [42]. The Z_{pd} was derived from the satellite-derived $k_d(490)$ using Equation (2).

$$Z_{pd} = \frac{1}{k_d(490)} \qquad (2)$$

- **Temporal components:** The temporal components are the day of the year (DOY) and the local time (LT) of the sampling profile. These components follow periodic evolution within certain time windows (0 to 365 days for DOY; 0 to 24 h for LT). The cyclic transformations (sine and cosine) of radian-transformed DOY and LT were used as temporal components (Equation (3) and (4)):

$$DOY_{rad} = \frac{DOY \times \pi}{182.625} \qquad (3)$$

$$LT_{rad} = \frac{LT \times \pi}{12} \tag{4}$$

The SOCA-light model outputs are the four light variables (PAR, ED380, ED412, and ED490) at 51 vertical levels from the surface to 250 m depth at every 5 m interval. The use of an ensemble of MLPs proved effective in improving the robustness and reliability of predictions compared to the use of a single MLP [43]. For this reason, as a first step, several MLPs were created, each with a unique architecture incorporating the hyperbolic tangent (tanh) as the activation function and adaptive moment estimation (ADAM) [44] as the solver. The ADAM solver streamlines the conclusion of iterations upon reaching model convergence, speeding up the process. At the same time, we identify the optimal number of epochs to ensure effective learning and prevent overfitting. The key distinction among these models lies in the varying number of neurons distributed across each hidden layer, with the intention of capturing diverse patterns and representations inherent in the data. We chose two hidden layers from the considered options of one, two, and three hidden layers for these light models. Notably, models with two hidden layers consistently outperformed, with the number of neurons in the second hidden layer always being fewer than or equal to that in the first hidden layer. The models were trained by changing the neuron numbers between 5 and 150 with an increment of one (altogether 10,585 iterations). The second step was then to select, from all these iterations, an ensemble of the 10 best MLPs based on minimum statistical metrics obtained from training and validation datasets (root mean square error (RMSE) and the median absolute percent error (MAPE)). Through this selection, the ensemble model aimed to capture diverse representations while ensuring the sound performance and consistency of individual MLPs.

2.2.3. Statistical Analyses

The performance of the model was evaluated by comparison between the modeled variable values (Y-axis) and the actual values used as references (X-axis). Two statistical criteria were used: the RMSE as well as the MAPE that were computed as in the equations below (Equations (5) and (6)):

$$RMSE = \sqrt{\frac{\sum_{i=1}^{n}(Obs_i - Pred_i)^2}{n}} \tag{5}$$

$$MAPE(\%) = median\left[\frac{|Obs_i - Pred_i|}{Obs_i}\right] \times 100 \tag{6}$$

where n, Obs, and $Pred$ correspond to the number of points, the observed value, and the predicted value, respectively.

3. Results

3.1. Validation of SOCA-Light Models

A rigorous set of validation protocols was adopted to assess the accuracy of the four light models. In this way, the model results were validated against the validation database (Section 3.1.1), then against the four independent BGC-Argo floats from four distinct oceanic basins (Section 3.1.2), as well as against the independent SeaBASS database (Section 3.1.3). Finally, proxies derived from SOCA-light products were further used to evaluate the prediction capability of the model (Section 3.1.4).

3.1.1. Validation of SOCA-Light Models Using 20% of the Global Database

The SOCA-light models were validated using 20% of the dataset randomly extracted from the BGC-Argo database, originating from a large diversity of oceanic regions. The comparison between modeled SOCA-light variables and BGC-Argo measurements (PAR, ED380, ED412, and ED490) is presented in Figure 5. Overall, there is a very good agreement between the predicted and the measured light variables. The density scatterplot reveals a close clustering of points along the identity line over more than five orders of magnitude. Statistical metrics extracted from linear regression between the modeled and observed PAR values reveal slope, r^2, RMSE and MAPE values of 1.01, 0.96, 76.42 µmol quanta $m^{-2}\ s^{-1}$, and 37.41%, respectively (Figure 5A). The validation for the three ED models shows satisfactory performances. The modeled ED380 profiles exhibit RMSE and MAPE values of 0.04 W $m^{-2}\ nm^{-1}$, and 39.01%, respectively, when compared with their measured counterparts (Figure 5B). Similarly, the MAPE values for ED412 and ED490 were 39.47% and 37.05%, respectively (Figure 5C,D).

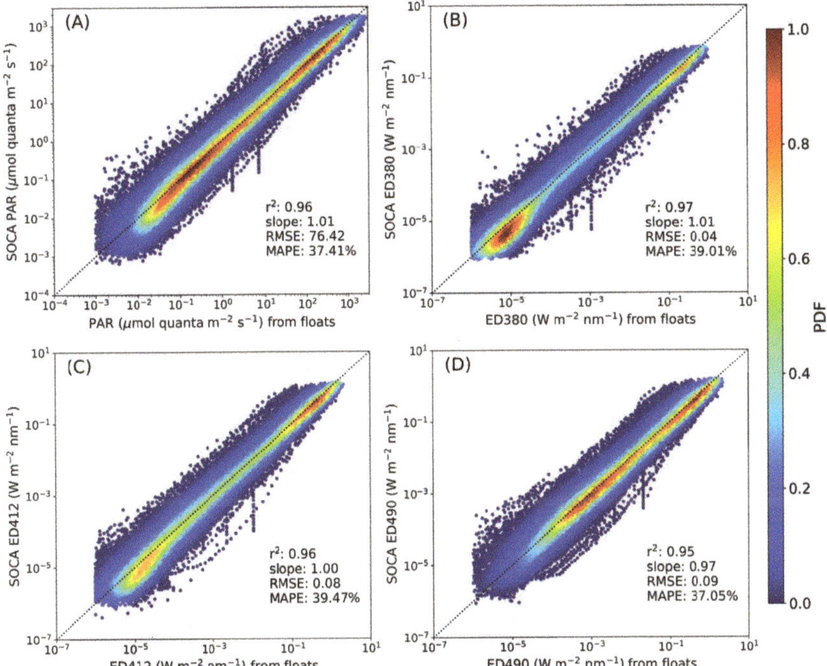

Figure 5. Scatterplots between light variables (PAR, ED380, ED412, and ED490) modeled by the SOCA-light models versus their corresponding BGC-Argo measurements: PAR (**A**); ED380 (**B**); ED412 (**C**); ED490 (**D**). This validation was performed using 20% of profiles randomly selected from the total database. The color code scales the probability density function (PDF). The identity line is represented by the 1:1 black dotted line.

3.1.2. Validation of SOCA-Light Models Using Four Independent BGC-Argo Floats from Different Oceanic Regions

An independent validation was performed using four BGC-Argo floats from distinct oceanic regions, namely the North Atlantic Subtropical Gyre (NASTG), the Eastern Mediterranean Sea (EMS), the Southern Ocean (SO), and the North Atlantic Subpolar Gyre (NASPG). The profiles for each region originated from a single float with a unique WMO, none of which were included in the training and validation databases. The validation results for each oceanic region are presented in Figure 6.

The scatterplot of PAR derived by the model shows strong agreement with PAR measured by the BGC-Argo floats for all four regions (Figure 6A). Statistical error estimators computed between modeled and observed PAR profiles for all four regions together show slope, r^2, RMSE, and MAPE values of 1.03, 0.96, 72.86 µmol quanta $m^{-2}\ s^{-1}$, and 30.50%, respectively. These statistical error estimators of PAR are comparable with the statistics obtained on 20% of the validation database (Figure 5A). For ED380 (Figure 6B), the slope, r^2, RMSE, and MAPE values are 1.04, 0.97, 0.034 W $m^{-2}\ nm^{-1}$, and 40.99%, respectively. For

ED412 (Figure 6C), the same statistical metrics yield values of 1.03, 0.97, 0.070 W m^{-2} nm^{-1}, and 36.67%, respectively. Finally, for ED490 (Figure 6D), the metrics take values of 1.03, 0.95, 0.087 W m^{-2} nm^{-1}, and 29.86%, respectively.

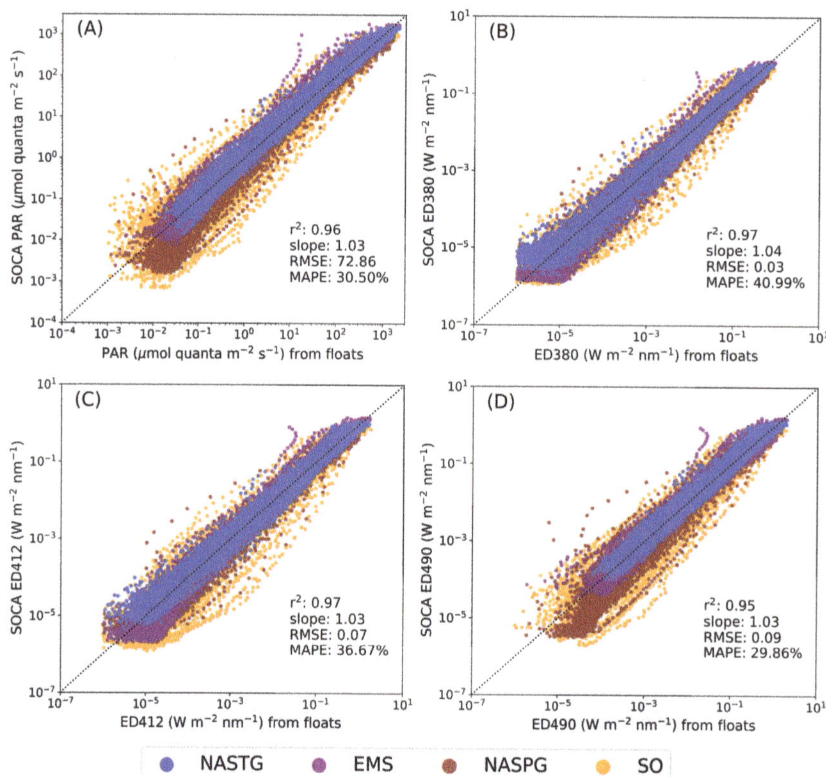

Figure 6. Scatterplots illustrating the comparison between SOCA-light modeled variables (PAR, ED380, ED412, and ED490) and their corresponding BGC-Argo measurements collected by the four independent floats. The subplots display: PAR (**A**), ED380 (**B**), ED412 (**C**), ED490 (**D**). Each color represents a specific float: blue for NASTG, purple for EMS, brown for NASPG, orange for SO. The identity line is represented by the 1:1 black dotted line.

North Atlantic Subtropical Gyre

The NASTG is an oligotrophic environment characterized by low surface nutrients, low Chla, and the presence of a permanent deep chlorophyll maximum (DCM), generally found below 100 m [45,46]. The multi-year time series (more than 6 years of measurement) of the vertical distribution of light variables (PAR, ED380, ED412, and ED490) measured by the NASTG BGC-Argo float (WMO = 6901472) and modeled by SOCA-light are presented in Figure 7 for a direct comparison. Overall, the SOCA-light models clearly reproduce, in a smoother way, the seasonal and vertical trends revealed by the float measurements. The SOCA-light models capture even subtle changes in the general trends of light variables, as evidenced by the less pronounced light penetration observed and reproduced by the model at the end of 2015. As well as reproducing the trends satisfactorily, the magnitude of the signals is retrieved well by the models for the four variables. The statistical metrics between the modeled and the observed PAR profiles show (Figure S4) slope, r^2, RMSE, and MAPE values of 0.99, 0.98, 73.09 µmol quanta m^{-2} s^{-1}, and 21.50%, respectively. For ED380, these metrics are, respectively, 0.96, 0.98, 0.04 W m^{-2} nm^{-1}, and 28.72%. For ED412, they are 0.95, 0.98, 0.08 W m^{-2} nm^{-1}, and 26.66%. Finally, for ED490, they are 0.98, 0.98, 0.09 W m^{-2} nm^{-1}, and 21.53%.

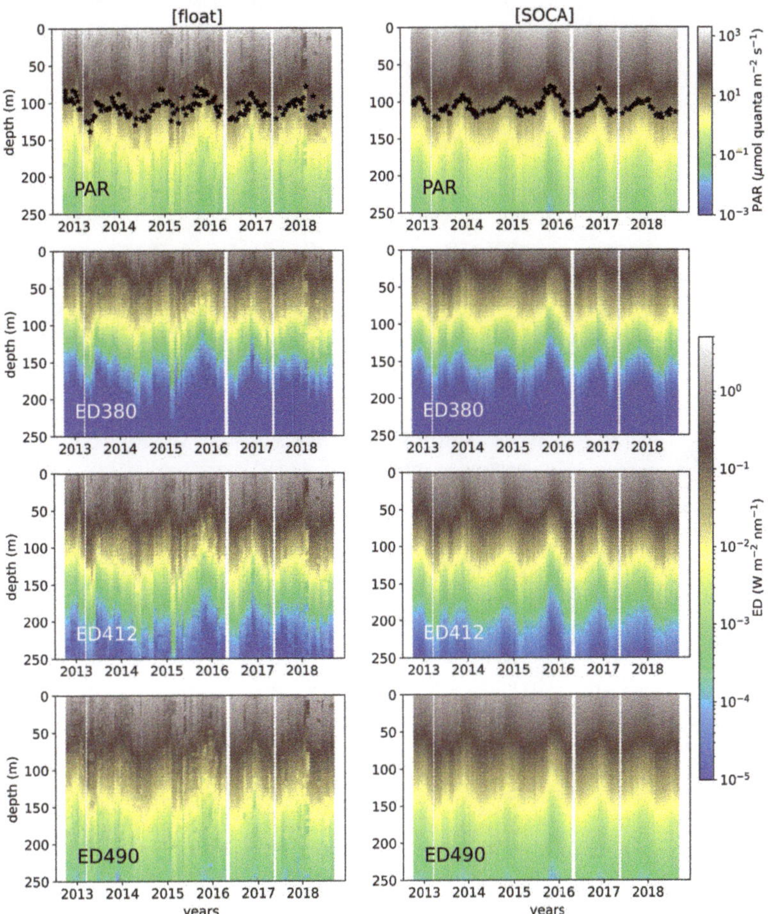

Figure 7. Time series of the vertical distribution of the four light variables in the North Atlantic Subtropical Gyre (NASTG), as measured by BGC-Argo float with WMO 6901472 (**left column**) and modeled by SOCA-light (**right column**). The variables in each subplot are indicated by text in the corresponding subplots. The black stars indicate the depth at which instantaneous PAR value = 15 µmol quanta m^{-2} s^{-1}.

Eastern Mediterranean Sea

The EMS is also a permanent oligotrophic system at temperate latitudes. The float selected (WMO = 6901773) measured all four light variables (PAR, ED380, ED412, and ED490) for nearly four years (Figure S5). Again, the multi-year vertical sections of these variables from this region show very good agreement between the measured and modeled values. The modeled variables exhibited seasonal fluctuations in their magnitude across different years, similar to those observed. The surface incoming solar radiation shows larger seasonal variability than the variability observed in the subtropical oligotrophic regime (NASTG, Figure 7), yet it is well captured by the model. As for the NASTG, the models reproduce light variables with much less noise compared to their corresponding BGC-Argo measurements. The statistical metrics between the modeled and the measured variables from the EMS for all four light variables are highly comparable with the global

20% validation metrics (Figure S6). The statistical metrics between the modeled and the observed PAR profiles display slope, r^2, RMSE, and MAPE values of 1.02, 0.98, 63.20 µmol quanta m^{-2} s^{-1} and 21.42%, respectively. For ED380, these metrics were, respectively, 1.08, 0.98, 0.03 W m^{-2} nm^{-1}, and 39.89%. For ED412, they were 1.06, 0.98, 0.06 W m^{-2} nm^{-1}, and 29.04%. Lastly, for ED490, they were 1.01, 0.98, 0.08 W m^{-2} nm^{-1}, and 22.78%. All the derived error estimators show comparable values (some even better, such as the RMSE and MAPE) with those obtained for the 20% global validation database. These results depict the robustness of the SOCA-light models for deriving light variables over several years of observation.

Southern Ocean

Over four years, the BGC-Argo float (WMO = 6901493) traveling eastwards (from 5°E to 83°E) and between 40°S and 50°S in the SO underwent the typical bio-physical conditions prevailing in the area. Overall, it captured four phytoplankton blooms and was regularly trapped or influenced by mesoscale features or fronts. The multi-year time series of the vertical distribution of light variables (PAR, ED380, ED412, and ED490) measured by this float and the SOCA-light modeled light variables are presented and compared in Figure 8. The gaps in the time series during the southern-hemisphere winter months are due to the unavailability of ocean color matchups resulting from cloud coverage during this period. In general, as for the NASTG and EMS, the SOCA-light models reproduce the seasonal and vertical trends of the float measurements in a smoother way. In addition to reproducing the seasonal trends, the magnitude of the retrieved light variables is in order with the measurements for the four light variables. The statistical metrics between the modeled and measured PAR profiles (Figure S7) show slope, r^2, RMSE, and MAPE values of 0.99, 0.91, 88.63 (µmol quanta m^{-2} s^{-1}), and 54.37%, respectively. For ED380, these metrics were 1.03, 0.95, 0.04 (W m^{-2} nm^{-1}), and 51.52%. For ED412, they were 1.03, 0.93, 0.08 (W m^{-2} nm^{-1}), and 54.33%. Finally, for ED490, the metrics were 0.99, 0.91, 0.10 (W m^{-2} nm^{-1}), and 51.47%. The statistical estimators from the SO, namely the RMSE and MAPE, are slightly larger than the global 20% validation metrics. These uncertainties could possibly originate from the highly dynamic nature of the area associated with the ocean color matchups of the closest pixel of the temporal (±5 days) and spatial (5×5 pixels) matchups. They may also be attributed to the higher level of this dataset's independence, thus providing a more rigorous test of the model's generalization capabilities. Indeed, a higher level of errors can be expected in a highly variable environment such as the SO. Nevertheless, the fact that errors from this dataset are only marginally greater than those from the 20% validation dataset suggests the model's robustness without signs of overfitting.

North Atlantic Subpolar Gyre

The data acquired by the float (WMO = 6901523) over its two years of exploration are representative of the diversity of the North Atlantic Subpolar Gyre conditions. In particular, it encountered intense convection periods (>1000 m) as well as intense spring phytoplankton blooms. Due to a lack of ocean color matchups, the NASPG region experienced similar problems as the SO region in obtaining SOCA-light variables during the winter. The two-year time series of the vertical distribution of light variables (PAR, ED380, ED412, and ED490) measured by this float and modeled by SOCA-light are presented and compared in Figure S8. Essentially, the SOCA-light models reproduce the seasonal trends in float measurements in a smoother way. The statistical metrics between the modeled and the observed PAR profiles manifest slope, r^2, RMSE, and MAPE values of 1.08, 0.93, 64.83 µmol quanta m^{-2} s^{-1}, and 58.79%, respectively (Figure S9). For ED380, these metrics were subsequently 1.02, 0.96, 0.02 W m^{-2} nm^{-1}, and 50.82%. For ED412, they were 1.02, 0.95, 0.04 W m^{-2} nm^{-1}, and 51.75%. Finally, for ED490, these metrics were 1.09, 0.92, 0.06 W m^{-2} nm^{-1}, and 57.63%. Similarly to the SO float, the statistical estimators, mainly the RMSE and MAPE, from the NASPG float are slightly larger than the global 20% validation metrics. This could mainly be due to the uncertainties associated with the retrieval of ocean color matchups from the closest pixel of the temporal (±5 days) and spatial (5 × 5 pixels) matchups in such a highly dynamic high-latitude environment, which seems less the case for low and temperate latitudes (see Figures 7 and S5 and the associated metrics in Figures S4 and S6).

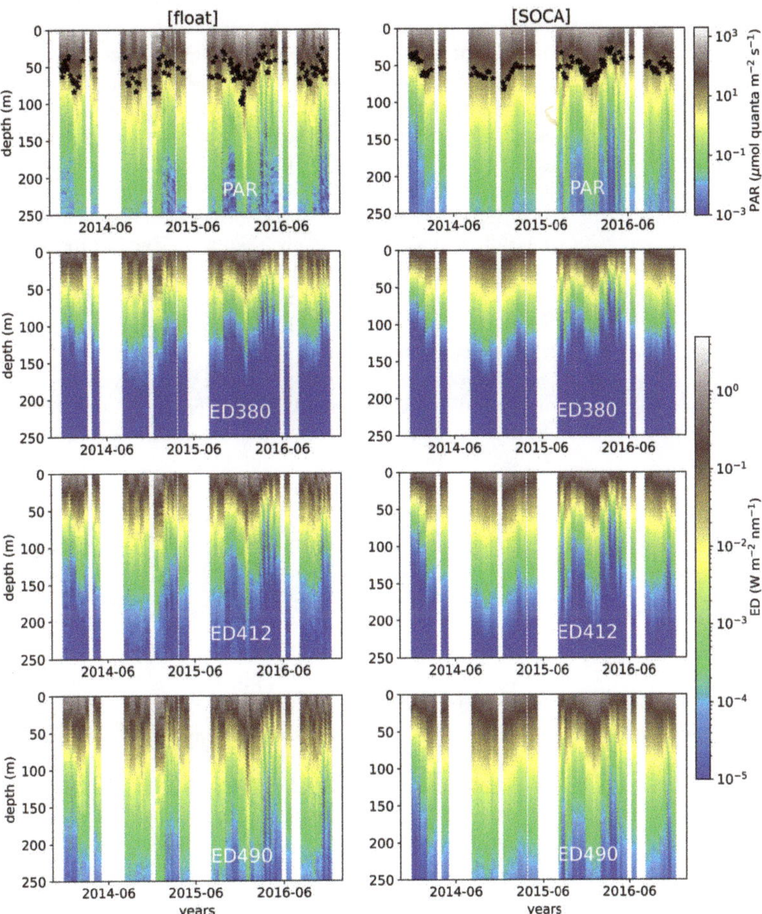

Figure 8. Time series of the vertical distribution of the four light variables in the Southern Ocean (SO) measured by BGC-Argo float WMO 6901493 (**left column**) and modeled by SOCA-light (**right column**). The variables in each subplot are specified by text in the corresponding subplots. The black stars indicate the depth at which instantaneous PAR value = 15 µmol quanta m^{-2} s^{-1}.

3.1.3. Validation of SOCA Light Models with the Independent Global SeaBASS Database

As well as validating SOCA-light models against a 20% subset of the BGC-Argo dataset or against the data of selected BGC-Argo floats not included in either the initial training or the 20% validation procedures, validation against datasets not acquired by BGC-Argo offered an informative complementary exercise. For this purpose, we used the global SeaBASS light database whose measurements originate from various cruises and field campaigns. It should be noted that, contrary to the BGC-Argo light measurements performed under any sky conditions, measurements from ships, which are more operator-dependent, are essentially conducted under a clear sky.

The input matchups were taken from the weekly binned files of ARMOR3D and GlobColour data that corresponded to each SeaBASS in situ station. The physical variables (temperature, salinity, and MLD) were extracted from the macro-pixel ($0.25° \times 0.25°$) nearest the in situ station and ocean color matchups from the mean of the 3×3 micro-pixels (4 km $\times$ 4 km) box centered at each in situ station. It should be noted that only a restricted number of stations from the original SeaBASS database were used for this validation exercise, as more than 90% of the stations (including coastal stations with bathymetric depths less than 1500 m) lacked a corresponding satellite ocean color matchup, mainly due to the contamination of signals, probably by clouds or sea ice.

The scatterplots of the light variables derived by the SOCA-light models compared with those measured in situ within the SeaBASS database are presented in Figure 9. The same metrics are reported as those in the other validation exercises based on float data (Sections 3.1.1 and 3.1.2). Figure 9 shows that the retrieval by SOCA-light systematically underestimates the SeaBASS measurements for each light variable. This bias is the same over the whole water column as the slopes between the modeled light and corresponding measurements are close to one (Figure 9). The fact that the SeaBASS database is essentially populated by data obtained under clear-sky conditions at a given time could explain this bias. By way of contrast, the weekly matchups of GlobColour products used as input for the SOCA-light models likely do not correspond to clear-sky conditions over such an extended temporal window.

The scatterplot of PAR produced by the model exhibits notable consistency with PAR measured in situ by SeaBASS data (Figure 9A). Statistical error metrics were extracted from linear regression between the modeled and observed PAR profiles, showing slope, r^2, RMSE, and MAPE values of 1.00, 0.88, 101.25 μmol quanta m^{-2} s^{-1}, and 65.48%, respectively. For ED380, these metrics were 1.00, 0.82, 0.11 (W m^{-2} nm^{-1}), and 76.30% (Figure 9B). For ED412, they were 1.00, 0.81, 0.18 W m^{-2} nm^{-1}, and 76.07% (Figure 9C). Finally, for ED490, they were 0.99, 0.85, 0.21 W m^{-2} nm^{-1}, and 62.32% (Figure 9D). These four light models (PAR, ED380, ED412, and ED490) were validated independently, and the extracted error metrics are quite satisfactory, even if these statistical estimators are slightly larger compared with the error metrics of both the global 20% validation database and four independent BGC-Argo floats. These larger error estimators could be because of the uncertainty associated with the physical and ocean color data considered as inputs (as well as the nature of the data in SeaBASS, essentially acquired under clear-sky conditions).

3.1.4. Additional Validation with iPAR_15

An alternative to validating the SOCA-light model results against light data from various databases (previous sections), that also allows gauging the model's prediction capabilities, is to quantify and assess the quality of model-derived products that are essential for certain applications. This is the case for the depth of iPAR_15 (Z_iPAR_15) [47], a variable that corresponds to the depth at which the instantaneous PAR, iPAR, equals 15 μmol quanta m^{-2} s^{-1}. This quantity is required for the correction of non-photochemical quenching (NPQ) that affects the chlorophyll-a fluorescence profiles. NPQ is a photo-physiological mechanism whereby the signal of chlorophyll-a fluorescence is depressed under high irradiances (maximal at noon). The method proposed by [47] and further improved by [48] uses Z_iPAR_15 as a depth threshold under which no NPQ is expected. In a way, Z_iPAR_15 can be considered as a proxy for water clarity with high values corresponding to the clearest waters, where the NPQ effect can be observed at the deepest depths. The present study extracted Z_iPAR_15 from PAR measured by the BGC-Argo floats and PAR derived using the SOCA-light PAR model for the validation database of 20% of the global database and for the four independent floats (Figure 10). Overall, the results are satisfactory with respect to the retrieval of Z_iPAR_15 by the SOCA-light PAR model. Furthermore, the range of values of Z_iPAR_15 for the four floats (Figure 10B) is equivalent to that for the 20% validation database (Figure 10A). This demonstrates that the four floats cover the entire range of trophic status currently detected by the BGC-Argo database throughout the global ocean.

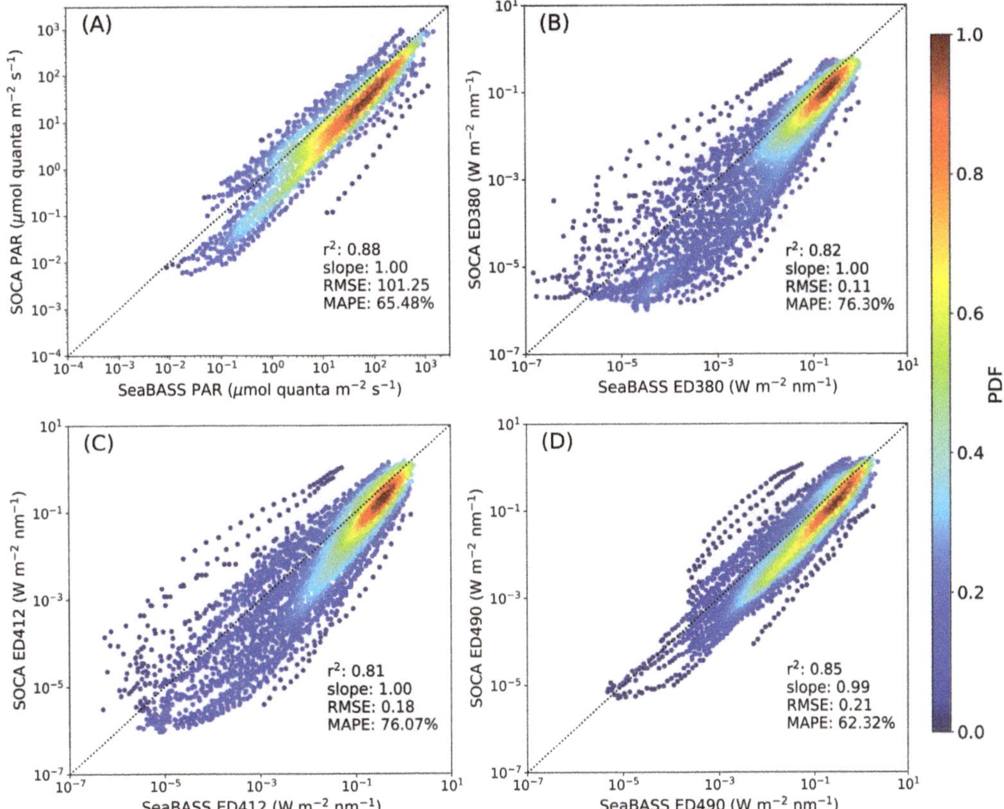

Figure 9. Scatterplots between light variables (PAR, ED380, ED412, and ED490) derived using SOCA-light models and SeaBASS in situ measurements. The subplots display: PAR (**A**), ED380 (**B**), ED412 (**C**), ED490 (**D**). The color code scales the PDF. The identity line is represented by the 1:1 black dotted line.

To illustrate a potential application of the SOCA-light models, we extracted global 3D multi-year monthly averaged climatologies of light variables at local noon at a 5 m resolution from the surface to 250 m depth. The inputs used to generate the climatologies were multi-year monthly averaged GlobColour data and ARMOR3D physical data. The satellite data were averaged ($0.25° \times 0.25°$) at the same spatial resolution as the physical ARMOR3D data. As an example, the extracted Z_iPAR_15 from these seasonal climatology fields is presented in Figure 11, and shows well-characterized latitudinal and seasonal variations.

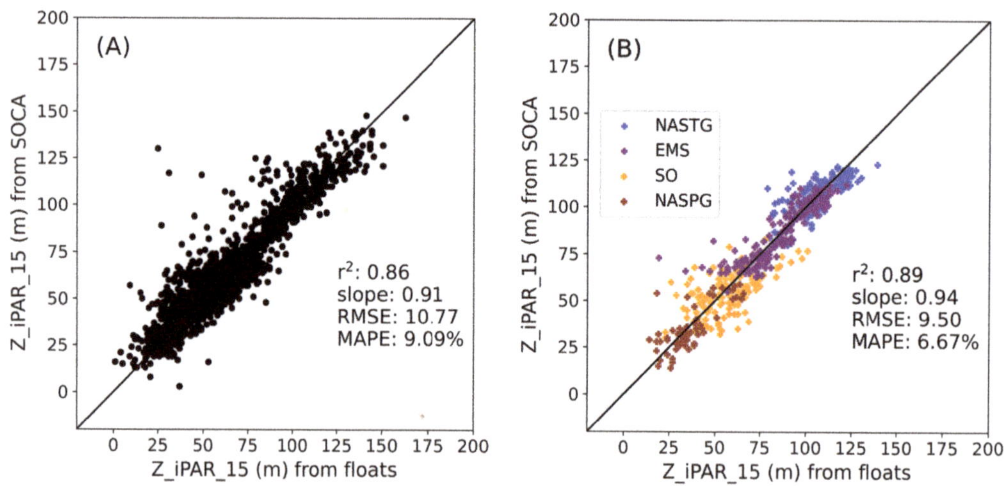

Figure 10. Comparisons of Z_iPAR_15 derived by the SOCA-light PAR model versus Z_iPAR_15 estimated by BGC-Argo float measurements for the 20% validation database (**A**) and for the 4 independent floats (**B**).

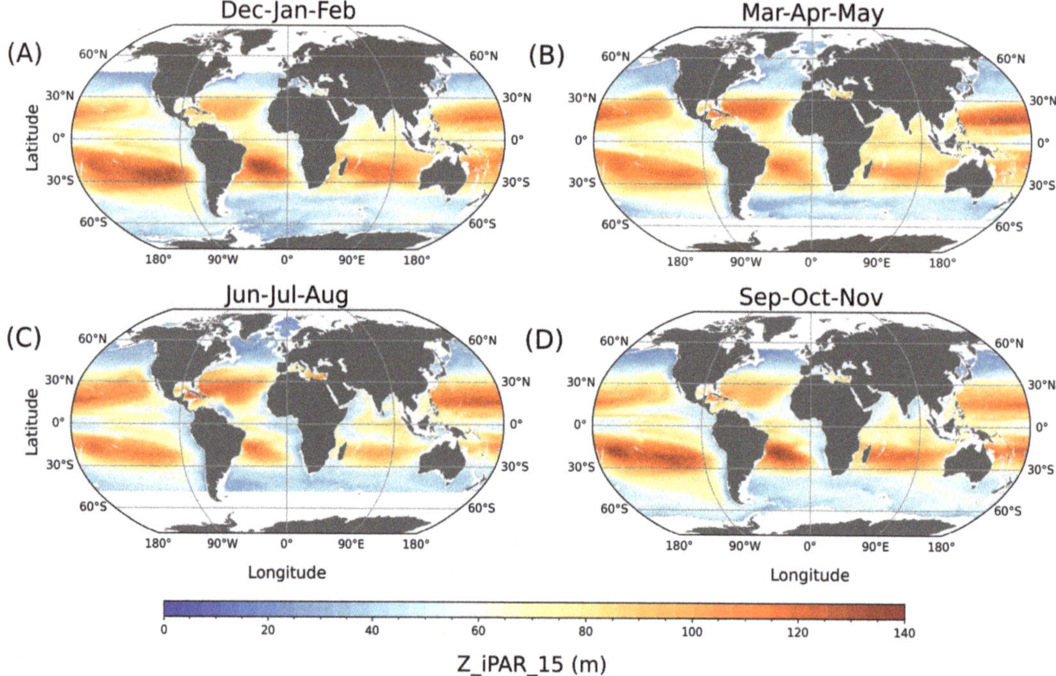

Figure 11. Seasonal climatology of Z_iPAR_15 derived at local noon using the SOCA-light PAR model applied to monthly climatological fields of inputs: Z_iPAR_15 averaged for the months of December, January, and February in (**A**); March, April, and May in (**B**); June, July, and August in (**C**); September, October, and November in (**D**).

4. Discussion and Conclusions

Knowledge about the irradiance vertical distribution is essential for improved understanding and quantification of many oceanic processes in the upper water column. Over time, a variety of bio-optical models has been developed to better predict light fields, particularly at the ocean surface. These models have drawn from various complex relationships, spanning from purely empirical to fully analytical algorithms. They have served as fundamental building blocks for various biogeochemical applications, including the retrieval of IOPs [49,50], biogeochemical quantities such as Chla [51–53], and POC [54], as well as the quantification of the oceanic heating rate [3,4] and the modeling of oceanic primary production [1,2]. These models serve as the foundation for bio-optical oceanography and satellite ocean color science.

The development of bio-optical models has, however, been constrained by the limited availability of in situ data, either for model construction (for empirical models) or model validation (in the case of analytical models). Substantial gaps in the acquisition of bio-optical data have resulted in limited coverage and sparse datasets, especially in remote open ocean areas. Additionally, the databases containing these measurements often exhibit heterogeneity in terms of acquisition modes, involving different platforms and sensors. These variations lead to consistency and interoperability issues, increasing the uncertainties of models relying on these data.

More recently, the prospect of developing more accurate bio-optical models for irradiance vertical distribution has emerged for two main reasons. The first one relates to the massive availability of the various oceanic properties, including optically significant substances and light variables. This availability is largely due to the extensive data-collection capacity of BGC-Argo, which has contributed to a rich and dense database of ED and PAR profiles. Importantly, in addition to being publicly and openly accessible, this database offers the advantage of being homogeneous and interoperable thanks to the development of dedicated methods to ensure its qualification [29,55,56]. Moreover, this database has proven instrumental in validating bio-optical models [57,58] and models based on Chla for estimating PAR [22]. The second reason is due to the increasing adoption of machine learning techniques that take advantage of data availability, which results in a strong improvement in the predictive capability of these purely empirical approaches. Pioneering work by [26] showed that global 3D reconstruction of the b_{bp} could be performed thanks to the development of the first SOCA model. More recently, subsets of irradiance data (ED380, ED412, and ED490) acquired by BGC-Argo floats have been used to predict PAR either through statistical approaches [59] or the use of neural networks [60].

The present study represents, to our best knowledge, the first attempt to develop a predictive model for the vertical profiles of light, encompassing both PAR and irradiance at three different wavelengths, thanks to the application of machine learning using the extensive BGC-Argo light database. This model rests on the initial SOCA methodology, which has been carefully refined to accommodate the specificity of light-related variables. While the model exhibits significant accuracy and potential, it does have some limitations that should be acknowledged. Certainly, the prediction of light profiles becomes challenging in the absence of R_{rs} data (e.g., due to cloud coverage), a situation particularly critical in high-latitude environments during the winter. Moreover, the majority of SOCA-light training involved local noon data (97% of profiles gathered between 10 and 13 h local time), suggesting a potential decrease in accuracy for predictions at other times of the day. Nevertheless, as more data from BGC-Argo become available at various times, this limitation could be easily addressed in the near future. In the meantime, it is recommended to preferentially use SOCA-light around noon local time.

The predictive power of SOCA-light appears to be robust (Figures 5, 6, 9 and 10). Until now, efforts to characterize vertical light profiles in oceanic waters have relied on various approaches, involving numerical models [17–21] and a combination of analytical, semi-analytical, and empirical relationships [11,20,22,33]. However, these models heavily rely on specific parameters, including AOPs, IOPs, and Chla resolved over the vertical dimension. The incorporation of these precise vertically resolved inputs presents challenges when attempting to compare such models with SOCA-light ones. The lack of these crucial input data poses a well-acknowledged challenge in the field of oceanography, especially within the marine optics and ocean color remote sensing communities. Consequently, this data gap potentially translates into more uncertainty over depths. The machine learning approaches proposed here potentially circumvent this weakness. Furthermore, assuming the model inputs are at the right resolution, SOCA-light can easily extract 3D global ocean light maps at any temporal resolution (daily, weekly, and monthly).

Due to their solidity and versatility, SOCA-light models offer great potential for supporting many applications for which light profiles are key variables but are unfortunately not measured. For instance, several applications for improving the BGC-Argo database can already be envisioned. At present, light profiles are not acquired from all BGC-Argo floats. Indeed, less than 45% of the ≈118,000 chlorophyll-a fluorescence profiles so far acquired have concurrent light measurements. Yet, light profiles are required for a more accurate estimation of Chla from chlorophyll-a fluorescence measured from floats. First, the correction of NPQ fluorescence is more accurate with the use of instantaneous PAR profiles [47,48] compared to former methods which do not rely on light [61]. Secondly, as the relation between Chla and chlorophyll-a fluorescence varies regionally and seasonally, methods have been proposed that rely on concurrent profiles of ED490 and chlorophyll-a fluorescence to estimate the slope correction to apply to the fluorescence profile in order to retrieve more accurate Chla [6]. The estimation of this slope correction relies on a bio-optical relationship linking $k_d(490)$ (derived from the ED490 profile) to Chla [11]. Having the whole BGC-Argo fleet delivering light profiles (either measured or modeled) would guarantee an overall more consistent and interoperable Chla dataset. Similar methods would allow the derivation of profiles of CDOM absorption at 412 nm from profiles of CDOM fluorescence, calibrated chlorophyll-a fluorescence (slope correction applied), and irradiance (ED412) [8]. Therefore, the potential of SOCA-light already appears enormous when simply considering its possible applications in relation to the BGC-Argo database alone.

Recently, floats have begun to acquire hyperspectral radiometric measurements [62,63]. New perspectives that consequently open up include refinements in the characterization of optically active substances, such as CDOM or phytoplankton community structure at large scale [63]. The SOCA-light method presented here has the potential to accommodate any increase in the spectral domain and resolution once sufficient data have been acquired to support training. The availability of such modeled data could represent a new step towards a better understanding of various components of biogeochemical cycles at a global scale.

Supplementary Materials: The following supporting information can be downloaded at: https://www.mdpi.com/article/10.3390/rs15245663/s1, Figure S1: Geographical distribution of BGC-Argo profiles used for the development and validation of the SOCA-light model for ED380, Figure S2: Geographical distribution of BGC-Argo profiles used for the development and validation of the SOCA-light model for ED412, Figure S3: Geographical distribution of BGC-Argo profiles used for the development and validation of the SOCA-light model for ED490, Figure S4: Scatter-plots between light variables (PAR, ED380, ED412, and ED490) modeled by SOCA light models versus their corresponding BGC-Argo measurements from NASTG. PAR (A); ED380 (B); ED412 (C); ED490 (D), Figure S5: Time series of the vertical distribution of the four light variables in the Easter Mediterranean Sea (EMS) measured by BGC-Argo float WMO 6901773 (left column) and modeled by SOCA-light (right column). The variables in each subplot are indicated by text in the corresponding subplots. The black stars indicate the depth at which instantaneous PAR value $=15$ µmol quanta m^{-2} s^{-1}, Figure S6: Scatter-plots between light variables (PAR, ED380, ED412, and ED490) modeled by SOCA light models versus their corresponding BGC-Argo measurements from EMS. PAR (A); ED380 (B); ED412 (C); ED490 (D), Figure S7: Scatter-plots between light variables (PAR, ED380, ED412, and ED490) modeled by SOCA light models versus their corresponding BGC-Argo measurements from SO. PAR (A); ED380 (B); ED412 (C); ED490 (D), Figure S8: Time series of the vertical distribution of the four light variables in the North Atlantic Subpolar Gyre (NASPG) measured by BGC-Argo float WMO 6901523 (left column) and modeled by SOCA-light (right column). The variables in each subplot are specified by text in the corresponding subplots. The black stars indicate the depth at which instantaneous PAR value =15 µmol quanta m^{-2} s^{-1}, Figure S9: Scatter-plots between light variables (PAR, ED380, ED412, and ED490) modeled by SOCA light models versus their corresponding BGC-Argo measurements from NASPG. PAR (A); ED380 (B); ED412 (C); ED490 (D). All of the models and functions (Jupyter Notebook) are open source and can be accessed via our GitHub page: https://github.com/renoshpr/SOCA-LIGHT-MODELS, (accessed on 10 November 2023).

Author Contributions: All the authors (P.R.R., J.Z., R.S. and H.C.) contributed to conceptualizing the methodology; P.R.R. implemented the methodology, the validation and formal analysis of results; P.R.R. wrote the initial draft; P.R.R., R.S. and H.C. revised and finalized the manuscript. All authors have read and agreed to the published version of the manuscript.

Funding: This study is a contribution to the following projects: REFINE (European Research Council, Grant agreement 834177), BGC-Argo-France (CNES-TOSCA), and the Copernicus Marine Environment Monitoring System (CMEMS) Ocean Multi-Observation TAC.

Data Availability Statement: The BGC-Argo data used in this study are publicly available at ftp://ftp.ifremer.fr/ifremer/argo, accessed on 17 February 2023, and supplied by the international Argo program. The ocean color (except surface PAR) and ARMOR3D data are publicly available at https://data.marine.copernicus.eu/products, (accessed on 17 February 20233) delivered by the Copernicus Marine Service. The ocean surface PAR data can be downloaded from the GlobColour portal from the following link https://hermes.acri.fr/?class=archive, (accessed on 17 February 2023) provided by ACRI-ST. The SeaBASS data are publicly available from the following link https://seabass.gsfc.nasa.gov/, (accessed on 17 February 2023) provided by the National Aeronautics and Space Administration (NASA).

Acknowledgments: These light profiles and physical variables (P, T, and S) were collected in the framework of the BGC-Argo program and made freely available by the International Argo program. The authors are also thankful to CMEMS for ocean color and physical data. The authors are grateful to NASA for the SeaBASS data.

Conflicts of Interest: The authors declare no conflicts of interest relevant to this study.

Abbreviations

The following abbreviations are used in this manuscript:

ADAM	Adaptive moment estimation
ANN	Artificial neural network
AOP	Apparent optical property
ARMOR3D	A 3D multi-observations T, S, U, V product of the ocean
b_{bp}	Particulate backscattering coefficient
BGC-Argo	BioGeoChemical Argo
CDOM	Colored dissolved organic matter
Chla	Chlorophyll-a concentration
CMEMS	Copernicus Marine Environment Monitoring System
DCM	Deep chlorophyll maxima
DOC	Dissolved organic carbon
DOY	Day of the year
ED	Downwelling irradiance
EMS	Eastern Mediterranean Sea
GOOS	Global Ocean Observing System
IOP	Inherent optical property
k_d	Diffuse attenuation coefficient
LT	Local time
LU	Upwelling radiance
MAPE	Median absolute percent error
MLD	Mixed layer depth
MLP	Multilayer perceptron
NASPG	North Atlantic Subpolar Gyre

NASTG	North Atlantic Subtropical Gyre
NN	Neural network
PAR	Photosynthetically available radiation
PDF	Probability density function
POC	Particulate organic carbon
RMSE	Root mean squared error
R_{rs}	Remote sensing reflectance
SeaBASS	SeaWiFS Bio-Optical Archive and Storage System
SLA	Sea-level anomaly
SO	Southern Ocean
SOCA	Satellite Ocean Color merged with Argo data
tanh	Hyperbolic tangent
WMO	World Meteorological Organization
Z_iPAR_15	The depth at which instantaneous PAR value $=15$ μmol quanta m^{-2} s^{-1}

References

1. Antoine, D.; Morel, A. Oceanic primary production: 1. Adaptation of a spectral light-photosynthesis model in view of application to satellite chlorophyll observations. *Glob. Biogeochem. Cycles* **1996**, *10*, 43–55. [CrossRef]
2. Westberry, T.; Behrenfeld, M.J.; Siegel, D.A.; Boss, E. Carbon-based primary productivity modeling with vertically resolved photoacclimation. *Glob. Biogeochem. Cycles* **2008**, *22*, GB003078. [CrossRef]
3. Ohlmann, J.C.; Siegel, D.A.; Mobley, C.D. Ocean Radiant Heating. Part I: Optical Influences. *J. Phys. Oceanogr.* **2000**, *30*, 1833–1848. [CrossRef]
4. Ohlmann, J.C.; Siegel, D.A. Ocean Radiant Heating. Part II: Parameterizing Solar Radiation Transmission through the Upper Ocean. *J. Phys. Oceanogr.* **2000**, *30*, 1849–1865. [CrossRef]
5. Tedetti, M.; Sempéré, R. Penetration of Ultraviolet Radiation in the Marine Environment. A Review. *Photochem. Photobiol.* **2006**, *82*, 389–397. [CrossRef] [PubMed]
6. Xing, X.; Morel, A.; Claustre, H.; Antoine, D.; D'Ortenzio, F.; Poteau, A.; Mignot, A. Combined processing and mutual interpretation of radiometry and fluorimetry from autonomous profiling Bio-Argo floats: Chlorophyll a retrieval. *J. Geophys. Res. Ocean.* **2011**, *116*, 6899. [CrossRef]
7. Roesler, C.; Uitz, J.; Claustre, H.; Boss, E.; Xing, X.; Organelli, E.; Briggs, N.; Bricaud, A.; Schmechtig, C.; Poteau, A.; et al. Recommendations for obtaining unbiased chlorophyll estimates from in situ chlorophyll fluorometers: A global analysis of WET Labs ECO sensors. *Limnol. Oceanogr. Methods* **2017**, *15*, 572–585. [CrossRef]
8. Xing, X.; Morel, A.; Claustre, H.; D'Ortenzio, F.; Poteau, A. Combined processing and mutual interpretation of radiometry and fluorometry from autonomous profiling Bio-Argo floats: 2. Colored dissolved organic matter absorption retrieval. *J. Geophys. Res. Ocean.* **2012**, *117*, 7632. [CrossRef]
9. Vodacek, A.; Blough, N.V.; DeGrandpre, M.D.; DeGrandpre, M.D.; Nelson, R.K. Seasonal variation of CDOM and DOC in the Middle Atlantic Bight: Terrestrial inputs and photooxidation. *Limnol. Oceanogr.* **1997**, *42*, 674–686. [CrossRef]
10. Morel, A. Optical modeling of the upper ocean in relation to its biogenous matter content (case I waters). *J. Geophys. Res. Oceans* **1988**, *93*, 10749–10768. [CrossRef]
11. Morel, A.; Maritorena, S. Bio-optical properties of oceanic waters: A reappraisal. *J. Geophys. Res. Oceans* **2001**, *106*, 7163–7180. [CrossRef]
12. Morel, A. Are the empirical relationships describing the bio-optical properties of case 1 waters consistent and internally compatible? *J. Geophys. Res. Oceans* **2009**, *114*, 4803. [CrossRef]
13. Morel, A.; Gentili, B. A simple band ratio technique to quantify the colored dissolved and detrital organic material from ocean color remotely sensed data. *Remote Sens. Environ.* **2009**, *113*, 998–1011. [CrossRef]
14. Scott, J.P.; Werdell, P.J. Comparing level-2 and level-3 satellite ocean color retrieval validation methodologies. *Opt. Express* **2019**, *27*, 30140–30157. [CrossRef] [PubMed]
15. Claustre, H.; Johnson, K.S.; Takeshita, Y. Observing the Global Ocean with Biogeochemical-Argo. *Annu. Rev. Mar. Sci.* **2020**, *12*, 23–48. [CrossRef] [PubMed]
16. Biogeochemical-ArgoPlanningGroup. *The Scientific Rationale, Design and Implementation Plan for a Biogeochemical-Argo Float Array*; Report; Ifremer: Plouzané, France, 2016. [CrossRef]
17. Gordon, H.R.; Brown, O.B.; Jacobs, M.M. Computed Relationships Between the Inherent and Apparent Optical Properties of a Flat Homogeneous Ocean. *Appl. Opt.* **1975**, *14*, 417–427. [CrossRef]
18. Morel, A.; Gentili, B. Diffuse reflectance of oceanic waters: Its dependence on Sun angle as influenced by the molecular scattering contribution. *Appl. Opt.* **1991**, *30*, 4427–4438. [CrossRef]
19. Lee, Z.; Du, K.; Arnone, R.; Liew, S.; Penta, B. Penetration of solar radiation in the upper ocean: A numerical model for oceanic and coastal waters. *J. Geophys. Res. Oceans* **2005**, *110*. [CrossRef]
20. Liu, C.C.; Miller, R.L.; Carder, K.L.; Lee, Z.; D'Sa, E.J.; Ivey, J.E. Estimating the underwater light field from remote sensing of ocean color. *J. Oceanogr.* **2006**, *62*, 235–248. [CrossRef]

21. Mobley, C.D.; Sundman, L.K. *HYDROLIGHT 5 ECOLIGHT 5*; Sequoia Scientific Inc.: Bellevue, WA, USA, 2008, p. 16.

22. Xing, X.; Boss, E. Chlorophyll-Based Model to Estimate Underwater Photosynthetically Available Radiation for Modeling, In-Situ, and Remote-Sensing Applications. *Geophys. Res. Lett.* **2021**, *48*, e2020GL092189. [CrossRef]

23. Gregg, W.W.; Carder, K.L. A simple spectral solar irradiance model for cloudless maritime atmospheres. *Limnol. Oceanogr.* **1990**, *35*, 1657–1675. [CrossRef]

24. Organelli, E.; Barbieux, M.; Claustre, H.; Schmechtig, C.; Poteau, A.; Bricaud, A.; Boss, E.; Briggs, N.; Dall'Olmo, G.; D'Ortenzio, F.; et al. Two databases derived from BGC-Argo float measurements for marinebiogeochemical and bio-optical applications. *Earth Syst. Sci. Data* **2017**, *9*, 861–880. [CrossRef]

25. Organelli, E.; Claustre, H.; Bricaud, A.; Barbieux, M.; Uitz, J.; D'Ortenzio, F.; Dall'Olmo, G. Bio-optical anomalies in the world's oceans: An investigation on the diffuse attenuation coefficients for downward irradiance derived from Biogeochemical Argo float measurements. *J. Geophys. Res. Oceans* **2017**, *122*, 3543–3564. [CrossRef]

26. Sauzède, R.; Claustre, H.; Uitz, J.; Jamet, C.; Dall'Olmo, G.; D'Ortenzio, F.; Gentili, B.; Poteau, A.; Schmechtig, C. A neural network-based method for merging ocean color and Argo data to extend surface bio-optical properties to depth: Retrieval of the particulate backscattering coefficient. *J. Geophys. Res. Oceans* **2016**, *121*, 2552–2571. . [CrossRef]

27. Copernicus Marine Service. *Global Ocean 3D Chlorophyll-A Concentration, Particulate Backscattering Coefficient and Particulate Organic Carbon*; Copernicus Marine Service Information (CMEMS); Marine Data Store (MDS); Copernicus Marine Service: Ramonville-Saint-Agne, France, 2023. [CrossRef]

28. Bittig, H.; Wong, A.; Plant, J.; Carval, T.; Rannou, J.P. *BGC-Argo Synthetic Profile File Processing and Format on Coriolis GDAC, v1.3*; Report; Ifremer: Plouzané, France, 2022. [CrossRef]

29. Jutard, Q.; Organelli, E.; Briggs, N.; Xing, X.; Schmechtig, C.; Boss, E.; Poteau, A.; Leymarie, E.; Cornec, M.; D'Ortenzio, F.; et al. Correction of Biogeochemical-Argo Radiometry for Sensor Temperature-Dependence and Drift: Protocols for a Delayed-Mode Quality Control. *Sensors* **2021**, *21*, 6217. [CrossRef] [PubMed]

30. Copernicus Marine Service. *Global Ocean Colour (Copernicus-GlobColour), Bio-Geo-Chemical, L3 (Daily) from Satellite Observations (1997-Ongoing)*; Copernicus Marine Service Information (CMEMS); Marine Data Store (MDS); Copernicus Marine Service: Ramonville-Saint-Agne, France, 2023. [CrossRef]

31. Garnesson, P.; Mangin, A.; Fanton d'Andon, O.; Demaria, J.; Bretagnon, M. The CMEMS GlobColour chlorophyll *a* product based on satellite observation: Multi-sensor merging and flagging strategies. *Ocean Sci.* **2019**, *15*, 819–830. [CrossRef]

32. Gohin, F.; Druon, J.N.; Lampert, L. A five channel chlorophyll concentration algorithm applied to SeaWiFS data processed by SeaDAS in coastal waters. *Int. J. Remote Sens.* **2002**, *23*, 1639–1661. [CrossRef]

33. Morel, A.; Huot, Y.; Gentili, B.; Werdell, P.J.; Hooker, S.B.; Franz, B.A. Examining the consistency of products derived from various ocean color sensors in open ocean (Case 1) waters in the perspective of a multi-sensor approach. *Remote Sens. Environ.* **2007**, *111*, 69–88. [CrossRef]

34. Werdell, P.J.; Fargion, G.S.; McClain, C.R.; Bailey, S.W. *The SeaWiFS Bio-Optical Archive and Storage System (SeaBASS): Current Architecture and Implementation*; Report NASA/TM-2002-211617; NASA: Washington, DC, USA, 2002.

35. Werdell, P.J.; Bailey, S.; Fargion, G.; Pietras, C.; Knobelspiesse, K.; Feldman, G.; McClain, C. Unique data repository facilitates ocean color satellite validation. *Eos Trans. Am. Geophys. Union* **2003**, *84*, 377–387. [CrossRef]

36. Guinehut, S.; Dhomps, A.L.; Larnicol, G.; Le Traon, P.Y. High resolution 3-D temperature and salinity fields derived from in situ and satellite observations. *Ocean Sci.* **2012**, *8*, 845–857. [CrossRef]

37. Mulet, S.; Rio, M.H.; Mignot, A.; Guinehut, S.; Morrow, R. A new estimate of the global 3D geostrophic ocean circulation based on satellite data and in-situ measurements. *Deep Sea Res. Part II Top. Stud. Oceanogr.* **2012**, *77–80*, 70–81. [CrossRef]

38. Copernicus Marine Service. *Multi Observation Global Ocean 3D Temperature Salinity Height Geostrophic Current and MLD*; Copernicus Marine Service Information (CMEMS); Marine Data Store (MDS); Copernicus Marine Service: Ramonville-Saint-Agne, France, 2023.

39. Rumelhart, D.E.; Hinton, G.E.; Williams, R.J. Learning representations by back-propagating errors. *Nature* **1986**, *323*, 533–536. [CrossRef]

40. Bishop, C.M. *Neural Networks for Pattern Recognition*; Oxford University Press: New York, NY, USA, 1995.

41. Taud, H.; Mas, J., Multilayer perceptron (MLP). In *Geomatic Approaches for Modeling Land Change Scenarios*; Springer International Publishing: Cham, Switzerland, 2018; pp. 451–455. [CrossRef]

42. de Boyer Montégut, C.; Madec, G.; Fischer, A.S.; Lazar, A.; Iudicone, D. Mixed layer depth over the global ocean: An examination of profile data and a profile-based climatology. *J. Geophys. Res. Oceans* **2004**, *109*, 2378. [CrossRef]

43. Linares-Rodriguez, A.; Ruiz-Arias, J.A.; Pozo-Vazquez, D.; Tovar-Pescador, J. An artificial neural network ensemble model for estimating global solar radiation from Meteosat satellite images. *Energy* **2013**, *61*, 636–645. [CrossRef]

44. Kingma, D.P.; Ba, J. Adam: A method for stochastic optimization. In Proceedings of the 3rd International Conference on Learning Representations (ICLR 2015), San Diego, CA, USA, 7–9 May 2015. https://doi.org/10.48550/arXiv.1412.6980.

45. Cornec, M.; Claustre, H.; Mignot, A.; Guidi, L.; Lacour, L.; Poteau, A.; D'Ortenzio, F.; Gentili, B.; Schmechtig, C. Deep Chlorophyll Maxima in the Global Ocean: Occurrences, Drivers and Characteristics. *Glob. Biogeochem. Cycles* **2021**, *35*, e2020GB006759. [CrossRef]

46. Bock, N.; Cornec, M.; Claustre, H.; Duhamel, S. Biogeographical Classification of the Global Ocean From BGC-Argo Floats. *Glob. Biogeochem. Cycles* **2022**, *36*, e2021GB007233. [CrossRef]

47. Xing, X.; Briggs, N.; Boss, E.; Claustre, H. Improved correction for non-photochemical quenching of in situ chlorophyll fluorescence based on a synchronous irradiance profile. *Opt. Express* **2018**, *26*, 24734–24751. [CrossRef] [PubMed]

48. Terrats, L.; Claustre, H.; Cornec, M.; Mangin, A.; Neukermans, G. Detection of Coccolithophore Blooms with BioGeoChemical-Argo Floats. *Geophys. Res. Lett.* **2020**, *47*, e2020GL090559. [CrossRef]

49. Bricaud, A.; Morel, A.; Babin, M.; Allali, K.; Claustre, H. Variations of light absorption by suspended particles with chlorophyll a concentration in oceanic (case 1) waters: Analysis and implications for bio-optical models. *J. Geophys. Res. Oceans* **1998**, *103*, 31033–31044. [CrossRef]

50. Werdell, P.J.; Franz, B.A.; Lefler, J.T.; Robinson, W.D.; Boss, E. Retrieving marine inherent optical properties from satellites using temperature and salinity-dependent backscattering by seawater. *Opt. Express* **2013**, *21*, 32611–32622. [CrossRef] [PubMed]

51. O'Reilly, J.E.; Werdell, P.J. Chlorophyll algorithms for ocean color sensors -OC4, OC5 and OC6. *Remote. Sens. Environ.* **2019**, *229*, 32–47. [CrossRef] [PubMed]

52. Hu, C.; Lee, Z.; Franz, B. Chlorophyll aalgorithms for oligotrophic oceans: A novel approach based on three-band reflectance difference. *J. Geophys. Res. Oceans* **2012**, *117*. [CrossRef]

53. Gilerson, A.A.; Gitelson, A.A.; Zhou, J.; Gurlin, D.; Moses, W.; Ioannou, I.; Ahmed, S.A. Algorithms for remote estimation of chlorophyll-a in coastal and inland waters using red and near infrared bands. *Opt. Express* **2010**, *18*, 24109–24125. [CrossRef]

54. Stramski, D.; Reynolds, R.A.; Babin, M.; Kaczmarek, S.; Lewis, M.R.; Röttgers, R.; Sciandra, A.; Stramska, M.; Twardowski, M.S.; Franz, B.A.; et al. Relationships between the surface concentration of particulate organic carbon and optical properties in the eastern South Pacific and eastern Atlantic Oceans. *Biogeosciences* **2008**, *5*, 171–201. [CrossRef]

55. Organelli, E.; Claustre, H.; Bricaud, A.; Schmechtig, C.; Poteau, A.; Xing, X.; Prieur, L.; D'Ortenzio, F.; Dall'Olmo, G.; Vellucci, V. A Novel Near-Real-Time Quality-Control Procedure for Radiometric Profiles Measured by Bio-Argo Floats: Protocols and Performances. *J. Atmos. Ocean. Technol.* **2016**, *33*, 937–951. [CrossRef]

56. O'Brien, T.; Boss, E. Correction of Radiometry Data for Temperature Effect on Dark Current, with Application to Radiometers on Profiling Floats. *Sensors* **2022**, *22*, 6771. [CrossRef] [PubMed]

57. Xing, X.; Boss, E.; Zhang, J.; Chai, F. Evaluation of Ocean Color Remote Sensing Algorithms for Diffuse Attenuation Coefficients and Optical Depths with Data Collected on BGC-Argo Floats. *Remote Sens.* **2020**, *12*, 2367. [CrossRef]

58. Begouen Demeaux, C.; Boss, E. Validation of Remote-Sensing Algorithms for Diffuse Attenuation of Downward Irradiance Using BGC-Argo Floats. *Remote Sens.* **2022**, *14*, 4500. [CrossRef]

59. Stahl, F.T.; Nolle, L.; Jemai, A.; Zielinski, O. A Model for Predicting the Amount of Photosynthetically Available Radiation from BGC-Argo Float Observations in the Water Column. In Proceedings of the European Council for Modelling and Simulation, Alesund, Norway, 30 May–3 June 2022; Hameed, I.A., Hasan, A., Alaliyat, S.A.A., Eds.; ECMS Digital Library: Dudweiler, Germany, 2022; Volume 36, pp. 174–180. [CrossRef]

60. Kumm, M.M.; Nolle, L.; Stahl, F.; Jemai, A.; Zielinski, O. On an Artificial Neural Network Approach for Predicting Photosynthetically Active Radiation in the Water Column. In Proceedings of the Artificial Intelligence XXXIX, Cambridge, UK, 13–15 December 2022; Bramer, M., Stahl, F., Eds.; Springer International Publishing: Cham, Switzerland, 2022; pp. 112–123.

61. Xing, X.; Claustre, H.; Blain, S.; D'Ortenzio, F.; Antoine, D.; Ras, J.; Guinet, C. Quenching correction for in vivo chlorophyll fluorescence acquired by autonomous platforms: A case study with instrumented elephant seals in the Kerguelen region (Southern Ocean). *Limnol. Oceanogr. Methods* **2012**, *10*, 483–495. [CrossRef]

62. Jemai, A.; Wollschläger, J.; Voß, D.; Zielinski, O. Radiometry on Argo Floats: From the Multispectral State-of-the-Art on the Step to Hyperspectral Technology. *Front. Mar. Sci.* **2021**, *8*, 676537. [CrossRef]

63. Organelli, E.; Leymarie, E.; Zielinski, O.; Uitz, J.; D'ortenzio, F.; Claustre, H. Hyperspectral radiometry on biogeochemical-argo floats: A bright perspective for phytoplankton diversity. *Oceanography* **2021**, *34*, 90–91. [CrossRef]

Communication

Spatiotemporal Prediction of Monthly Sea Subsurface Temperature Fields Using a 3D U-Net-Based Model

Nengli Sun [1], Zeming Zhou [1], Qian Li [1,*] and Xuan Zhou [2]

1   The College of Meteorology and Oceanography, National University of Defense Technology, Changsha 410005, China
2   Independent Researcher, Mailbox No. 5111, Beijing 100094, China
*   Correspondence: liqian@nudt.edu.cn

Abstract: The ability to monitor and predict sea temperature is crucial for determining the likelihood that ocean-related events will occur. However, most studies have focused on predicting sea surface temperature, and less attention has been paid to predicting sea subsurface temperature (SSbT), which can reflect the thermal state of the entire ocean. In this study, we use a 3D U-Net model to predict the SSbT in the upper 400 m of the Pacific Ocean and its adjacent oceans for lead times of 12 months. Two reconstructed SSbT products are added to the training set to solve the problem of insufficient observation data. Experimental results indicate that this method can predict the ocean temperature more accurately than previous methods in most depth layers. The root mean square error and mean absolute error of the predicted SSbT fields for all lead times are within 0.5–0.7 °C and 0.3–0.45 °C, respectively, while the average correlation coefficient scores of the predicted SSbT profiles are above 0.96 for almost all lead times. In addition, a case study qualitatively demonstrates that the 3D U-Net model can predict realistic SSbT variations in the study area and, thus, facilitate understanding of future changes in the thermal state of the subsurface ocean.

Keywords: sea temperature prediction; reconstructed sea subsurface temperature data; 3D U-Net

Citation: Sun, N.; Zhou, Z.; Li, Q.; Zhou, X. Spatiotemporal Prediction of Monthly Sea Subsurface Temperature Fields Using a 3D U-Net-Based Model. *Remote Sens.* **2022**, *14*, 4890. https://doi.org/10.3390/rs14194890

Academic Editors: Ana B. Ruescas, Veronica Nieves and Raphaëlle Sauzède

Received: 25 August 2022
Accepted: 27 September 2022
Published: 30 September 2022

Publisher's Note: MDPI stays neutral with regard to jurisdictional claims in published maps and institutional affiliations.

1. Introduction

In oceanographic investigations, the sea temperature is a crucial measure that can indicate the thermal state of seawater [1]. Its variation strongly correlates with the global climate and meteorological events [2–5] and can affect the marine ecological environment, underwater acoustic communication, and commercial fisheries [6–8]. Therefore, sea temperature prediction is crucial for assisting in the early assessment of the likelihood of associated events. Two categories can be used to group the methods for predicting sea temperature. The first group is based on dynamical models to simulate atmosphere–ocean variables with physical constraints and then make forecasts [9–11]. The second group attempts to capture the relationships between past observations and future sea temperature through data analysis. This group includes statistical model-based approaches [12,13] and machine learning-based approaches [14–17]. In the last two decades, machine learning-based approaches have been increasingly adopted to predict sea temperature due to lower computational costs and higher flexibility in comparison with numerical model-based approaches [18,19]. For example, to estimate the time series of sea surface temperature (SST) in isolated locations, Zhang et al. [17] used long short-term memory (LSTM). The gated recurrent unit model, which has less trainable parameters compared with LSTM, was used by Zhang et al. [20] for SST time series prediction. Yang et al. [21] used a fully connected LSTM (FC-LSTM) layer and a convolution layer to predict the SST of an area of nearby points to incorporate the temporal and spatial information. However, the abovementioned models cannot capture the spatial linkage of the sea temperature values in a large region, thus limiting their prediction performance. Researchers have paid close attention to the performance of the convolutional LSTM (ConvLSTM) model in precipitation nowcasting [22].

This model replaces the matrix operations of the FC-LSTM with convolutions, making it a powerful tool that can extract expressive spatial feature maps of input images and model their time evolution. The ConvLSTM model has been used in some studies to successfully estimate sea temperature [23–25].

Sea subsurface temperature (SSbT), which is a 3D sea temperature field, is essential for understanding the mechanisms and processes in the ocean as a whole [26] and is thus preferred in oceanographic studies [27]. A review of previous studies revealed that the prediction of the SSbT has not been investigated as extensively as that of the SST. Liu et al. [28] employed LSTM to predict the mean seawater temperature at various depths at each observation point for the following month. To predict the SSbT for 3-day and 5-day lead times at an observation point, Patil and Iiyama [29] first used a ConvLSTM network to extract the spatiotemporal information of the past SST around this point and then used a multilayer perceptron to analyze the past observed SSbT profiles at this point. The results of the prediction were then produced by combining the outputs of these two networks. Zhang et al. [25] used a multilayer ConvLSTM (M-ConvLSTM) model to predict the mean SSbT field of the upcoming month in a subarea of the Pacific Ocean in terms of the spatiotemporal prediction of the SSbT in a region rather than at specific points. To solve the extraction problem of 3D spatial correlation of the SSbT field, a 4D convolutional neural network (CNN) was designed by Zuo et al. [30] for SSbT horizontal and profile prediction of the next day, respectively, in which a 4D convolution operation was implemented by linearly adding the results of several 3D convolution operations. Although deep learning (DL) has significantly improved SSbT prediction, there are still few studies that perform SSbT field prediction for lead times greater than 1 month, which limits their usefulness as a reference for longer-term ocean-related studies. Moreover, an insufficient amount of monthly observation data limits the generalization ability of the network during training.

This study investigates the DL-based long-term SSbT field prediction. Utilizing the SSbT fields from the previous 12 months as the input, a 3D U-Net-based model was constructed to perform the prediction of the monthly SSbT fields, mostly in the Pacific Ocean, for lead times up to 12 months. This model is capable of extracting the spatiotemporal features of historical SSbT fields and mapping them into future SSbT fields. In addition, two SSbT products that have been recreated based on objective analysis have been added to the training dataset to address the issue of insufficient monthly mean observation data. The rest of this manuscript is organized as follows. Section 2 briefly introduces some related works about SSbT prediction. Section 3 describes the data used in this study. The suggested approach is illustrated in Section 4. Section 5 presents the experimental findings. Finally, Section 6 provides the conclusions of this study.

2. Related Works

Liu et al. [28] considers the spatiotemporal SSbT field prediction as a combination of the independent time series prediction of each observation point in a 3D grid region. They first use a matrix fusion approach to capture the features of the closeness and period in the temperature time series and then leverage LSTM to conduct the temperature prediction for the next month. This method is evaluated at different depth levels of three regions, including the Coral Sea, the equatorial Pacific Ocean, and the South China Sea. Experimental results show that the optimal parameters of the fusion matrix are different for different depth levels. Comparable or even better overall performance is achieved by using this method compared to support vector regression and a multilayer perceptron. However, the spatiotemporal relationship between observation points is ignored in the modeling process, which limits the prediction performance.

Instead of using a DL model to predict the temperature of each depth layer separately, Patil and Iiyama [29] investigate the DL-based SSbT profile prediction for 3-day and 5-day lead times at a specific location in the eastern Indian Ocean. In their developed model, a ConvLSTM network is first adopted to extract the spatiotemporal features of the past SST around this location. Then, a multilayer perceptron is used to extract the patterns of the past observed SSbT profiles at this location. Finally, these two types of features are fused by another multilayer perceptron to generate the prediction result. Experimental results show that compared with using SST only, the prediction accuracy can be significantly improved by incorporating the past SSbT profiles into the model. In addition, it is found that the proposed model produces higher prediction errors for the intermediate depth levels from 100 m to 300 m compared to other depth levels.

To model the spatiotemporal relationship of the sea temperature in the whole 3D grid region, Zhang et al. [25] and Zuo et al. [30] propose the M-ConvLSTM model and 4D-CNN model, respectively. Zhang et al. [25] focus on predicting the mean SSbT field of the next month based on monthly mean SSbT fields of the previous 28 months. The input SSbT field at each time step is considered as a multi-channel image. The M-ConvLSTM model, which consists of multiple ConvLSTM layers, is used to transmit and update the inner states along the time direction of the input sequence of SSbT fields. The prediction result is generated at the last time step. The M-ConvLSTM model achieves improvements over the FC-LSTM model [21] for SST prediction. The coefficient of determination for most depth layers exceeds 0.95. Zuo et al. [30] develop a 4D-CNN-based model to perform SSbT horizontal and profile prediction of the next day, respectively. The 4-D convolution module is the core part of this model. When conducting SSbT horizontal prediction, they first divide the SSbT fields into several temporal sequences of horizontal fields according to depth levels. Then, different 3D convolutional layers are applied to the temporal sequences of different depths, and their outputs are added up to enrich feature representation. As for the SSbT profile prediction, the SSbT fields are divided into several temporal sequences of profiles along the latitude, and then the 4-D convolution module is used to extract their features. Experiments show that this model achieves competitive performance when predicting the sea temperature for different depth levels, locations and seasonal thermocline. This method requires a lot of parameters and computation when there are many depth levels for SSbT horizontal prediction or many latitudes for SSbT profile prediction.

3. Data

The study area, which includes the Pacific Ocean, the eastern Indian Ocean, and the western Atlantic Ocean, is located between latitudes 59.5°S and 59.5°N and longitudes 95.5°E and 25.5°W, as shown in Figure 1.

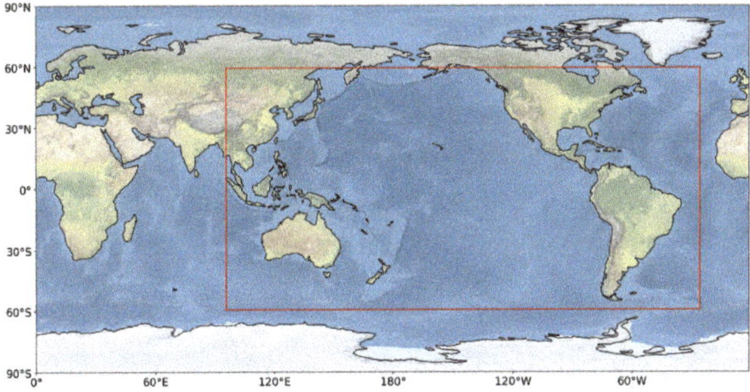

Figure 1. Selected study area.

The SSbT data used in this investigation are monthly mean sea temperature in the upper 400 m over the study area. The depth levels are 0, 10, 20, 30, 50, 75, 100, 150, 200, 250, 300, and 400 m. The amount of data available for neural network training is insufficient since accurate SSbT observations with full spatial and temporal coverage are only available after the early 1980s. Therefore, the reconstructed historical SSbT products provided by the Institute of Atmospheric Physics (IAP) [27] and the Research Data Archive (RDA) at the National Center for Atmospheric Research [31] are added to the training set to increase the sample size and avoid the overfitting problem in the training process. As for SSbT observations, the SST data are from the National Oceanic and Atmospheric Administration (NOAA) Optimum Interpolation SST V2 (OISST V2) dataset [32], and data of seawater temperature at depths of 10 to 400 m are from the Global Ocean Data Assimilation System (GODAS) at the National Centers for Environmental Prediction (NCEP) [33]. To make the GODAS data consistent with the reconstructed products, bilinear interpolation (a widely accepted and often used interpolation algorithm that can resample the grid products to a new resolution [34,35]) was utilized to adjust the horizontal spatial (latitude–longitude) resolution from $0.333° \times 1°$ to $1° \times 1°$. In addition, if one product does not have the data for one selected depth level, the sea temperature fields at close levels would be linearly interpolated to this level. After the data processing was complete, spatiotemporal sequences were produced for each type of monthly SSbT product using a 24-month sliding window with a sliding step of 1 month.

Regarding the division of these generated sequences, all reconstructed SSbT sequences and most of the observation sequences were used for training, and the remaining observation sequences were used for validation and testing. In addition, we prevented time overlap between the target SSbT fields in the training, validation, and test set sequences. For details, see Table 1.

Table 1. Organization of the built dataset.

Subset	Product	Period	Number of Sequences
Training	IAP RDA OISST V2 and GODAS	January 1956–December 2007 January 1945–December 2007 January 1982–December 2007	1623
Validation	OISST V2 and GODAS	January 2008–December 2012	60
Test	OISST V2 and GODAS	January 2013–May 2022	102

4. Method

When modeling, the shape of the input data or feature map tensors is (T, H, W, C). Here, T, H, W, and C refer to the time, height, width, and channel dimensions, respectively. Specifically, for the input sequence of SSbT fields for the previous 12 months, T is 12 (number of timesteps), H is 120 (latitude grid size), W is 240 (longitude grid size), and C is 12 (number of depth levels).

The created model, which is shown in Figure 2, is an end-to-end trainable model-based on the 3D U-Net model [36], which was designed initially for volumetric segmentation. It is composed of multiple 3D convolutional layers, each of which is followed by a rectified linear unit activation function, except for the final convolutional layer. This model, like U-Net, has a downsampling path, a symmetrical upsampling path, and skip connections.

In the downsampling path, the temporal and spatial sizes of the input sequence are progressively halved by using two 3D convolutional layers with strides of 2, each of which is followed by two 3D convolutional layers, and spatiotemporal features with different representation levels are extracted. Two transposed 3D convolutional layers, each of which is followed by two more 3D convolutional layers, are used in the upsampling process to gradually restore the high-level features to their original size. Furthermore, low-level features are received from the downsampling path through skip connections, delivering detailed information to generate more comprehensive representations. To speed up model training, batch normalization (BN) [37] is performed after the second-to-last convolutional layer. The predicted monthly SSbT fields are output through the final convolutional layer with a kernel size of $1 \times 1 \times 1$.

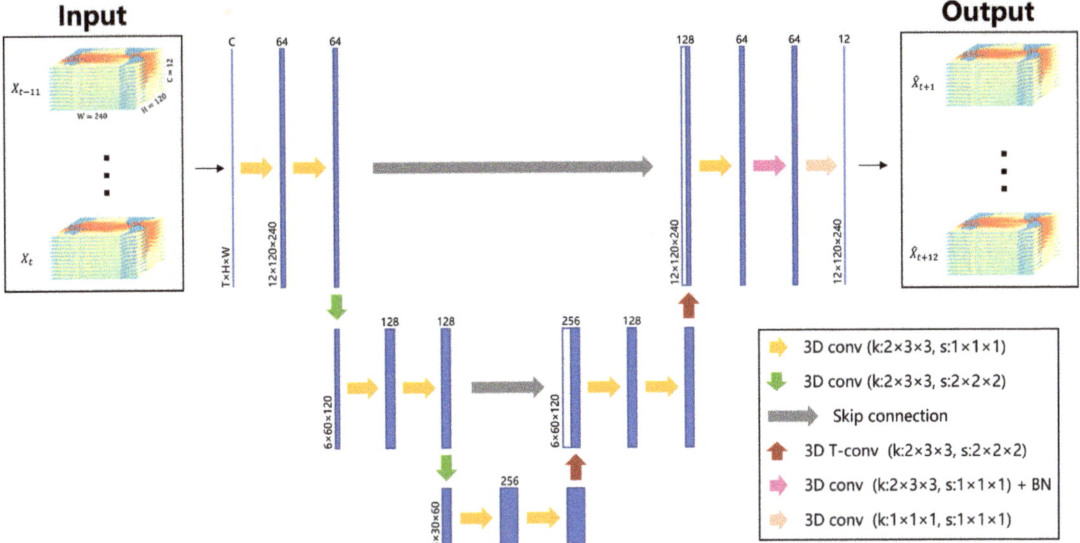

Figure 2. The architecture of the proposed model. k and s indicate the size and stride of convolution kernels, respectively.

The root mean square error (RMSE) is utilized as the loss function to guide the training of the developed model in this study, which is defined as

$$L_{RMSE} = \sum_{m=1}^{12} \sqrt{\frac{1}{N} \sum_{n=1}^{N} \left(\hat{X}_{t+m}^{(n)} - X_{t+m}^{(n)} \right)^2} \tag{1}$$

where $\hat{X}_{t+m}^{(n)}$ is the value of the n grid point of the predicted SSbT field for the time step $t + m$, and $X_{t+m}^{(n)}$ is the corresponding ground truth value.

5. Experiments

5.1. Experimental Settings

A seasonal naïve model that copies the input SSbT fields from the past 12 months unaltered to its outputs (assuming the interannual variation of SSbT is zero), M-ConvLSTM model [25], and a simple 3D-CNN (S3D-CNN) [38] are three baselines against which the 3D U-Net model is compared. The temperature values are normalized by dividing by 30 before being input into the models. All the DL-based models are trained using the RMSE loss function defined in Section 4 for a fair comparison. The learning rate is set at 0.0001, and the batch size is 4. When the validation loss does not decrease for 12 epochs, the training process is terminated to prevent overfitting. The training plots of the 3D U-Net model are shown in Figure 3. These DL-based models are implemented using TensorFlow [39] and run on a TITAN RTX GPU (24 GB).

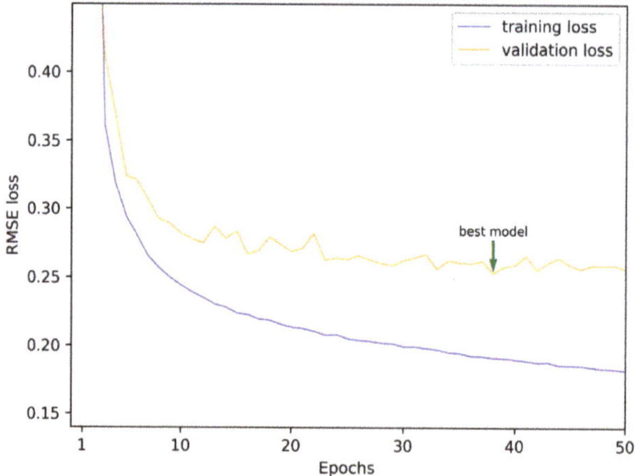

Figure 3. RMSE loss against epochs for the 3D U-Net model.

5.2. Evaluation Metrics

RMSE and mean absolute error (MAE) are used to measure the SSbT prediction error, which is defined as

$$RMSE = \sqrt{\frac{1}{N}\sum_{n=1}^{N}(\hat{X}^{(n)} - X^{(n)})^2}$$ (2)

$$MAE = \frac{1}{N}\sum_{n=1}^{N}\left|\hat{X}^{(n)} - X^{(n)}\right|$$ (3)

A correlation coefficient (CC) is adopted to evaluate the consistency between the observed SSbT profile s and the predicted SSbT profile $\hat{s}$ at the (i, j) observation point, which is defined as

$$CC_{i,j} = \frac{Cov(\hat{s}, s)}{\sqrt{Var(\hat{s}) \cdot Var(s)}}$$ (4)

The average CC of all observation points of the predicted SSbT field can be calculated as

$$CC = \frac{1}{HW}\sum_{i=1}^{H}\sum_{j=1}^{W} CC_{i,j}$$ (5)

5.3. Results

We compared the prediction performance of the model between training on observation-only data and training on observation and reconstruction data. The results are shown in Table 2, in which 12-month average RMSE, MAE, and CC are adopted as indicators. Using additional reconstruction data resulted in relative improvements of 7.5% and 9.3% in terms of RMSE and MAE, while the CC value increased from 0.9506 to 0.9616. This proves that the reconstructed historical data can help improve the overall performance of the model for SSbT prediction.

Table 2. The effect of adding reconstruction data during model training.

Dataset	RMSE	MAE	CC
Observation data	0.6857	0.4227	0.9506
Reconstructed data + observation data	0.6343	0.3832	0.9616

The verification of depth levels for lead times of 1, 3, 6, 9, and 12 months is shown in Figure 4. For all depth levels, the 3D U-Net model's temperature prediction error increased with growing lead times. The prediction performance of the model varied at different depths. In general, the prediction error of SST and 50–150 m SSbT is larger than that at other depths. Relatively large prediction error obtained at 0 m may be due to the presence of more erratic elements, such as solar radiation and sea surface winds. To investigate the reasons for the large prediction error at the depth of 50–150 m, we calculated the average root-mean-square (RMS) values of the temperature inter-annual variations for different depth levels, as shown in Figure 5a. It can be seen that the subsurface temperature inter-annual variability peaks in the range 50–150 m, which is consistent with the findings in [40], making it difficult to accurately model the temporal variations at these depth levels. In addition, the vertical temperature gradients for different depth levels was computed. As shown in Figure 5b, the vertical temperature gradients are higher at the depth of 50–150 m than those at other depth levels, which also causes a significant prediction error.

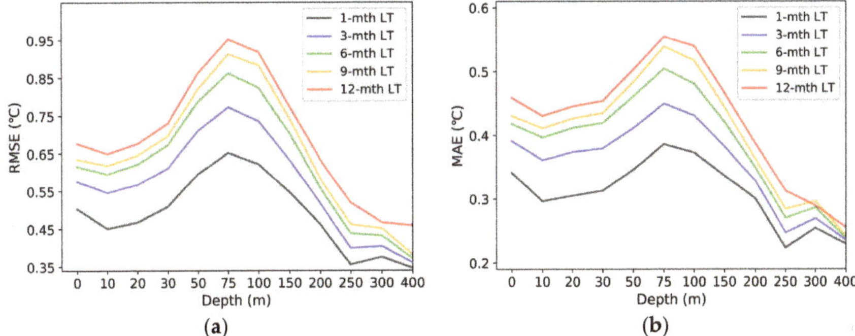

Figure 4. RMSE (**a**) and MAE (**b**) of the prediction results at different depth levels for selected lead times (LTs).

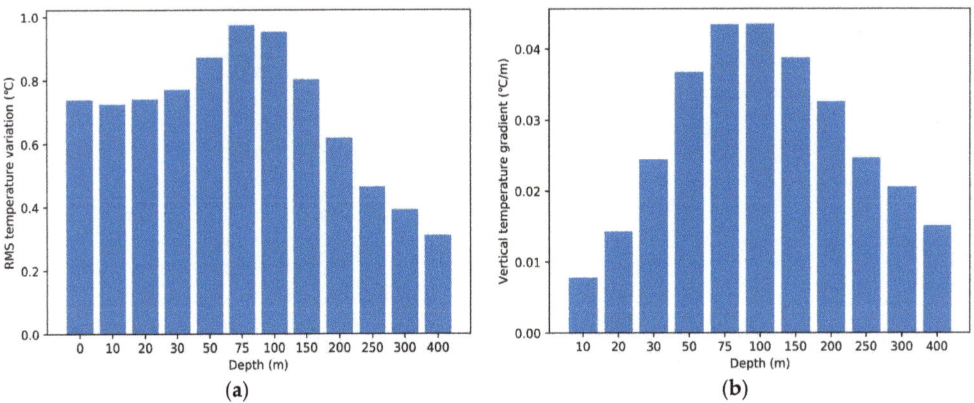

Figure 5. Average root-mean-square (RMS) temperature variations (**a**) and average vertical temperature gradients (**b**) at different depth levels in the study area.

We then examined the prediction performance of this method for SSbT in different geographical parts of the study area, including north temperate (between 35.5°N–59.5°N), north subtropics (between 23.5°N–35.5°N), tropics (between 23.5°N–23.5°S), south subtropics (between 23.5°S–35.5°S), and south temperate (between 35.5°S–59.5°S). First, the prediction performance for different depth levels over these regions were examined. For simplicity and intuition, six depth levels at appropriate intervals were selected within 0–400 m, including 0, 20, 50, 100, 200, and 400 m. The 12-month average RMSE values are shown in Table 3. The SST prediction error in tropics is lower than in other regions. However, the highest RMSE values are obtained for the prediction of temperature at depths of 50, 100, and 200 m over this region. The changes in prediction error with depth in north temperate, north subtropics and tropics are generally consistent with the trend of the whole study area (shown in Figure 4), while the prediction error in south subtropics and south temperate increases with depth. Beyond that, the overall prediction performance for all 12 depth levels in the upper 400 m over the six regions was evaluated using 12-month average RMSE, MAE, and CC as indicators. As shown in Table 4, the RMSE and MAE values are higher in tropics than in other regions, in which the error produced in middle layers contributes a lot. It seems that the prediction performance for SSbT in the Southern Hemisphere is better than that in the Northern Hemisphere, and the prediction error in subtropical zones is lower than that in temperate zones. It is noteworthy that the CC values of the predicted SSbT profiles in subtropical and tropical zones are above 0.99, while those of north temperate and south temperate are around 0.91. This indicates that the predicted SSbT profiles in temperate zones has poorer consistency with the observations compared to that in other regions.

Table 3. RMSE values of the predicted SSbT for selected depth levels over different geographical zones.

Depth	North Temperate	North Subtropics	Tropics	South Subtropics	South Temperate
0	0.7323	0.6200	0.5363	0.5468	0.5876
20	0.7194	0.5811	0.6003	0.4648	0.5256
50	0.7754	0.6334	0.9259	0.4568	0.5182
100	0.6494	0.6109	1.0686	0.4424	0.5179
200	0.5163	0.4657	0.6488	0.4376	0.4540
400	0.4180	0.4123	0.3924	0.3407	0.3147

Table 4. Evaluation of the overall prediction performance for the SSbT fields in different geographical zones.

Region	RMSE	MAE	CC
North temperate	0.6391	0.3330	0.9177
North subtropics	0.5565	0.3718	0.9935
Tropics	0.7345	0.4454	0.9960
South subtropics	0.4492	0.3130	0.9955
South temperate	0.4893	0.3499	0.9081

For lead times of 1–12 months, the SSbT prediction performance of different methods was examined at depths of 0, 20, 50, 100, 200, and 400 m. As shown in Figure 6, the prediction error of the seasonal naïve model for a specific depth is positively correlated with the inter-annual variation of temperature at that depth. Three DL models outperformed the seasonal naïve model for most of the selected depths. However, they are worse than the seasonal naïve model for 400 m where the inter-annual variation of temperature at 400 m is much smaller compared to those at other depth levels. This is probably because redundant layers in these DL models cause a loss of input information. The use of residual blocks [41] may alleviate this problem. In comparison to the M-ConvLSTM model, it was found that the S3D-CNN and 3D U-Net models had reduced RMSE values for all the chosen depth levels at all lead times, demonstrating their potential in modeling spatiotemporal data with seasonal periodicity. In addition, the designed 3D U-Net model is superior to the M-ConvLSTM and S3D-CNN models for these depth levels.

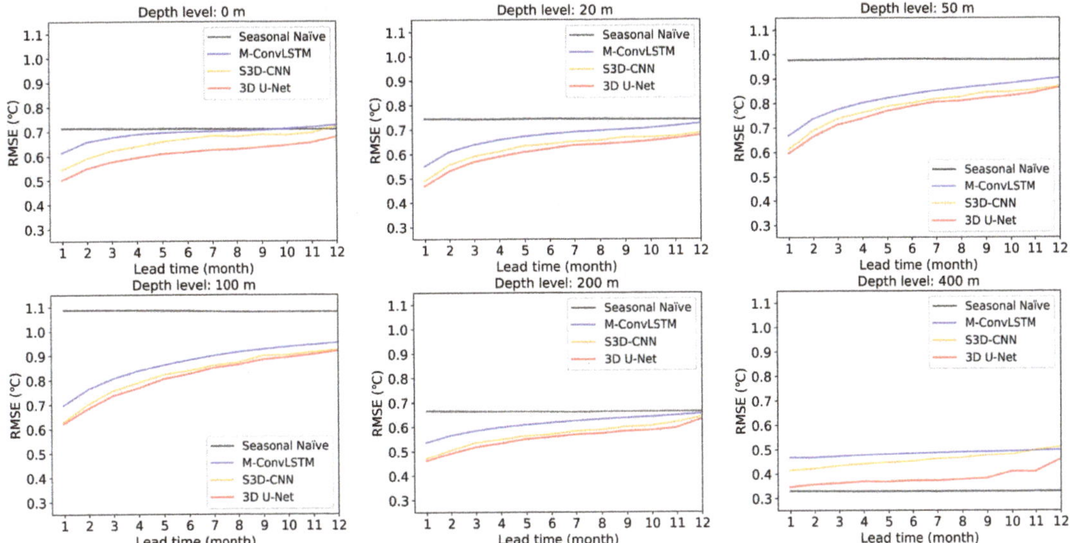

Figure 6. RMSE curves of the compared methods against lead times of 1–12 months for selected depth levels.

For all 12 depth levels, the total prediction performance of the comparative approaches was assessed. The results are displayed in Figure 7. It can be seen that the 3D U-Net model has a lower prediction error than the other methods for all lead times. The CC values obtained by the four approaches fall between 0.94 and 0.97, demonstrating that the projected SSbT profiles of the four methods are consistent with the observed SSbT profiles. It is noteworthy that although the seasonal naïve model achieved the highest prediction error, it performs even better than the 3D U-Net model in terms of CC. This indicates that there is still room for improvement in retaining spatial information in the vertical direction of input SSbT fields.

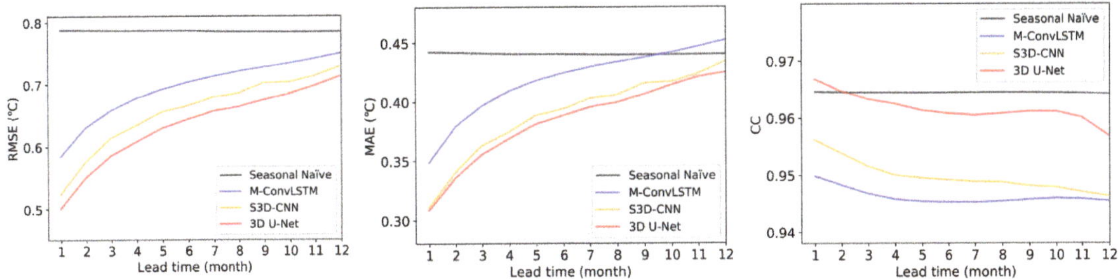

Figure 7. Evaluation of the overall prediction performance of different models.

A case study was then conducted to qualitatively assess the 3D U-Net model's performance in terms of prediction. The SSbT fields observed in the past 12 months from February 2020 to January 2021 were input into the model, and the prediction results for the next 12 months from February 2021 to January 2022 were obtained. Figure 8a–d show the prediction results for the lead times of 3, 6, 9, and 12 months together with the associated ground truth SSbT fields for the selected depths of 0, 100, 200, and 400 m. Maps of the relative error between the predictions and observations are also displayed. The SST shows an obvious seasonal variation, and the prediction results of the network well reflect this temporal characteristic. The distribution of the predicted sea temperature is comparable to that of the ground truth observations in the other three depth layers. In addition, the relative error in most regions is less than 10%. However, large relative error of more than 20% is found in temperate zones, mainly at the boundary of the study area and the regions close to the land. This may be due to the lack of utilization of temperature outside the study area and the complexity of SSbT variations in coastal seas. In general, the 3D U-Net model can predict realistic SSbT variations in the study area.

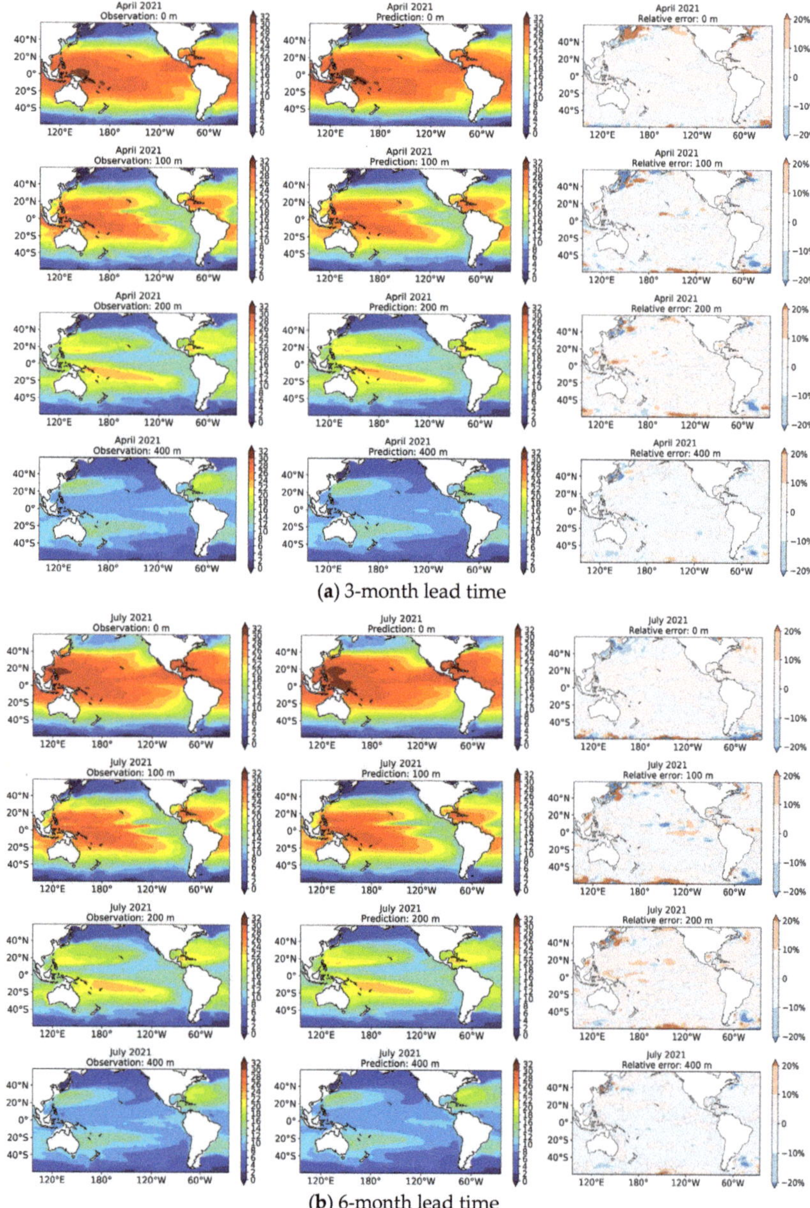

(a) 3-month lead time

(b) 6-month lead time

Figure 8. *Cont.*

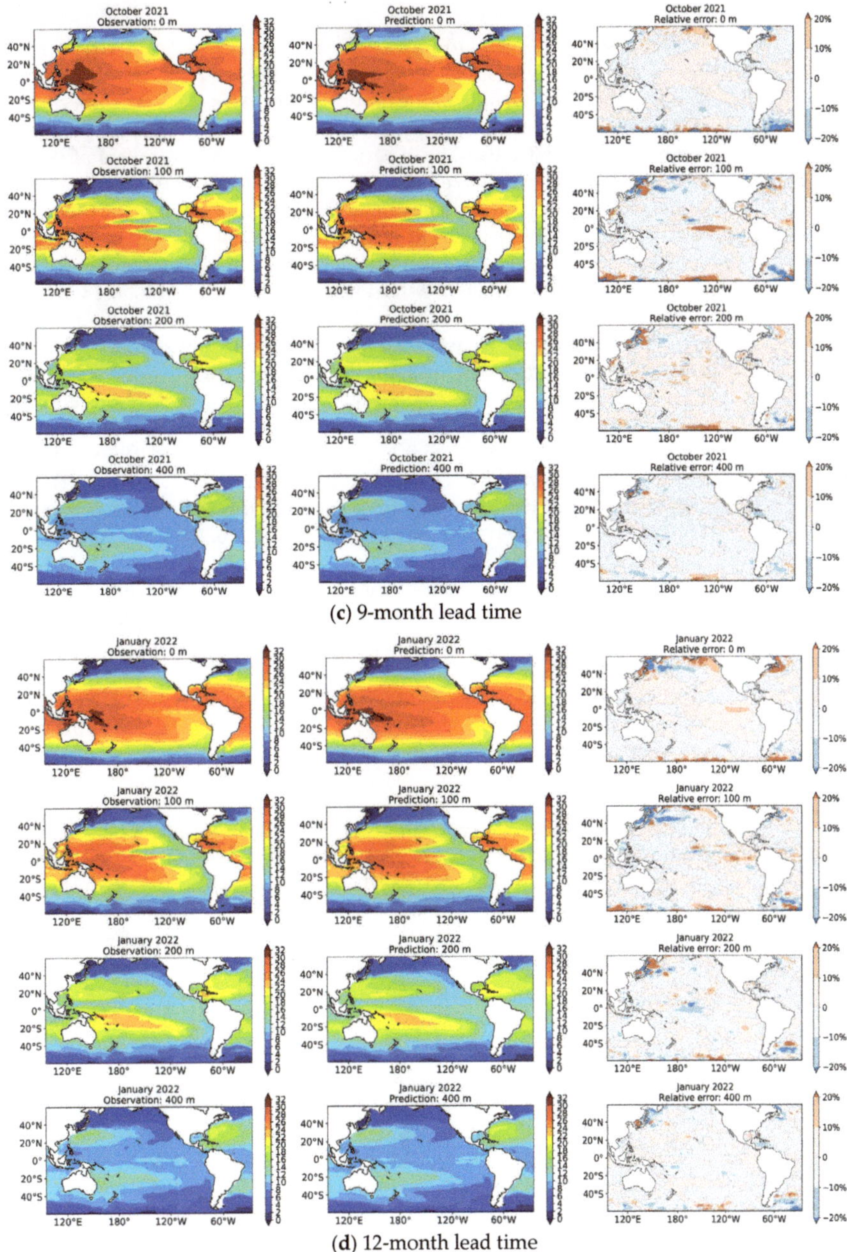

Figure 8. A prediction case starting from January 2021 for lead times of 3 (**a**), 6 (**b**), 9 (**c**), and 12 (**d**) months. In each subfigure, from left to right are the observations, prediction results, and maps of the relative error between them, and from top to bottom, subfigures correspond to depth levels of 0, 100, 200, and 400 m.

6. Conclusions

This study attempts to predict the SSbT over the Pacific Ocean and its surrounding oceans in the upcoming 12 months. We designed a 3D U-Net model to extract the spatiotemporal correlation of the input SSbT fields in the past 12 months and map them to the prediction results. The issue of insufficient training data was resolved using the reconstructed SSbT data. The experimental results indicate that after using additional reconstruction data, relative improvements of 7.5% and 9.3% are achieved in terms of RMSE and MAE, and a higher average CC value is obtained. The prediction of 50–150 m SSbT is more difficult than that at other depths, possibly because the inter-annual variations and vertical temperature gradients at these layers are larger than those at other layers. In addition, the SSbT prediction performance of the 3D U-Net model varies across different geographical parts of the study area. For the prediction of SST, the model has the lowest prediction error in tropics, while for the prediction of mesosphere temperature, the model has the largest prediction error in the tropical region. The overall prediction performance for SSbT in the Southern Hemisphere is better than that in the Northern Hemisphere, and the model performance in subtropical zones are better than that in temperate zones. The prediction error of this method is lower at most depth levels compared with that of other methods. It achieves better overall and longer-term prediction performance. The RMSE and MAE of the predicted SSbT fields for all lead times are in the range 0.5–0.7 °C and 0.3–0.45 °C, respectively, and the average CC scores of the predicted SSbT profiles exceed 0.96 for almost all lead times. A prediction case starting from January 2021 over the study area demonstrates that the 3D U-Net model is capable of simulating the temporal variations of the SSbT fields, and its predictions were in line with the observations, which can facilitate understanding of future changes in the thermal state of the subsurface ocean.

However, there is still a problem of input information loss in the 3D U-Net model, which makes its prediction accuracy for 400 m and the CC values of the predicted sea temperature profiles inferior to those of the seasonal naïve model. In the future, we will try to address this problem and incorporate more oceanic parameters such as ocean currents into the network to further improve the SSbT prediction performance.

Author Contributions: Conceptualization, N.S. and Q.L.; methodology, N.S.; software, N.S.; validation, N.S., Q.L., Z.Z. and X.Z.; formal analysis, N.S.; investigation, N.S.; resources, Q.L.; data curation, N.S.; writing—original draft preparation, N.S.; writing—review and editing, Q.L. and Z.Z.; visualization, N.S.; supervision, Q.L., Z.Z. and X.Z.; project administration, Q.L.; funding acquisition, Q.L. All authors have read and agreed to the published version of the manuscript.

Funding: This research was funded by the National Natural Science Foundation of China (Grant No. 42075139, 41305138), the China Postdoctoral Science Foundation (Grant No. 2017M621700), Hunan Province Natural Science Foundation (Grant No. 2021JJ30773), and Fengyun Application Pioneering Project (FY-APP).

Data Availability Statement: The IAP-reconstructed historical SSbT data can be downloaded from http://www.ocean.iap.ac.cn/ftp/cheng/CZ16_v3_IAP_Temperature_gridded_1month_netcdf/ (accessed on 4 July 2022); the RDA-reconstructed historical SSbT data are available from https://rda.ucar.edu/datasets/ds285.3/ (accessed on 4 July 2022); and the NOAA Optimum Interpolation (OI) SST V2 data and NCEP Global Ocean Data Assimilation System (GODAS) data are provided by the NOAA PSL, Boulder, Colorado, USA, from their website at https://psl.noaa.gov (accessed on 4 July 2022).

Acknowledgments: The authors would like to thank the anonymous reviewers for providing professional and insightful comments on the previous article. We offer our deepest thanks to the data contributors for sharing their data freely.

Conflicts of Interest: The authors declare no conflict of interest.

References

1. Wentz, F.J.; Gentemann, C.; Smith, D.; Chelton, D. Satellite measurements of sea surface temperature through clouds. *Science* **2000**, *288*, 847–850. [CrossRef] [PubMed]
2. Yao, S.L.; Luo, J.J.; Huang, G.; Wang, P. Distinct global warming rates tied to multiple ocean surface temperature changes. *Nat. Clim. Change* **2017**, *7*, 486–491. [CrossRef]
3. Kao, H.Y.; Yu, J.Y. Contrasting eastern-Pacific and central-Pacific types of ENSO. *J. Clim.* **2009**, *22*, 615–632. [CrossRef]
4. Anderson, B.T. On the joint role of subtropical atmospheric variability and equatorial subsurface heat content anomalies in initiating the onset of ENSO events. *J. Clim.* **2007**, *20*, 1593–1599. [CrossRef]
5. Cione, J.J.; Uhlhorn, E.W. Sea surface temperature variability in hurricanes: Implications with respect to intensity change. *Mon. Weather. Rev.* **2003**, *131*, 1783–1796. [CrossRef]
6. Belkin, I.M. Rapid warming of large marine ecosystems. *Prog. Oceanogr.* **2009**, *81*, 207–213. [CrossRef]
7. Gou, Y.; Zhang, T.; Liu, J.; Wei, L.; Cui, J.H. DeepOcean: A general deep learning framework for spatio-temporal ocean sensing data prediction. *IEEE Access* **2020**, *8*, 79192–79202. [CrossRef]
8. Klemas, V. Fisheries applications of remote sensing: An overview. *Fish. Res.* **2013**, *148*, 124–136. [CrossRef]
9. Barnston, A.G.; Glantz, M.H.; He, Y. Predictive skill of statistical and dynamical climate models in SST forecasts during the 1997–98 El Niño episode and the 1998 La Niña onset. *Bull. Am. Meteorol. Soc.* **1999**, *80*, 217–244. [CrossRef]
10. Stockdale, T.N.; Balmaseda, M.A.; Vidard, A. Tropical Atlantic SST prediction with coupled ocean–atmosphere GCMs. *J. Clim.* **2006**, *19*, 6047–6061. [CrossRef]
11. Johnson, S.J.; Stockdale, T.N.; Ferranti, L.; Balmaseda, M.A.; Molteni, F.; Magnusson, L.; Monge-Sanz, B.M. SEAS5: The new ECMWF seasonal forecast system. *Geosci. Model Dev.* **2019**, *12*, 1087–1117. [CrossRef]
12. Kondrashov, D.; Kravtsov, S.; Robertson, A.W.; Ghil, M. A hierarchy of data-based ENSO models. *J. Clim.* **2005**, *18*, 4425–4444. [CrossRef]
13. Shirvani, A.; Nazemosadat, S.M.J.; Kahya, E. Analyses of the Persian Gulf sea surface temperature: Prediction and detection of climate change signals. *Arab. J. Geosci.* **2015**, *8*, 2121–2130. [CrossRef]
14. Lins, I.D.; Araujo, M.; das Chagas Moura, M.; Silva, M.A.; Droguett, E.L. Prediction of sea surface temperature in the tropical Atlantic by support vector machines. *Comput. Stat. Data Anal.* **2013**, *61*, 187–198. [CrossRef]
15. Corchado, J.M.; Fyfe, C. Unsupervised neural method for temperature forecasting. *Artif. Intell. Eng.* **1999**, *13*, 351–357. [CrossRef]
16. Wei, L.; Guan, L.; Qu, L. Prediction of sea surface temperature in the South China Sea by artificial neural networks. *IEEE Geosci. Remote Sens. Lett.* **2019**, *17*, 558–562. [CrossRef]
17. Zhang, Q.; Wang, H.; Dong, J.; Zhong, G.; Sun, X. Prediction of sea surface temperature using long short-term memory. *IEEE Geosci. Remote Sens. Lett.* **2017**, *14*, 1745–1749. [CrossRef]
18. Ham, Y.G.; Kim, J.H.; Luo, J.J. Deep learning for multi-year ENSO forecasts. *Nature* **2019**, *573*, 568–572. [CrossRef]
19. Haghbin, M.; Sharafati, A.; Motta, D.; Al-Ansari, N.; Noghani, M.H.M. Applications of soft computing models for predicting sea surface temperature: A comprehensive review and assessment. *Prog. Earth Planet. Sci.* **2021**, *8*, 1–19. [CrossRef]
20. Zhang, Z.; Pan, X.; Jiang, T.; Sui, B.; Liu, C.; Sun, W. Monthly and quarterly sea surface temperature prediction based on gated recurrent unit neural network. *J. Mar. Sci. Eng.* **2020**, *8*, 249. [CrossRef]
21. Yang, Y.; Dong, J.; Sun, X.; Lima, E.; Mu, Q.; Wang, X. A CFCC-LSTM model for sea surface temperature prediction. *IEEE Geosci. Remote Sens. Lett.* **2017**, *15*, 207–211. [CrossRef]
22. Shi, X.; Chen, Z.; Wang, H.; Yeung, D.Y.; Wong, W.K.; Woo, W.-C. Convolutional LSTM network: A machine learning approach for precipitation nowcasting. *Adv. Neural Inf. Process. Syst.* **2015**, *28*, 802–810.
23. Xiao, C.; Chen, N.; Hu, C.; Wang, K.; Xu, Z.; Cai, Y.; Gong, J. A spatiotemporal deep learning model for sea surface temperature field prediction using time-series satellite data. *Environ. Model. Softw.* **2019**, *120*, 104502. [CrossRef]
24. Gupta, M.; Kodamana, H.; Sandeep, S. Prediction of ENSO beyond spring predictability barrier using deep convolutional LSTM networks. *IEEE Geosci. Remote Sens. Lett.* **2020**, *19*, 1501205. [CrossRef]
25. Zhang, K.; Geng, X.; Yan, X.H. Prediction of 3-D ocean temperature by multilayer convolutional LSTM. *IEEE Geosci. Remote Sens. Lett.* **2020**, *17*, 1303–1307. [CrossRef]
26. Wu, X.; Yan, X.H.; Jo, Y.H.; Liu, W.T. Estimation of subsurface temperature anomaly in the North Atlantic using a self-organizing map neural network. *J. Atmos. Ocean. Technol.* **2012**, *29*, 1675–1688. [CrossRef]
27. Cheng, L.; Zhu, J. Benefits of CMIP5 multimodel ensemble in reconstructing historical ocean subsurface temperature variations. *J. Clim.* **2016**, *29*, 5393–5416. [CrossRef]
28. Liu, J.; Zhang, T.; Han, G.; Gou, Y. TD-LSTM: Temporal dependence-based LSTM networks for marine temperature prediction. *Sensors* **2018**, *18*, 3797. [CrossRef]
29. Patil, K.R.; Iiyama, M. Deep Neural Networks to Predict Sub-surface Ocean Temperatures from Satellite-Derived Surface Ocean Parameters. In *Soft Computing for Problem Solving*; Springer: Singapore, 2021; pp. 423–434.
30. Zuo, X.; Zhou, X.; Guo, D.; Li, S.; Liu, S.; Xu, C. Ocean temperature prediction based on stereo spatial and temporal 4-D convolution model. *IEEE Geosci. Remote Sens. Lett.* **2021**, *19*, 1–5. [CrossRef]
31. Ishii, M.; Kimoto, M.; Sakamoto, K.; Iwasaki, S. *Subsurface Temperature and Salinity Analyses*; Research Data Archive at the National Center for Atmospheric Research. Accessed 4 June 2022; Computational and Information Systems Laboratory: Boulder, CO, USA, 2005. [CrossRef]

32. Reynolds, R.W.; Rayner, N.A.; Smith, T.M.; Stokes, D.C.; Wang, W. An improved in situ and satellite SST analysis for climate. *J. Clim.* **2002**, *15*, 1609–1625. [CrossRef]
33. Behringer, D.W.; Ji, M.; Leetmaa, A. An improved coupled model for ENSO prediction and implications for ocean initialization. Part I: The ocean data assimilation system. *Mon. Weather. Rev.* **1998**, *126*, 1013–1021. [CrossRef]
34. Yang, K.; Cai, W.; Huang, G.; Wang, G.; Ng, B.; Li, S. Oceanic processes in ocean temperature products key to a realistic presentation of positive Indian Ocean Dipole nonlinearity. *Geophys. Res. Lett.* **2020**, *47*, e2020GL089396. [CrossRef]
35. Tang, Y.; Duan, A. Using deep learning to predict the East Asian summer monsoon. *Environ. Res. Lett.* **2021**, *16*, 124006. [CrossRef]
36. Çiçek, Ö.; Abdulkadir, A.; Lienkamp, S.S.; Brox, T.; Ronneberger, O. U-net: Learning dense volumetric segmentation from sparse annotation. In Proceedings of the International Conference on Medical Image Computing and Computer-Assisted Intervention, Athens, Greece, 17–21 October 2016; Springer: Cham, Switzerland, 2016; pp. 424–432.
37. Ioffe, S.; Szegedy, C. Batch normalization: Accelerating deep network training by reducing internal covariate shift. In Proceedings of the International Conference on Machine Learning, Lille, France, 7–9 July 2015; pp. 448–456.
38. Tran, D.; Bourdev, L.; Fergus, R.; Torresani, L.; Paluri, M. Learning spatiotemporal features with 3d convolutional networks. In Proceedings of the IEEE International Conference on Computer Vision, Santiago, Chile, 7–13 December 2015; pp. 4489–4497.
39. Abadi, M.; Agarwal, A.; Barham, P.; Brevdo, E.; Chen, Z.; Citro, C.; Zheng, X. Tensorflow: Large-Scale Machine Learning on Heterogeneous Distributed Systems. *arXiv* **2016**, arXiv:1603.04467.
40. Doney, S.; Yeager, S.; Danabasoglu, G.; Large, W.; McWilliams, J. *Modeling Global Oceanic Interannual Variability (1958–1997): Simulation Design and Model-Data Evaluation*; Tech. Note NCAR/TN-452+ STR; National Center for Atmospheric Research: Boulder, CO, USA, 2003.
41. Zhang, Y.; Li, K.; Li, K.; Wang, L.; Zhong, B.; Fu, Y. Image super-resolution using very deep residual channel attention networks. In Proceedings of the European Conference on Computer Vision (ECCV), Munich, Germany, 8–14 September 2018; pp. 286–301.

remote sensing

MDPI

Article

Deep Learning to Near-Surface Humidity Retrieval from Multi-Sensor Remote Sensing Data over the China Seas

Rongwang Zhang [1,2,3] , Weihao Guo [1,4] and Xin Wang [1,2,3,*]

1   State Key Laboratory of Tropical Oceanography, South China Sea Institute of Oceanology, Chinese Academy of Sciences, Guangzhou 510301, China
2   Southern Marine Science and Engineering Guangdong Laboratory (Guangzhou), Guangzhou 511458, China
3   Innovation Academy of South China Sea Ecology and Environmental Engineering, Chinese Academy of Sciences, Guangzhou 511458, China
4   University of Chinese Academy of Sciences, Beijing 100049, China
*   Correspondence: wangxin@scsio.ac.cn

Abstract: Near-surface humidity (Q_a) is a key parameter that modulates oceanic evaporation and influences the global water cycle. Remote sensing observations act as feasible sources for long-term and large-scale Q_a monitoring. However, existing satellite Q_a retrieval models are subject to apparent uncertainties due to model errors and insufficient training data. Based on in situ observations collected over the China Seas over the last two decades, a deep learning approach named Ensemble Mean of Target deep neural networks (EMTnet) is proposed to improve the satellite Q_a retrieval over the China Seas for the first time. The EMTnet model outperforms five representative existing models by nearly eliminating the mean bias and significantly reducing the root-mean-square error in satellite Q_a retrieval. According to its target deep neural network selection process, the EMTnet model can obtain more objective learning results when the observational data are divergent. The EMTnet model was subsequently applied to produce 30-year monthly gridded Q_a data over the China Seas. It indicates that the climbing rate of Q_a over the China Seas under the background of global warming is probably underestimated by current products.

Keywords: near-surface humidity; remote sensing; deep learning; China Seas

Citation: Zhang, R.; Guo, W.; Wang, X. Deep Learning to Near-Surface Humidity Retrieval from Multi-Sensor Remote Sensing Data over the China Seas. *Remote Sens.* **2022**, *14*, 4353. https://doi.org/10.3390/rs14174353

Academic Editors: Ana B. Ruescas, Veronica Nieves and Raphaëlle Sauzède

Received: 4 July 2022
Accepted: 26 August 2022
Published: 2 September 2022

Publisher's Note: MDPI stays neutral with regard to jurisdictional claims in published maps and institutional affiliations.

1. Introduction

As the primary source of global evaporation and precipitation, the ocean plays an important role in the transportation and redistribution of water resources on Earth [1–3]. On this basis, the near-surface humidity (Q_a) over global oceans is crucial, as it modulates oceanic evaporation and influences the global water cycle [4–6]. Nevertheless, there are non-negligible uncertainties in Q_a estimates in satellite-derived products [7,8] and reanalysis products [9,10]. The imperfection of Q_a data quality has been reported as one of the leading error sources of uncertainties in freshwater exchanges across the air–sea interface and in global water budgets [11–14]. Even in coupled general circulation models, the performance of Q_a is highly related to simulations of oceanic evaporation [15]. Accurate estimates of Q_a are thus necessary for studies on the global water cycle, air–sea interactions, and climate change [16].

The measurements of Q_a can be generally divided into two approaches: in situ observations and remote sensing observations. The former are direct measurements of Q_a and have relatively high credibility, but these observations are subject to poor continuity in time and space. The latter have the advantage of long-term and large-scale Q_a monitoring. Still, remote sensing is an indirect approach that requires a relevant retrieval model to convert satellite measurements into Q_a. With the development of space-borne technology and microwave radiometers, the last several decades have experienced the prosperity of investigations in model development for satellite Q_a retrieval. Considering that a large portion of the total column precipitable water (TPW) is confined in the atmospheric surface layer, pioneering work by [17,18] (hereafter L86) linked the TPW to Q_a in light of the Q_a–W relation. It was reported that the Q_a–W relationship has excellent performance with training data on a monthly scale [18] and can also work well with synoptic-scale data [19]. Considering the decoupling of the atmospheric boundary layer from the upper troposphere, Ref. [20] proposed replacing the TPW data in the Q_a–W relation with the precipitable water constrained in the lowest 500 m, which can be derived from brightness temperature (TB) measurements. To reduce the propagation of uncertainties within input data, Ref. [21] established a direct linear regression between Q_a and TB. Under the scheme of the Q_a–W relation, Ref. [22] developed an empirical orthogonal function method for satellite Q_a retrieval. A neural network combining TPW and sea surface temperature (SST) was first developed to estimate Q_a by [23]. Subsequently, estimates of Q_a with multichannel TBs as input data by multivariate linear regression [24–27] or nonlinear regression [16,28,29] prevailed in the last two decades.

The models above were primarily designed for global oceans. However, Ref. [28] reported that satellite Q_a retrieval differed in regions of the tropics and high latitudes, and a high-latitude enhancement was considered in their model. It indicates that attention should be given to different regional features of satellite Q_a retrieval. The China Seas, consisting of the Bohai Sea, Yellow Sea (YS), East China Sea (ECS), and South China Sea (SCS), are the largest marginal group of seas in the northwestern Pacific and are strongly influenced by complex continental environments. Previous investigations have pointed out that the Q_a data in this region suffer from significant uncertainties and are the leading error source of air–sea heat fluxes [30,31].

In recent years, machine learning, especially deep learning, has been widely used to provide new insights into traditional and/or emerging research in earth science [32–36]. With the accumulation of high-quality in situ observations of Q_a over the China Seas in the last two decades, the main objective of this study is to develop a deep-learning-based model to improve the satellite Q_a retrieval over the China Seas. The data and methods are introduced in Section 2. The main results are presented in Section 3. Section 4 discusses the interpretability of the deep-learning-based model. Finally, Section 5 draws the main conclusion of this study.

2. Data and Methods

2.1. In Situ Observations

In situ observation information from this study is listed in Table 1. There are 20 observational stations in the coastal and open oceans (Figure 1a), including 18 buoys, 1 offshore platform, and 1 flux tower on an island. Compared to ship observations, these fixed-point observations usually have more stable data quality performance. The data collected in coastal areas are valuable touchstones to validate the performance of remote sensing observations, including Q_a, surface wind, and sea surface temperature. The data span from 1998 to 2018, and the sampling intervals vary from 1 min to 30 min. All the raw data were processed with quality control procedures as suggested by [31,37,38]. For all stations, the Q_a and surface wind data were adjusted to standard heights of 2 m and 10 m, respectively, according to the COARE 3.0 algorithm [39].

Table 1. Information on in situ observations collected in this study. The station names with prefixes "DH" and "HH" are located in the ECS and the YS, respectively. The remaining 13 stations are located in the SCS. The data span from 1998 to 2018 and the sampling intervals vary from 1 min to 30 min.

Name	Location	Ocean Depth	Type	Sampling Interval	Period
Maoming	111.66°E, 20.75°N	~100 m	buoy	1 min	26 May 2010–28 September 2011
Shantou	117.34°E, 22.33°N	~100 m	buoy	1 min	16 October 2010–16 May 2011
Bohe	111.32°E, 21.46°N	~15 m	offshore platform	10 min	26 November 2009–15 May 2010 4 January 2011–28 April 2011 13 March 2012–3 June 2012
Xisha flux tower	112.33°E, 16.83°N	island	tower	1~10 min	26 April 2008–6 October 2008 19 July 2013–31 January 2017
Xisha buoy	112.33°E, 16.86°N	~1000 m	buoy	10 min	19 September 2009–7 April 2013 14 May 2018–12 June 2018
Kexue 1	110.26°E, 6.41°N	~1300 m	buoy	15 min	7 May 1998–20 June 1998
Shiyan 3	117.40°E, 20.60°N	~1000 m	buoy	15 min	6 May 1998–23 June 1998
SCS1	115.60°E, 8.10°N	~3000 m	buoy	15 min	19 April 1998–29 April 1998
SCS3	114.41°E, 12.98°N	~4500 m	buoy	15 min	8 June 1998–16 June 1998
SCS3[+]	114.00°E, 13.00°N	~4000 m	buoy	15 min	13 April 1998–29 May 1998
QF301	115.59°E, 22.28°N	~100 m	buoy	30 min	1 March 2011–31 May 2011
QF302	114.00°E, 21.50°N	~100 m	buoy	30 min	1 March 2011–31 May 2011
QF303	112.83°E, 21.12°N	~100 m	buoy	30 min	1 March 2011–31 May 2011
DH06	123.13°E, 30.72°N	<100 m	buoy	30 min	29 March 2012–30 December 2013
DH10	122.00°E, 31.37°N	<100 m	buoy	30 min	1 September 2013–2 December 2015
DH11	122.82°E, 31.00°N	<100 m	buoy	30 min	1 January 2014–30 December 2016
DH20	122.75°E, 29.75°N	<100 m	buoy	30 min	6 November 2014–1 November 2016
HH07	122.58°E, 37.01°N	<100 m	buoy	30 min	29 March 2012–31 December 2013
HH09	120.27°E, 35.90°N	<100 m	buoy	30 min	1 January 2014–31 December 2016
HH19	119.60°E, 35.42°N	<100 m	buoy	30 min	6 November 2014–31 December 2016

The basic characteristics of in situ observations of Q_a are examined to check their representativeness, considering many missing data and the uneven sampling in time and space. As shown in Figure 1b–d, Q_a varies from 5.5 to 24.0, 1.5 to 24.5, and 0.8 to 24.8 g/kg in the SCS, ECS, and YS. Q_a's mean values plus/minus one standard deviation (STD) are 16.6 ± 4.1, 10.1 ± 5.5, and 8.5 ± 5.7 g/kg in the SCS, ECS, and YS. For the probability density distributions (PDDs) of Q_a, a left-skewed distribution in the SCS and right-skewed distributions in the ECS and YS can be observed. These lower limits, mean values, and PDDs of Q_a in the three seas coincide well with the latitudes they locate in, as Q_a usually decreases from low to high latitudes. With the data in the three seas considered as a whole, Q_a presents a fairly even distribution in the range of 2~22 g/kg, which varies basically around a steady density across each bin (Figure 1e). The highly uniform PDD of Q_a shows an acceptable representativeness of in situ observations collected here. Therefore, it is expected that the analyses based on those data could be relatively objective and with high significance.

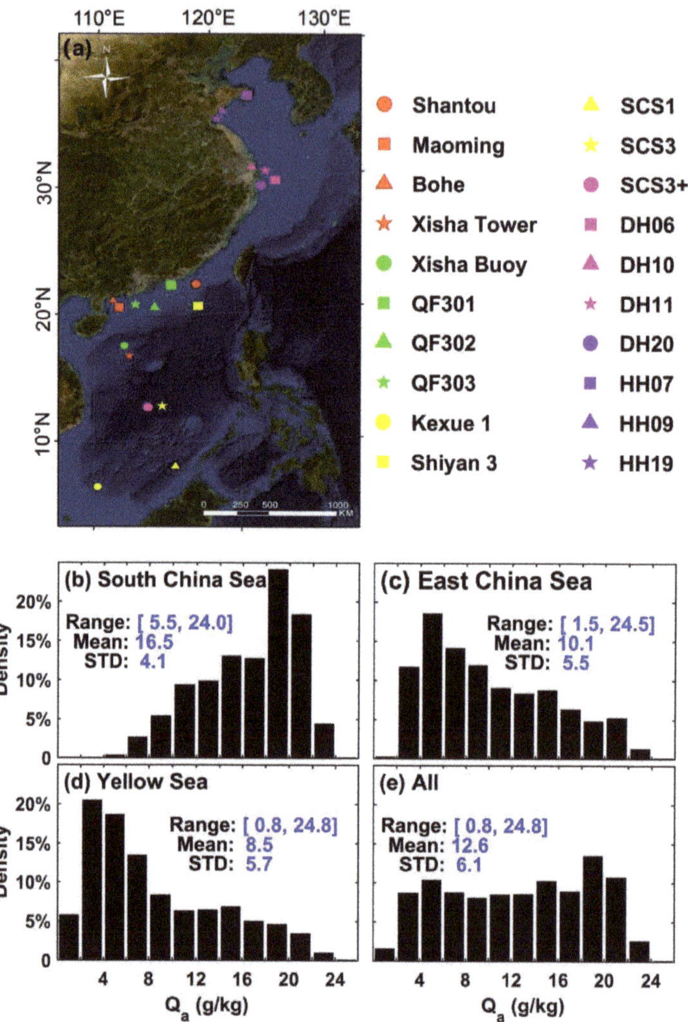

Figure 1. (**a**) Geographical distribution of observational stations. Shading denotes the ocean depth. (**b–d**) The PDDs of in situ observations of Q_a over the SCS, ECS, and YS. (**e**) The mean results for all data. The range, mean value, and STD of the data in each panel are shown in blue text with unit of g/kg.

2.2. Remote Sensing Data

Remote sensing observations of TPW, wind speed (U), cloud liquid water (CLW), and SST from various satellite microwave radiometers are utilized in this study. The sensors include the Special Sensor Microwave Imager (SSM/I), the Special Sensor Microwave Imager Sounder (SSMIS), the Advanced Microwave Scanning Radiometer series (AMSR-E and AMSR-2), the WindSat Polarimetric Radiometer, the Tropical Rainfall Measuring Mission (TRMM) Microwave Imager (TMI), and the Global Precipitation Measurement Microwave Imager (GMI). For SSM/I and SSMIS, the instruments are referred to by satellite number starting with F08. Here, F15 from SSMI/I and F16 to F18 from SSMIS are employed because of their relatively long time coverage. Detailed descriptions for each sensor can be found at the Remote Sensing Systems (RSS; www.remss.com, accessed on 30 June 2022). Due to the satellite swath width and orbit seam, banded gaps exist in the ascending and descending daily maps of satellited-derived variables. To facilitate the matching of satellite data and in situ data, the products incorporating both the ascending and descending data by using 3-day running average are utilized, which can achieve more homogeneous spatial distributions for these variables. Note that there exist differences in data from various sensors (Figure 2), and uncertainties in results may be caused by a single data source. Consequently, multi-sensor inputs from SSM/I F15, SSMIS F16, SSMIS F17, SSMIS F18, AMSR-E, AMSR-2, WindSat, TMI, and GMI are utilized in the following study.

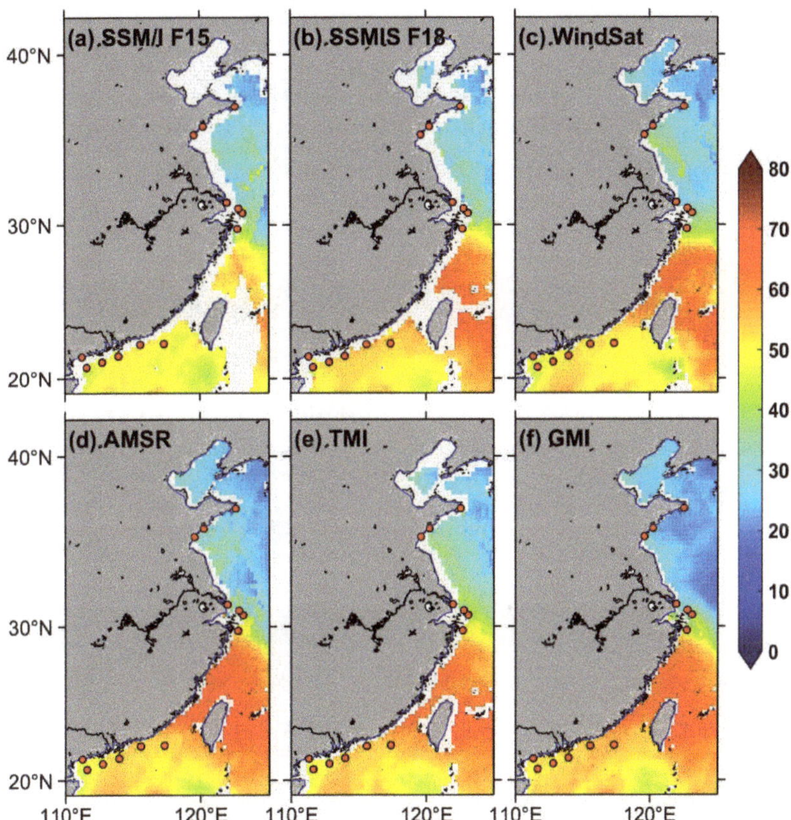

Figure 2. Comparison of 3-day running averaged TPW data on 1 June 2014, from various types of sensors. (**a–f**) Sensors SSM/I F15, SSMIS F18, WindSat, AMSR, TMI, and GMI, respectively. Red dots are stations located in coastal areas. Note that different sensors have different scopes of data coverage in coastal areas.

TB data from six channels are used for the models listed in Table 2 that directly retrieve Q_a from multichannel TBs. These channels are 19 H, 19 V, 22 V, and 37 V GHz from SSMIS F17 and 52 V GHz from Advanced Microwave Sounding Unit-A (AMSU-A), where H and V denote the horizontal and vertical polarizations, respectively. As underscored by RSS, TBs from the SSMIS are produced using uniform processing techniques. They are intercalibrated by considering the differences in sensor frequencies, channel resolutions, instrument operation, and other radiometer characteristics [40]. The AMSU-A is a multichannel microwave radiometer that performs atmospheric sounding of temperature and moisture by passively recording atmospheric microwave radiation at multiple wavelengths. Detailed descriptions of how TBs from the AMSU-A are processed can be found in [41,42].

The collocating strategy between satellite data and in situ observations for each station is as follows:

(i) For variables TPW, *U*, and CLW, which are already in daily values, first, average the high-frequency in situ observations to daily values with the local standard time adjusted to Coordinated Universal Time and then apply a 3-day running average. Second, locate the $0.25° \times 0.25°$ box in the satellite data where the corresponding observational station lies. Subsequently, the mean value of the four corners of that box is used as a proxy for satellite data. Attempts such as extending the search area to $1° \times 1°$ box and/or applying an inverse-distance-weighted average present similar results.

(ii) For TBs with ascending and descending measurements per day, temporal and spatial windows of 90 min and 50 km following Yu and Jin (2018) are used. If multiple points of satellite data meet the criterion, the average of those points is taken. If no satellite data match the in situ observations, a missing value is set.

Table 2. Summary of the methods of surface humidity retrieval validated in this study. Here, the *W* denotes the parameter TPW in the main text.

Algorithm	Equation	RMSE (g/kg)
Liu et al. (1986) [18]	$Q_a = C_1 \times W + C_2 \times W^2 + C_3 \times W^3 + C_4 \times W^4 + C_5 \times W^5,$ where $C_1 = 0.006088244$, $C_2 = 0.1897219$, $C_3 = 0.1891893$, $C_4 = -0.07549036$, and $C_5 = 0.006088244$.	0.40 in tropics and 0.80 in globe
Jones et al. (1999) [23]	$Q_a = C_0 + C_1 \times SST + C_2 \times SST^2 + C_3 \times W^1 + C_4 \times W^2,$ where $C_0 = 2.1052$, $C_1 = -0.0551$, $C_2 = 0.0138$, $C_3 = 0.2435$, and $C_4 = -0.0019$.	0.77 ± 0.39
Bentamy et al. (2003) [24]	$Q_a = C_0 + C_1 T_{19V} + C_2 T_{19H} + C_3 T_{22V} + C_4 T_{37V}$, where $C_0 = -55.9227$, $C_1 = 0.4035$, $C_2 = -0.2944$, $C_3 = 0.3511$, and $C_4 = -0.2395$.	1.40
Jackson et al. (2006) [25]	$Q_a = C_0 + C_1 T_{52V} + C_2 T_{19V} + C_3 T_{19H} + C_4 T_{37V}$, where $C_0 = -105.117$, $C_1 = 0.31743$, $C_2 = 0.62754$, $C_3 = -0.12056$, and $C_4 = -0.33940$.	0.83
Yu and Jin (2018) [28]	$Q_a = a_0 + a_1 T_{19v} + a_2 T_{22v} + a_3 T_{37v} + a_4 T_{52v} + b_1 T_{19v}{}^2 + b_2 T_{22v}{}^2 + b_3 T_{37v}{}^2 + b_4 T_{52v}{}^2$, where $a_0 = 1423.34$, $a_1 = 0.46967$, $a_2 = 0.43401$, $a_3 = -0.92292$, $a_4 = -11.494$, $b_1 = -0.00071$, $b_2 = -0.00072$, $b_3 = 0.00155$, and $b_4 = 0.02336$ for the global model, $a_0 = -127.10$, $a_1 = -0.21113$, $a_2 = 0.71712$, $a_3 = -0.78268$, $a_4 = 1.1918$, $b_1 = 0.00062$, $b_2 = -0.00139$, $b_3 = 0.00153$, and $b_4 = -0.00222$ for the high-latitude model.	0.82

2.3. Reanalysis Data

Two reanalysis products are employed to make comparisons with the Q_a data derived from the model proposed in this study. They are the European Centre for Medium Range Weather Forecast (ECWMF) fifth generation (ERA5) reanalysis product [43] and the National Centers for Environmental Prediction/Department of Energy Global Reanalysis 2 (NCEP2) product [44]. Both products are the latest versions of their corresponding series and improvements have been made in their data assimilations and model physics. Monthly Q_a data from 1990 to 2019 are extracted from the two reanalyses and are interpolated onto $1° \times 1°$ grid maps.

2.4. Existing Satellite Q_a Retrieval Models

Five representative satellite Q_a retrieval models are employed to intercompare with the deep-learning-based model proposed in this study. Table 2 summarizes the basic information of these five models. The L86 model uses a fifth-order polynomial regression approach to estimate Q_a with TPW data. The model reported in [23] (hereafter J99) uses a nonlinear neural network approach to estimate Q_a with TPW and SST. The models reported in [24] (hereafter B03) and [25] (hereafter J06) use multivariate linear regression to estimate Q_a with multichannel TBs. Recently, the model reported in [28] (hereafter Y18) uses multivariate nonlinear regression to estimate Q_a with multichannel TBs and considers enhancement in high latitudes.

2.5. Ensemble Mean of Target Deep Neural Network Development

A model named Ensemble Mean of Target deep neural networks (EMTnet) was proposed to improve the satellite Q_a retrieval over the China Seas (Figure 3). The tool to build and perform the EMTnet model is TensorFlow (https://tensorflow.org/, accessed on 30 June 2022), an open-source machine learning library. The EMTnet model is generated from a large number of deep neural networks (DNNs). Each DNN is based on the error backpropagation (BP) algorithm [45] and consists of an input layer, three hidden layers, and an output layer. The BP algorithm takes advantage of the gradient descent and error backpropagation methods to adjust the connection weights of corresponding neurons to achieve its nonlinear learning ability. The EMTnet model works via the following steps:

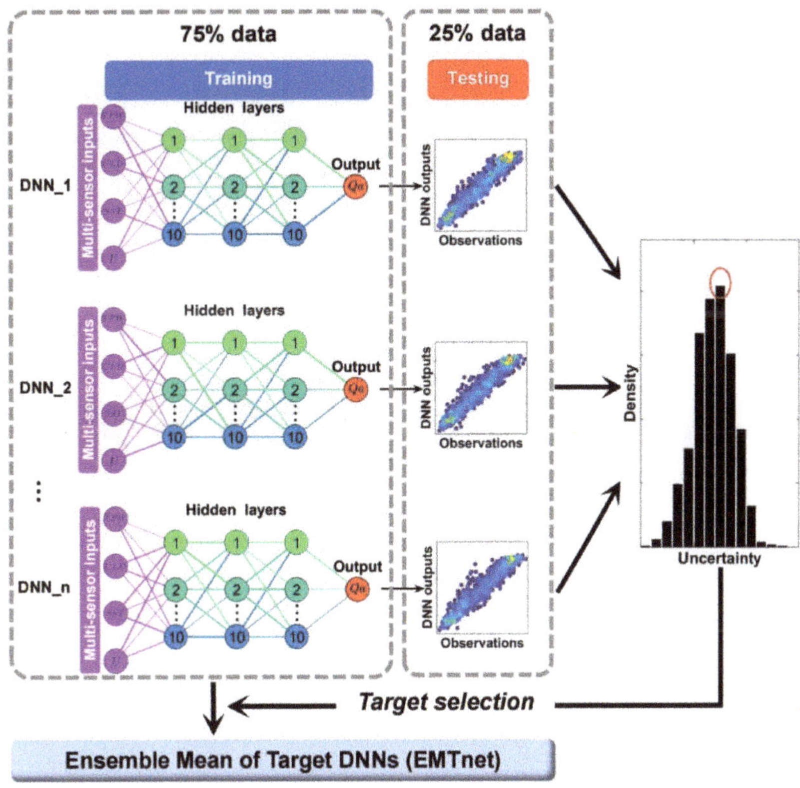

Figure 3. Architecture of the EMTnet. Each DNN_n ($n = 1, 2, 3, \ldots$) uses 75% randomly sampled data from all in situ observations as training data, while the remaining 25% are used as testing data.

(i) Four satellite-derived variables, TPW, CLW, U, and SST, are put in the input layer of each DNN. Note that different combinations of input variables can lead to different learning abilities of the EMTnet model, which are discussed in Section 4. Attempts using pure multichannel TBs or the mix of level-3 variables and multichannel TBs show similar or even slightly worse results.

(ii) Normalize these four input variables to the range between 0 and 1 according to their maximum and minimum values. It is known that the DNNs are quite sensitive to the magnitude difference in various input variables. Therefore, normalizing the input variables in advance can lead to better computational efficiency and results.

(iii) Determine the specific configuration of each DNN. The EMTnet model does not necessarily use the DNN approach. We have also tried other machine learning approaches such as the support vector machine (SVM) and the random forest (RF). It is found that the results of SVM and RF were comparable with those of the DNN, or sometimes slightly worse. The DNN approach is eventually employed to build the EMTnet model considering its better abilities in big data processing efficiency and nonlinear learning ability. Critical parameters are eventually determined through a number of tests to make the DNNs suitable for the learning task here. For instance, three hidden layers, each with ten neurons, are used. The number of iterations is set to 5000, and the learning rate is set to 0.005. In addition, the activation function is essential in forming nonlinear learning abilities for the DNNs. Three widely used activation functions, sigmoid, rectified linear units (ReLU), and hyperbolic tangent function (tanh), have been tested and compared. These activation functions have some defects, for example, the neuronal death for the ReLU activation function and a vanishing gradient for the sigmoid activation function. Activation function tanh is free of neuronal death, and the vanishing gradient problem has been alleviated to some extent. In addition, tanh has a faster convergence speed and a lower number of iterations. Preliminary tests show that the sigmoid and tanh activation functions perform better than the ReLU activation function in this study. Further inspections find that the accuracy of the result using the tanh activation function is 2~4% higher than that using the sigmoid activation function. Therefore, the tanh activation function is employed in this study. Note the current hyper-parameter tuning for each DNN is not unique. It only aims to make each DNN suitable for the learning task here. For those who are interested in the EMTnet model, they can adjust these hyper-parameter tunings according to specific tasks.

(iv) A total of 75% of Q_a observations are randomly sampled as training data, while the remaining 25% are used as independent testing data. As for traditional DNN training and testing, this operation is usually conducted once. However, it is found that samplings of different training and testing data can result in different uncertainty levels for the DNNs. To reduce the uncertainty of learning results caused by man-made operations in setting training and testing data, the Ensemble Mean approach is adopted. The EMTnet model trains n different DNNs with randomly sampled n sets of training and testing data to produce an ensemble of DNNs. In this study, the n is set to 1000. For each DNN, the testing data are used to compute the mean bias and root-mean-square error (RMSE). The sum of absolute values of mean bias and RMSE, that is, the absolute error, is taken as the uncertainty of each DNN.

According to the PDD of uncertainties constructed by the DNN ensemble, the top 10% DNNs with uncertainties falling into the highest density intervals are selected as target DNNs. The Ensemble Mean of Target DNNs is then used to produce the EMTnet model outputs.

3. Results

3.1. EMTnet Model Validation

The Q_a predictions from the EMTnet model and five existing models are intercompared with respect to Q_a observations. Three representative stations from the three China seas are selected according to their data quality, continuity, and integrity to facilitate the intercomparison. They are the Xisha Tower station in the SCS, the DH11 station in the ECS, and the HH09 station in the YS. A whole year of data from 2016 is used for each station to reduce the possible seasonal dependence of the results.

Figure 4 shows the scatter diagram between the Q_a predictions and observations and the corresponding correlation coefficient (CC), mean bias, and RMSE for each model in each sea. The CCs all exceed the 99% confidence level, varying from 0.59–0.91, 0.89 to 0.98, and 0.92 to 0.98 in the SCS, ECS, and YS, respectively. Among them, the EMTnet model has the highest CCs at each station. The mean biases and RMSEs present a large spread in different models and stations. In the SCS, the EMTnet model slightly overestimates Q_a by 0.06 g/kg, while the rest of the models underestimate Q_a from 1.07 (L86 model) to 7.33 (Y18 model) g/kg. In the ECS, except for the EMTnet model and the L86 model which overestimate Q_a by 0.13 and 0.78 g/kg, all the models underestimate Q_a by 0.41 (B03 model) to 4.12 (J06 model) g/kg. In the YS, except for the J99 and B03 models which underestimate Q_a by 3.22 and 1.37 g/kg, all the models overestimate Q_a by 0.06 (EMTnet model) to 4.32 (J06 model) g/kg. The RMSEs of these models in the SCS, ECS, and YS are varying from 1.10 (EMTnet model) to 2.72 (Y18 model) g/kg, 1.17 (EMTnet model) to 3.36 (Y18 model) g/kg, and 1.22 (EMTnet model) to 3.08 (J99 model) g/kg.

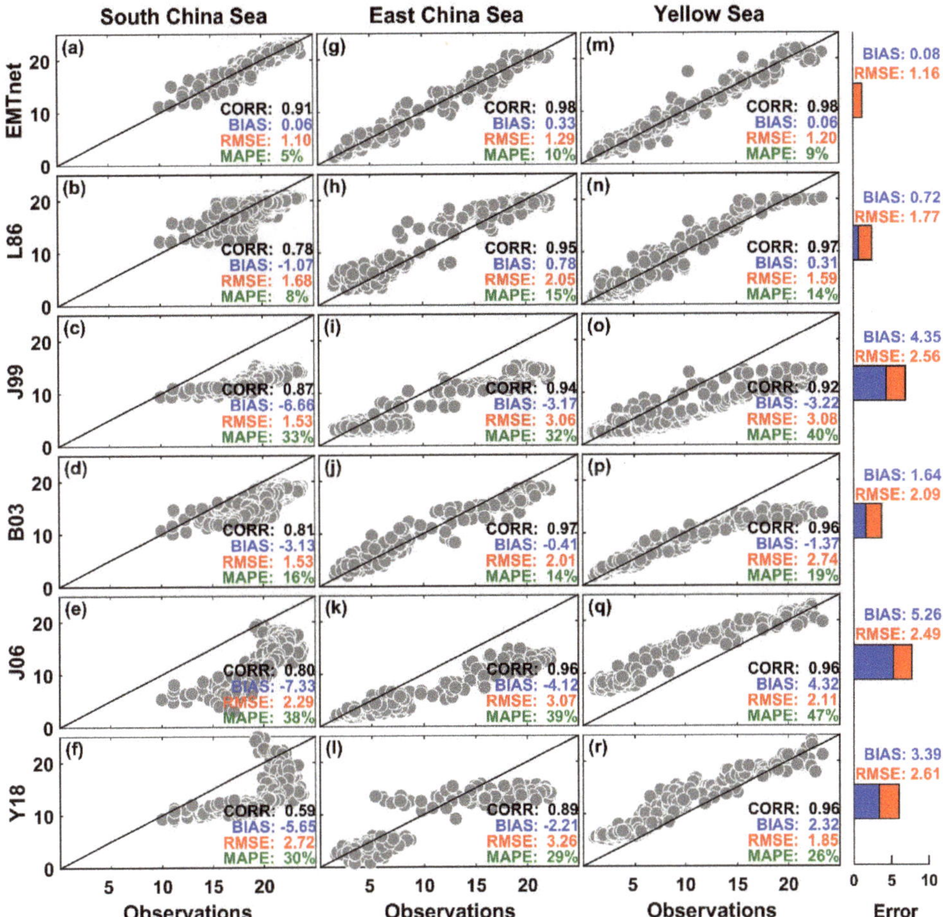

Figure 4. Comparisons between Q_a predictions (ordinate) and observations (abscissa). (**a–f**) The results for the Xisha buoy station in the SCS using the EMTnet model and five existing models summarized in Table 2. (**g–l**) and (**m–r**) The same as (**a–f**) but for the DH11 station in the ECS and the HH09 station in the YS, respectively. Bars on the rightmost side show the mean results of the three stations for each model. The units of mean bias and RMSE are g/kg.

The absolute values of the mean bias and RMSEs from the three seas are averaged for each model to compare their overall uncertainty level. The mean biases plus/minus RMSEs are 0.08 ± 1.16, 0.72 ± 1.77, 4.35 ± 2.56, 1.64 ± 2.09, 5.26 ± 2.49, and 3.39 ± 2.61 for the EMTnet, L86, J99, B03, J06, and Y18 models. Quantitatively, the EMTnet model has the lowest mean bias and RMSE on average. The EMTnet model also shows the least mean absolute percentage error (MAPE) at each station.

3.2. EMTnet Model Application

All the Q_a observations collected here are subsequently used to fully train the EMTnet model. Q_a predictions of the L86 model, which has the best performance among these five existing models, are used as a reference here. It is noted that both the EMTnet model and the L86 model take TPW as an input variable, which confirms the good relationship between TPW and Q_a over the China Seas. Figure 5a compares the Q_a predictions of the EMTnet model and the L86 model in the form of the Q_a–W relation. The dots determined by Q_a and TPW data cluster around the classical curve of the L86 model. The data density distribution shows that the majority of the data coincide well with the L86 model. Compared to the medians of Q_a observations, however, biases of the L86 model occur primarily under moderate Q_a values. For example, in the range of 10~20 g/kg, the L86 model overestimates Q_a from 0.44 to 1.98 g/kg, while the biases of the EMTnet model are almost negligible.

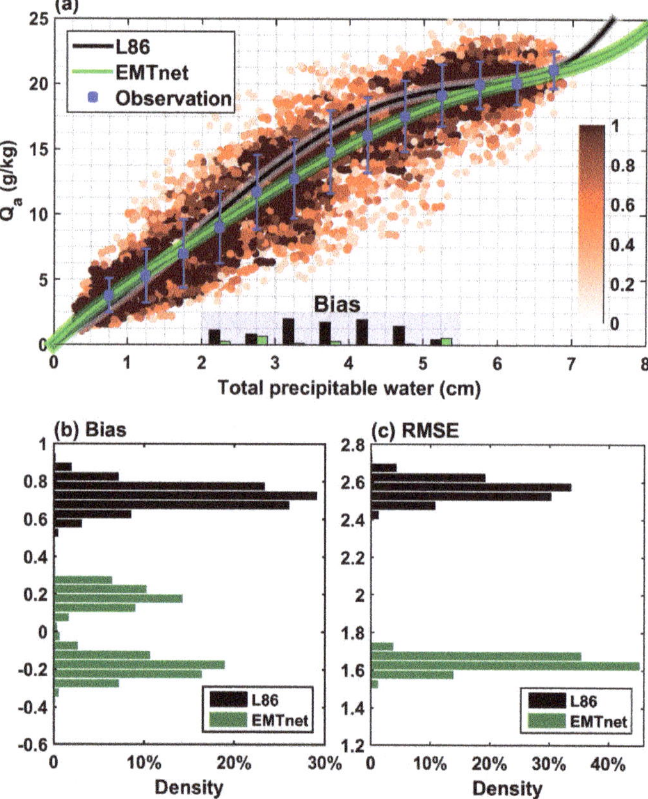

Figure 5. Comparisons of the EMTnet model and the L86 model. (**a**) The scatter diagram of Q_a (ordinate) and TPW (abscissa), while the bars show the model biases. In (**a**), the red dot denotes the data density in each 0.5 cm bin of TPW and 0.5 g/kg bin of Q_a with units of ‰. The blue square and error bar are the median and one STD of Q_a observations in each 0.5 cm bin of TPW. (**b**,**c**) The PDDs of the mean bias and RMSE in 1000 sets of DNN computations. Black and green lines and bars are the results of the L86 model and the EMTnet model. The units of mean bias and RMSE are g/kg.

Figure 5b,c depict the PDDs of the mean biases and RMSEs of the EMTnet model and the L86 model according to the 1000 samples of testing data used in the EMTnet model.

The mean biases of the L86 model are concentrated from 0.60 to 0.80 (87% data). The mean biases of the EMTnet model have two peaks, which are around 0.10 to 0.30 g/kg and −0.10 to −0.30 g/kg. On average, the mean biases of the L86 model and the EMTnet model are 0.72 ± 0.06 and -0.02 ± 0.19 g/kg. The RMSEs of the L86 and EMTnet models are concentrated from 2.45 to 2.65 (94% data) and 1.55 to 1.70 (95% data) g/kg, which are on average 2.56 ± 0.05 g/kg and 1.64 ± 0.04 g/kg. Thus, the EMTnet model reduces the mean bias and RMSE of the L86 model by approximately 0.70 and 0.90 g/kg, respectively. The mean bias for the EMTnet model in satellite Q_a retrieval is almost zero, reducing the RMSE of the L86 model by 36%.

Monthly gridded Q_a data over the China Seas were produced with satellite multi-sensor inputs by applying the fully trained EMTnet model. Both the input and output data are on $0.25° \times 25°$ gridded maps and span from 1990 to 2019. Figure 6a–d show the climatologies of Q_a from two satellite Q_a retrieval models (EMTnet and L86) and two reanalyses (ERA5 and NCEP2). Except for some differences in detail, apparent gradients from south to north in the mean state and seasonal variation in Q_a can be observed in all four data sources, which is higher in the south and lower in the north. Here, the intensity of seasonal variations is defined by the standard deviation from January to December.

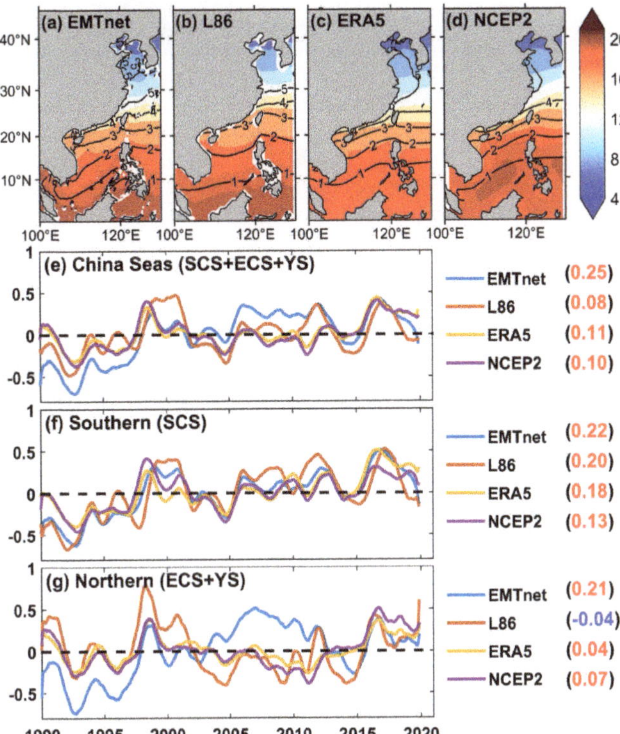

Figure 6. (**a–d**) Climatology of Q_a distributions (shading) over the China Seas from the EMTnet, L86, ERA5, and NCEP2. Contours denote the intensity of seasonal variation, which is defined by one standard deviation from January to December on each grid. (**e**) The time series of Q_a anomalies over the China Seas. (**f,g**) The same as (**e**) but for the southern (SCS) and northern (ECS and YS) sections of the China Seas. The time series in (**e–g**) have been applied to 13-point running average operations. The values in parentheses denote the long-term trend of Q_a during the period from 1990 to 2019 with unit of g/kg per decade. The units are g/kg.

The atmosphere's capacity to hold water vapor will increase in a warming climate according to the Clausius–Clapeyron relation [46]. The long-term trend of Q_a over the China Seas is depicted in Figure 6e–g. With the global warming in recent decades, the four data sources show consistent upward trends of Q_a (Figure 6e). However, the L86 model, ERA5, and NCEP2 have relatively lower climbing rates of Q_a, which are 0.08, 0.11, and 0.10 g/kg per decade, compared to the results of the EMTnet model (0.23 g/kg per decade). It is found that the long-term trends of Q_a are probably underestimated in both the southern (the SCS) and northern (the ECS and YS) sections, especially for the latter. In the southern section, the long-term trends of Q_a are 0.22, 0.20 0.18, and 0.13 g/kg per decade in the EMTnet model, L86 model, ERA5, and NCEP2 (Figure 6f). Except for NCEP2, all the data show quite similar climbing rates of Q_a. In contrast, the long-term trends of Q_a show larger spread in the northern section, which are 0.21, −0.04, 0.04, and 0.07 g/kg per decade in the EMTnet model, L86 model, ERA5, and NCEP2 (Figure 6g). Therefore, the possible underestimation of upward trends of Q_a in the L86 model, ERA5, and NCEP2 can be mainly attributed to their too-weak trends of Q_a variations in the northern section.

4. Discussions

The interpretability of deep learning is of great significance for its development and application. The EMTnet model and the L86 model, which take TPW as an input variable, are the top two best performing models investigated here. The possible reasons why the EMTnet model can further improve the satellite Q_a retrieval compared to the L86 model are discussed. Taking the result of the L86 model as a reference, eight sensitivity experiments (Exp1 to Exp8) are designed to examine whether the improvement of the EMTnet model is due to the model itself or the additional training data such as CLW, U, and SST compared to the L86 model. All the sensitivity experiments employ TPW as a fixed variable and adopt the eight combinations of CLW, U, and SST to construct their training data. Note that Exp1, including the full CLW, U, and SST information, is the result shown in Figure 5. The statistical results for Exp1 to Exp8 are shown in Table 3. As revealed in Table 3, all the experiments show improvements in satellite Q_a retrieval compared to the L86 model. They reduce the mean biases and RMSEs to varying extents. If only TPW data are used as training data as in the L86 model, the absolute error is reduced by 23% (Exp8). Taking into account CLW, U, and SST, the absolute errors are reduced by 35% (Exp5), 24% (Exp6), and 42% (Exp7). The three pairwise combinations of CLW, U, and SST are considered in Exp2 to Exp4. The reductions in absolute error in Exp2 to Exp4 are 42%, 47% and 48%.

Table 3. The mean bias and RMSE of each sensitivity experiment with EMTnet model. "Reference" refers to the result of the L86 model. In the nomenclature of Exp1 to Exp7, postfixes C, U, and S denote parameters CLW, U, and SST considered in the corresponding experiment, respectively. In Exp8, the postfix "none" means no additional information is considered. The percent change means the ratio of changes in absolute error compared to the reference value. The units of mean bias, RMSE, and absolute error are g/kg.

	Reference	Exp1_CUS	Exp2_CU	Exp3_CS	Exp4_US	Exp5_C	Exp6_U	Exp7_S	Exp8_None
Bias	0.72	−0.02	0.08	0.13	−0.05	−0.31	−0.22	−0.08	−0.18
RMSE	2.56	1.64	1.81	1.62	1.64	1.81	2.28	1.83	2.36
Absolute error	3.28	1.66	1.89	1.75	1.69	2.12	2.50	1.91	2.54
Percent change	-	−49%	−42%	−47%	−48%	−35%	−24%	−42%	−23%

The results of Exp2 to Exp7 suggest that factors CLW, U, and SST are helpful to improve the deep learning for satellite Q_u retrieval. If these three factors are superimposed together, a most significant improvement of 49% (Exp1) can be archived. The abilities of CLW, U, and SST in improving satellite Q_a retrieval are probably due to their roles in reflecting the environmental information. In the following, examples of the Q_a–W relation under different CLW, U, and SST conditions are shown in Figures 7–9, respectively.

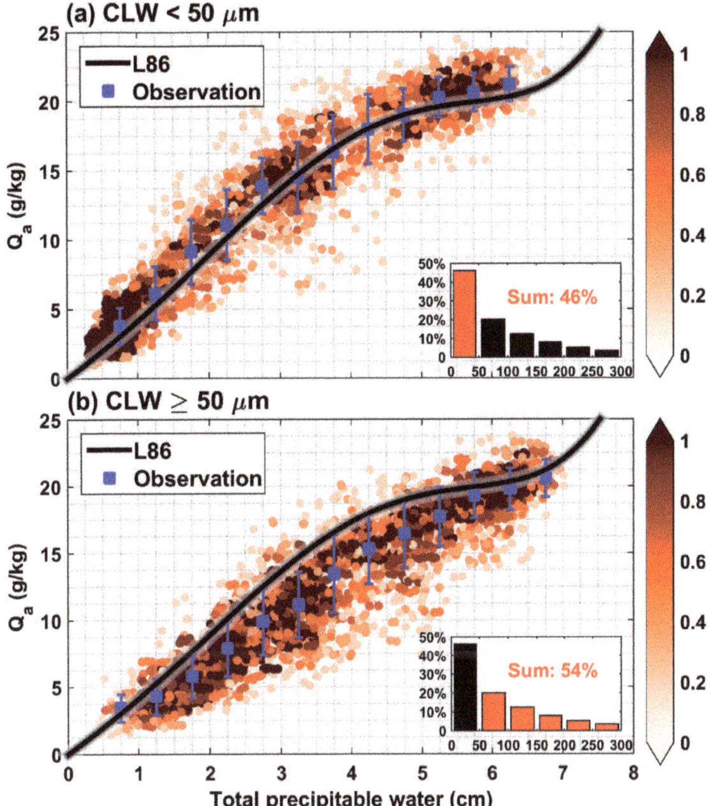

Figure 7. Scatter diagram of Q_a (ordinate) and TPW (abscissa) for data under conditions of CLW (**a**) below and (**b**) above 50 μm. The black line denotes the L86 model. The red dot denotes the data density in each 0.5 cm bin of TPW and 0.5 g/kg bin of Q_a with units of ‰. The blue square and error bar are the median and one STD of Q_a in each bin of 0.5 cm of TPW, respectively. The bar plot is the probability density distribution of CLW, with red bars representing the data range used in the corresponding panel.

The CLW is a measure of the total liquid water contained in a cloud in a vertical column of the atmosphere. As a component of TPW, the content of CLW will undoubtedly have an impact on the determination of the Q_a–W relation. However, none of the existing models for satellite Q_a retrieval incorporate cloud information. Figure 7 shows the Q_a–W relation under two conditions, one under CLW less than 50 μm (46% data) and the other greater than 100 μm (54% data). Note that the criterion of 50 μm here is only determined by the PDD of CLW, which ensures the data balance in both cases. Under a relatively low CLW (Figure 7a), the reference curve of the L86 model passes through most of the medians of Q_a observations, presenting high consistency with observations. Under a relatively high CLW (Figure 7b), however, the reference curve of the L86 model is nearly above all the medians of the Q_a observations. This result indicates that a relatively high CLW condition can interfere with the determination of the Q_a–W relation and lead to evident overestimations in satellite Q_a retrieval if no CLW information is considered.

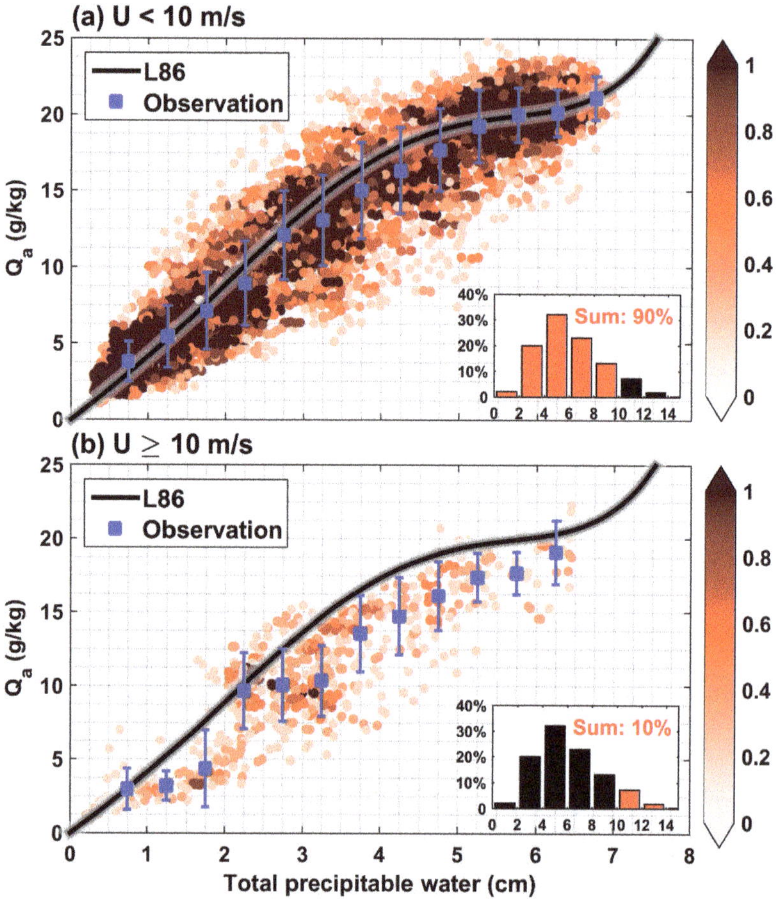

Figure 8. The same as Figure 7 but for two conditions of U.

The surface wind plays a vital role in reflecting the weather conditions near the sea surface and influencing the Q_a variations. Consequently, the surface wind is expected to be a potential factor that may improve the skill of satellite Q_a retrieval, which has not been considered in existing models. Figure 8 shows the Q_a–W relation with U less than 10 m/s (90% data) and greater than 10 m/s (10% data). It can be observed that the reference curve of the L86 model fits the Q_a observations well under lower to moderate U (Figure 8a). In contrast, the L86 model overestimates Q_a in almost all ranges of Q_a under relatively high U (Figure 8b). The different performances of the L86 model here imply that the Q_a–W relation is sensitive to surface wind conditions. One possible reason is that the water vapor distributions in the vertical column of the atmosphere are relatively stable under relatively weak U, which is conducive to the estimation of Q_a from TPW. As a portion of water vapor can be carried away by horizontal advection under relatively high U, the observed Q_a will be smaller than the model-predicted Q_a.

SST is an important variable that reflects information on the marine environment and underlying atmospheric surface. For example, under a relatively warm SST (Figure 9b), the predictions of the L86 model are more consistent with the observations. In contrast, Q_a is overestimated in the range of Q_a from 10 to 20 g/kg under rather cold SSTs (Figure 9a). A colder SST means weaker sea surface evaporation, which might lead to less actual moisture than predicted values. Therefore, it is suggested that attention should be given to the Q_a–W relation under different SST conditions.

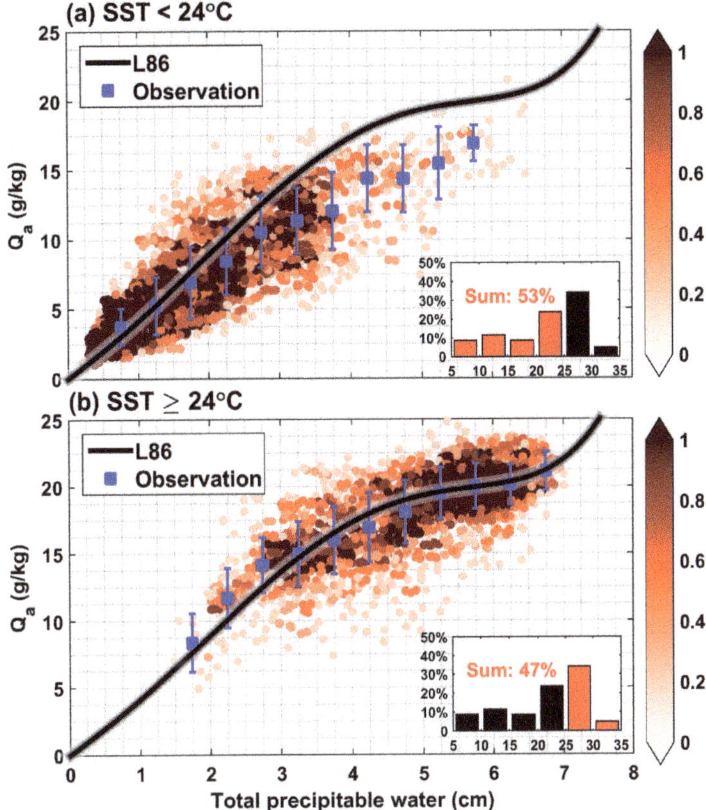

Figure 9. The same as Figure 7 but for two conditions of SST.

5. Conclusions

Previous satellite Q_a retrieval models suffer significant uncertainties due to factors such as model errors, scarce in situ observations, environmental interference, and so on. In this study, a deep learning approach, the EMTnet model, is proposed to improve the satellite Q_a retrieval over the China Seas. The EMTnet model is based on multiple DNNs, and the ensemble mean of target DNNs is used to produce output predictions, which can obtain more objective learning results when the observational data are quite divergent. The Q_a predictions from the EMTnet model outperform five existing models by nearly eliminating the mean bias and significantly reducing the RMSE. Compared to the L86 model, which has the best performance among five existing models, the outperformance of the EMTnet model can be attributed to two aspects. Firstly, if only TPW data are used as training data as in the L86 model, the EMTnet model reduces the absolute error by 23% (Table 3). This level of improvement can be attributed to the EMTnet model itself. Secondly, if CLW, U, and SST are added, the EMTnet model reduces the absolute error by 49% (Table 3). The approximately doubled increase in absolute error reduction benefits from the good interpretability of CLW, U, and SST on the determination of the Q_a–W relation. Note that the in situ observations are with uneven distribution in time and space, which could cause errors to the performance of the EMTnet model to a certain extent. The further development of the EMTnet model needs more in situ observations.

The fully trained EMTnet model has been applied to learn from remote sensing data to produce a 30-year monthly gridded Q_a data over the China Seas. It is found that current products perform well in depicting the mean state and seasonal variations in Q_a. However, they show much weaker upward trends of Q_a in the context of global warming, which are less than half of the EMTnet model result. As a locally well-trained and well-validated model, the different perspectives on the long-term variations in Q_a suggested by the EMTnet model may help to provide new understandings for humidity-related multi-disciplinary research over the China Seas. In addition, the EMTnet model is capable of merging Q_a observations from other regions as training data, which is to be applied to more oceans globally.

Author Contributions: Conceptualization, R.Z. and X.W.; methodology, R.Z. and W.G.; validation, R.Z. and W.G.; experiment, R.Z. and W.G.; formal analysis, R.Z. and W.G.; writing—original draft preparation, R.Z.; writing—review and editing, X.W.; visualization, R.Z.; supervision, X.W.; project administration, X.W. All authors have read and agreed to the published version of the manuscript.

Funding: This work was supported by the National Key R&D Program of China (Grant No. 2017YFA0603200), the Strategic Priority Research Program of Chinese Academy of Sciences (Grant Nos. XDB42000000), the National Natural Science Foundation of China (Grant Nos. 41925024, 41906178), the Key Special Project for Introduced Talents Team of Southern Marine Science and Engineering Guangdong Laboratory (Guangzhou) (GML2019ZD0306), the Innovation Academy of South China Sea Ecology and Environmental Engineering, the Chinese Academy of Sciences (ISEE2021ZD01), and the China-Sri Lanka Joint Center for Education and Research, Chinese Academy of Sciences. Rongwang Zhang was supported by the Independent Research Project Program of State Key Laboratory of Tropical Oceanography (LTOZZ2004).

Data Availability Statement: The remote sensing data are available at Remote Sensing Systems (www.remss.com, accessed on 30 June 2022). The ERA5 and NCEP2 data are available at https://www.ecmwf.int/en/forecasts/datasets/reanalysis-datasets/era5 (accessed on 30 June 2022) and https://psl.noaa.gov/data/gridded/data.ncep.reanalysis2.html (accessed on 30 June 2022), respectively. The in situ observational data can be acquired via the link https://pan.cstcloud.cn/s/4hLRaDMoRfM (accessed on 30 June 2022). The code and datasets of the EMTnet model can be acquired via the link https://pan.cstcloud.cn/s/YjAZM1vUQOA (accessed on 30 June 2022).

Acknowledgments: We thank Jun-Yue Yan for providing the in situ data collected by the program National Key Basic Research Program of China (Grant No. 2006CB403604). Acknowledgment is given for the data support of the Xisha Marine Environment National Observation and Research Station, the Marine Meteorological Science Experiment Base at Bohe of ITMM, CMA, the Yellow Sea Ocean Observation and Research Station of OMORN, and the East China Sea Ocean Observation and Research Station of OMORN.

Conflicts of Interest: The authors declare no conflict of interest.

References

1. Baumgartner, A.; Reichel, E. *The World Water Balance*; Elsevier: New York, NY, USA, 1975; p. 179.
2. Chahine, M.T. The hydrological cycle and its influence on climate. *Nature* **1992**, *359*, 373–380. [CrossRef]
3. Trenberth, K.E.; Smith, L.; Qian, T.; Dai, A.; Fasullo, J. Estimates of the global water budget and its annual cycle using observational and model data. *J. Hydrometeorol.* **2007**, *8*, 758–769. [CrossRef]
4. Yu, L. Global variations in oceanic evaporation (1958–2005): The role of the changing wind Speed. *J. Clim.* **2007**, *20*, 5376–5390. [CrossRef]
5. Lorenz, D.J.; DeWeaver, E.T.; Vimont, D.J. Evaporation change and global warming: The role of net radiation and relative humidity. *J. Geophys. Res.* **2010**, *115*, D20118. [CrossRef]
6. Jin, X.; Yu, L.; Jackson, D.L.; Wick, G.A. An improved near-surface specific humidity and air temperature climatology for the SSM/I satellite period. *J. Atmos. Ocean. Technol.* **2015**, *32*, 412–433. [CrossRef]
7. Tomita, H.; Kubota, M.; Cronin, M.F.; Iwasaki, S.; Konda, M.; Ichikawa, H. An assessment of surface heat fluxes from J-OFURO2 at the KEO and JKEO sites. *J. Geophys. Res.* **2010**, *115*, C03018. [CrossRef]
8. Kinzel, J.; Fennig, K.; Schröder, M.; Andersson, A.; Bumke, K.; Hollmann, R. Decomposition of random errors inherent to HOAPS-3.2 near-surface humidity estimates using multiple triple collocation analysis. *J. Atmos. Ocean. Tech.* **2016**, *33*, 1455–1471. [CrossRef]

9. Brunke, M.A.; Wang, Z.; Zeng, X.; Bosilovich, M.; Shie, C.-L. An assessment of the uncertainties in ocean surface turbulent fluxes in 11 reanalysis, satellite-derived, and combined global datasets. *J. Clim.* **2011**, *24*, 5469–5493. [CrossRef]

10. Kent, E.C.; Berry, D.I.; Prytherch, J.; Roberts, J.B. A comparison of global marine surface-specific humidity datasets from in situ observations and atmospheric reanalysis. *Int. J. Climatol.* **2014**, *34*, 355–376. [CrossRef]

11. Robertson, F.R.; Bosilovich, M.G.; Roberts, J.B.; Reichle, R.H.; Adler, R.; Ricciardulli, L.; Berg, W.; Huffman, G.J. Consistency of estimated global water cycle variations over the satellite era. *J. Clim.* **2014**, *27*, 6135–6154. [CrossRef]

12. Rodell, M.; Beaudoing, H.K.; L'Ecuyer, T.S.; Olson, W.S.; Famiglietti, J.S.; Houser, P.R.; Adler, R.; Bosilovich, M.G.; Clayson, C.A.; Chambers, D.; et al. The observed state of the water cycle in the early twenty-first century. *J. Clim.* **2015**, *28*, 8289–8318. [CrossRef]

13. Liman, J.; Schröder, M.; Fenning, K.; Andersson, A.; Hollmann, R. Uncertainty characterization of HOAPS 3.3 latent heat-flux-related parameters. *Atmos. Meas. Tech.* **2018**, *11*, 1793–1815. [CrossRef]

14. Roberts, J.B.; Clayson, C.A.; Robertson, F.R. Improving near-surface retrievals of surface humidity over the global open oceans from passive microwave observations. *Earth Space Sci.* **2019**, *6*, 1220–1233. [CrossRef]

15. Zhang, R.; Wang, X.; Wang, C. On the simulations of global oceanic latent heat flux in the CMIP5 multimodel ensemble. *J. Clim.* **2018**, *31*, 7111–7128. [CrossRef]

16. Tomita, H.; Hihara, T.; Kubota, M. Improved satellite estimation of near-surface humidity using vertical water vapor profile information. *Geophys. Res. Lett.* **2018**, *45*, 899–906. [CrossRef]

17. Liu, W.T.; Niiler, P.P. Determination of monthly mean humidity in the atmospheric surface layer over oceans from satellite data. *J. Phys. Oceanogr.* **1984**, *14*, 1451–1457. [CrossRef]

18. Liu, W.T. Statistical relation between monthly precipitable water and surface-level humidity over global oceans. *Mon. Weather Rev.* **1986**, *114*, 1591–1602. [CrossRef]

19. Hsu, S.A.; Blanchard, B.W. The relationship between total precipitable water and surface-level humidity over the sea surface: A further evaluation. *J. Geophys. Res.* **1989**, *94*, 14539–14545. [CrossRef]

20. Schulz, J.; Schluessel, P.; Grassl, H. Water vapour in the atmospheric boundary layer over oceans from SSM/I measurements. *Int. J. Remote Sens* **1993**, *14*, 2773–2789. [CrossRef]

21. Schlüssel, P.; Schanz, L.; Englisch, G. Retrieval of latent heat flux and longwave irradiance at the sea surface from SSM/I and AVHRR measurements. *Adv. Space Res.* **1995**, *16*, 107–116. [CrossRef]

22. Chou, S.-H.; Atlas, R.M.; Shie, C.-L.; Ardizzone, J. Estimates of surface humidity and latent heat fluxes over oceans from SSM/I Data. *Mon. Weather Rev.* **1995**, *123*, 2405–2425. [CrossRef]

23. Jones, C.; Peterson, P.; Gautier, C. A new method for deriving ocean surface specific humidity and air temperature: An artificial neural network approach. *J. Appl. Meteorol.* **1999**, *38*, 1229–1245. [CrossRef]

24. Bentamy, A.; Katsaros, K.B.; Mestas-Nunoz, A.M.; Drennan, W.M.; Forde, E.B.; Roquet, H. Satellite estimates of wind speed and latent heat flux over the global oceans. *J. Clim.* **2003**, *16*, 637–656. [CrossRef]

25. Jackson, D.L.; Wick, G.A.; Bates, J.J. Near-surface retrieval of air temperature and specific humidity using multi-sensor microwave satellite observations. *J. Geophys. Res.* **2006**, *111*, D10306. [CrossRef]

26. Kubota, M.; Hihara, T. Retrieval of surface air specifific humidity over the ocean using AMSR-E measurements. *Sensors* **2008**, *8*, 8016–8026. [CrossRef]

27. Jackson, D.L.; Wick, G.A.; Robertson, F.R. Improved multisensor approach to satellite-retrieved near-surface specific humidity observations. *J. Geophys. Res.* **2009**, *114*, D16303. [CrossRef]

28. Yu, L.; Jin, X. A regime-dependent retrieval algorithm for near-surface air temperature and specific humidity from multi-microwave sensors. *Remote Sens. Environ.* **2018**, *215*, 199–216. [CrossRef]

29. Gao, Q.; Wang, S.; Yang, X. Estimation of surface air specific humidity and air–sea latent heat flux using FY-3C microwave observations. *Remote Sens.* **2019**, *11*, 466. [CrossRef]

30. Wang, D.; Zeng, L.; Li, X.; Shi, P. Validation of satellite-derived daily latent heat flux over the South China Sea, compared with observations and five products. *J. Atmos. Ocean. Technol.* **2013**, *30*, 1820–1832. [CrossRef]

31. Wang, X.; Zhang, R.; Huang, J.; Zeng, L.; Huang, F. Biases of five latent heat flux products and their impacts on mixed-layer temperature estimates in the South China Sea. *J. Geophys. Res. Oceans* **2017**, *122*, 5088–5104. [CrossRef]

32. Roberts, J.B.; Clayson, C.A.; Robertson, F.R.; Jackson, D.L. Predicting near-surface atmospheric variables from Special Sensor Microwave/Imager using neural networks with a first-guess approach. *J. Geophys. Res.* **2010**, *115*, D19113. [CrossRef]

33. Reichstein, M.; Camps-Valls, G.; Stevens, B.; Juang, M.; Denzier, J.; Carvalhais, N. Prabhat Deep learning and process understanding for data-driven earth system science. *Nature* **2019**, *566*, 195–204. [CrossRef]

34. Liu, B.; Li, X.; Zheng, G. Coastal inundation mapping from bitemporal and dual-polarization SAR imagery based on deep convolutional neural networks. *J. Geophys. Res. Oceans* **2019**, *124*, 9101–9113. [CrossRef]

35. Li, X.; Liu, B.; Zheng, G.; Zhang, S.; Liu, Y.; Gao, L.; Liu, Y.; Zhang, B.; Wang, F. Deep-learning-based information mining from ocean remote-sensing imagery. *Nat. Sci. Rev.* **2020**, *7*, 1585–1606. [CrossRef]

36. Wang, Y.; Li, X.; Song, J.; Li, X.; Zhong, G.; Zhang, B. Carbon sinks and variations of pCO_2 in the Southern Ocean from 1998 to 2018 based on a deep learning approach. *IEEE J. Sel. Top. Appl. Earth Observ. Remote Sens.* **2021**, *14*, 3495–3503. [CrossRef]

37. Zhang, R.; Huang, J.; Wang, X.; Zhang, J.A.; Huang, F. Effects of precipitation on sonic anemometer measurements of turbulent Fluxes in the atmospheric surface layer. *J. Ocean Univ. China* **2016**, *15*, 389–398. [CrossRef]

38. Zhou, F.; Zhang, R.; Shi, R.; Chen, J.; He, Y.; Wang, D.; Xie, Q. Evaluation of OAFlux datasets based on in situ air–sea flux tower observations over the Yongxing Islands in 2016. *Atmos. Meas. Tech.* **2018**, *11*, 6091–6106. [CrossRef]
39. Fairall, C.W.; Bradley, E.F.; Hare, J.E.; Grachev, A.A.; Edson, J.B. Bulk parameterization of air-sea fluxes: Updates and verification for the COARE algorithm. *J. Clim.* **2003**, *16*, 571–591. [CrossRef]
40. Wentz, F.J. *SSM/I Version-7 Calibration Report (Report Number 011012)*; Remote Sensing Systems: Santa Rosa, CA, USA, 2013; 46p.
41. Zou, C.-Z.; Wang, W. *Climate Algorithm Theoretical Basis Document (C-ATBD)—AMSU Radiance Fundamental Climate Data Record Derived From Integrated Microwave Inter-calibration Approach*; Technical Report; NOAA: Asheville, NC, USA, 2013.
42. Zou, C.-Z.; Hao, X. AMSU-A Brightness Temperature FCDR—Climate Algorithm Theoretical Basis Document. NOAA Climate Data Record Program CDRP-ATBD-0345, Rev. 2.0. 2016. Available online: http://www.ncdc.noaa.gov/cdr/operationalcdrs.html (accessed on 30 June 2022).
43. Hersbach, H.; Bell, B.; Berrisford, P.; Hirahara, P.; Horanyi, S.; Munoz-Sabater, J.; Nicolas, J.; Peubey, C.; Radu, R.; Schepers, D.; et al. The ERA5 global reanalysis. *Quart. J. R. Meteor. Soc.* **2020**, *146*, 1999–2049. [CrossRef]
44. Kanamitsu, M.; Ebisuzaki, W.; Woollen, J.; Yang, S.-K.; Hnilo, J.J.; Fiorino, M.; Potter, G.L. NCEP–DOE AMIP-II reanalysis (R-2). *Bull. Am. Meteorol. Soc.* **2002**, *83*, 1631–1644. [CrossRef]
45. Rumelhart, D.E.; Hinton, G.E.; Williams, R.J. Learning internal representations by error propagation. *Read. Cogn. Sci.* **1988**, *323*, 399–421. [CrossRef]
46. Held, I.M.; Soden, B.J. Robust responses of the hydrological cycle to global warming. *J. Clim.* **2006**, *19*, 5686–5699. [CrossRef]

remote sensing

MDPI

Article

An Algorithm to Bias-Correct and Transform Arctic SMAP-Derived Skin Salinities into Bulk Surface Salinities

David Trossman [1,2,3,*] and Eric Bayler [4]

1 Global Science & Technology, NOAA/NESDIS Center for Satellite Applications and Research (STAR), Greenbelt, MD 20770, USA
2 Department of Oceanography & Coastal Sciences, Louisiana State University, Baton Rouge, LA 70803, USA
3 Center for Computation & Technology, Louisiana State University, Baton Rouge, LA 70803, USA
4 NOAA/NESDIS Center for Satellite Applications and Research (STAR), College Park, MD 20740, USA; eric.bayler@noaa.gov
* Correspondence: dtrossman@lsu.edu

Abstract: An algorithmic approach, based on satellite-derived sea-surface ("skin") salinities (SSS), is proposed to correct for errors in SSS retrievals and convert these skin salinities into comparable in-situ ("bulk") salinities for the top-5 m of the subpolar and Arctic Oceans. In preparation for routine assimilation into operational ocean forecast models, Soil Moisture Active Passive (SMAP) satellite Level-2 SSS observations are transformed using Argo float data from the top-5 m of the ocean to address the mismatch between the skin depth of satellite L-band SSS measurements ($\sim$1 cm) and the thickness of top model layers (typically at least 1 m). Separate from the challenge of Argo float availability in most of the subpolar and Arctic Oceans, satellite-derived SSS products for these regions currently are not suitable for assimilation for a myriad of other reasons, including erroneous ancillary air-sea forcing/flux products. In the subpolar and Arctic Oceans, the root-mean-square error (RMSE) between the SMAP SSS product and several in-situ salinity observational data sets for the top-5 m is greater than 1.5 pss (Practical Salinity Scale), which can be larger than their temporal variability. Thus, we train a machine-learning algorithm (called a Generalized Additive Model) on in-situ salinities from the top-5 m and an independent air-sea forcing/flux product to convert the SMAP SSS into bulk-salinities, correct biases, and quantify their standard errors. The RMSE between these corrected bulk-salinities and in-situ measurements is less than 1 pss in open ocean regions. Barring persistently problematic data near coasts and ice-pack edges, the corrected bulk-salinity data are in better agreement with in-situ data than their SMAP SSS equivalent.

Keywords: salinity; SMAP; skin-effect; bias; air-sea; Arctic; ocean; machine-learning

Citation: Trossman, D.; Bayler, E. An Algorithm to Bias-Correct and Transform Arctic SMAP-Derived Skin Salinities into Bulk Surface Salinities. *Remote Sens.* **2022**, *14*, 1418. https://doi.org/10.3390/rs14061418

Academic Editors: Veronica Nieves, Ana B. Ruescas, Raphaëlle Sauzède and Ali Khenchaf

Received: 3 January 2022
Accepted: 9 March 2022
Published: 15 March 2022

1. Introduction

In this paper, we present an algorithm to bias-correct and convert sea-surface salinity (SSS) fields from L-band passive microwave satellite retrievals at northern high latitudes into surface salinity fields that can be assimilated by ocean modeling systems. Satellite L-band passive microwave observations (Section 2.1) have demonstrated information potential for nearly global depiction of SSS. Satellite SSS derived from NASA's Aquarius mission and Argo products are generally consistent to about 60°N/S, particularly when compared over coarser resolutions [1]. The collected satellite observations from the European Space Agency (ESA) Soil Moisture and Ocean Salinity (SMOS) mission (greater than 10 years) and National Aeronautics and Space Administration (NASA) Soil Moisture Active Passive (SMAP) mission (greater than 5 years) have been summarized and shown to generally agree with in-situ observations [2]. However, monitoring SSS in the Arctic is more challenging. There are many challenges associated with using SMAP-derived SSS observations to monitor Arctic freshwater changes [3], including its accuracy in the colder waters at such high latitudes and utility to monitor variations in the vicinity of ice and/or coasts. The sources of these accuracy problems include systematic errors in the ancillary wind fields or the wind roughness model that are used in the surface roughness correction, ancillary sea surface temperature (SST) fields, and ancillary sea-ice products that are used in sea-ice contamination correction, all of which are used in the SSS retrievals. Adding to these challenges, temporally-varying SSS drift-like behavior exists in the SMOS data, which, at least partially, accounts for the inability of SMOS to characterize the annual cycle of SSS in the subpolar North Atlantic Ocean [4]. Because SMAP data suffer from fewer problems than SMOS data in the northern high latitudes and a previous study has achieved an improved surface salinity product based on SMOS data [5], we demonstrate the utility of our algorithmic approach using SMAP data.

One difficulty with assimilating SSS into ocean models is the mismatch between the depth levels that L-band satellites observe (top centimeter of the ocean) and the resolution of the top ocean model layer (typically at least one meter). This mismatch can be seen in the in-situ salinity data collected during multiple observational campaigns (e.g., [6]), including two NASA-sponsored Salinity Processes in the Upper-ocean Regional Study (SPURS) campaigns [7,8]: SPURS-1 in the subtropical North Atlantic Ocean and SPURS-2 in the eastern equatorial Pacific Ocean, with a planned third campaign in the Arctic Ocean. We examine whether differences can be reconciled between the "skin" (satellite-derived from the top centimeter) salinity and "bulk" (in-situ at 1–5 m) salinity at high latitudes using an algorithmic approach. If there is a difference in salinity between the top-centimeter and top-meter, or so, of the ocean due to evaporation, precipitation, runoff, ice melt, or freezing/brine rejection effects, then a correction is needed in order to assimilate the satellite SSS observations. Under evaporation, a theory [9] argues for the existence of a salty, cool, sea surface skin layer; however, this theory was revisited after the creation of an air-sea exchange data set [10]. The latter study found that the cooler and saltier skin layer is always statically unstable, and that cooling controls the tendency to overturn, after which it takes 90 times longer to reestablish the skin salinity than the skin temperature. The skin-effect from this theory depends on several air-sea forcing/flux fields, and SSS retrievals depend on some of the same fields using ancillary data. The availability of in-situ and air-sea forcing/flux data sets is, therefore, crucial to the conversion of skin salinity to bulk salinity for use in ocean data assimilation models. The in-situ data are available from several campaigns at high latitudes (Section 2.2), but remain sparse.

Cold-induced biases in satellite-derived skin SSS observations, however, present a problem at high latitudes for data-assimilating ocean models. Due to the strong dependency of density on salinity in the polar regions, SSS can have a significantly higher impact than SST on constraining the modeled circulation of the Arctic Ocean, notably the waters that overlie the warm, salty Atlantic water mass transiting the Norwegian Sea into the Arctic Ocean. Placing upstream constraints on this Atlantic water can significantly impact on the heat imported to the base of the mixed layer along the shelf-basin slopes in the Eastern Arctic, which subsequently impacts the mixed-layer salinity and sea-ice melting in this region [11,12]. Subsequent sea-ice melt, in turn, can influence the Arctic Ocean's salinity [13], which, when exported to the subpolar North Atlantic Ocean, can have consequences for the Atlantic Meridional Overturning Circulation (AMOC) [14–17]. Several theories have been developed to explain the complicated relationships between the sea-ice state, the ocean's salinity, and the circulation in the Arctic context [18–23]. Thus, if SSS can be better constrained in ocean models, then there is the potential to unravel and understand the relationships between sea ice and ocean properties in the Arctic Ocean.

In this study, our primary objectives are to: (1) assess associated biases in a SMAP-derived SSS product relative to in-situ observations in the top-5 m, (2) characterize the statistics of SMAP-derived SSS observations, and (3) assess whether an algorithm to correct for biases and convert the SMAP-derived (skin) salinities to near-surface (bulk) salinities improves their agreement with in-situ observations, thereby permitting ocean data assimilation systems to exploit northern high-latitude SMAP-derived skin SSS. Currently, SMOS and SMAP observations can be adjusted to correlate with Argo float salinity observations in the top-5 m; but, because there are no Argo floats in the Arctic Ocean, we need an alternative method to effectively correct for both satellite measurement biases, particularly in colder waters, and skin-induced effects that are inconsistent with the model's top-layer thickness. (We refer to the resulting bulk surface salinity product as "corrected" hereafter.) To achieve this objective, we first identify potential issues with Arctic SSS, characterizing the statistics of SMAP SSS and how those data compare with in-situ observations for 2015 through 2019. We then demonstrate the utility of our algorithm for the corrected bulk surface salinity by comparing the root-mean-square error (RMSE) of the algorithm's estimates, relative to the in-situ observations, with the RMSE of the SMAP SSS product, relative to the same in-situ observations. Finally, we characterize the statistics of the algorithm's corrected bulk salinity product derived from SMAP SSS data. The same exercises can be conducted using a SMOS data product. In demonstrating that our algorithm improves the validity of SMAP data, we note that the algorithm can easily be extended to SMOS data. We highlight differences between the original SMAP product and the corrected bulk salinity product derived from SMAP data.

2. Data and Methods

2.1. Satellite Data

Satellite passive microwave retrievals of ocean salinity exploit the L-Band (1.41 GHz), with SMOS employing a synthetic aperture interferometer and SMAP using a scanning radiometer. Because of the different instrumentation, they have different strengths and weaknesses. For example, SMOS retrievals are known to be challenged near land and the ice-pack edge due to greater off-viewing-angle sea-ice contamination of the salinity signal [24]. SMOS spatial resolution depends on incident angle, spanning from about 40 km near nadir to about 60 km near 55-degrees incident angle [25], whereas SMAP has a fixed incident angle of 40 degrees, with spatial resolution around 40 km [26]. Higher SSS accuracy can be achieved through spatial averaging. Global coverage from SMOS is approximately 3 days, with suborbital repeats being 23 days and the exact repeat period being 149 days [25]. Global coverage from SMAP is exactly 8-days, with nominal global coverage also every 3 days [26]. The 3-day periods mostly have a single data value for each data set so we choose to average over longer periods for each gridded file. Higher SSS accuracy can be achieved through spatial averaging.

In processing the Level-2 SMAP SSS data [27], we perform the following steps. First, we reduce latitude/longitude coordinates in numerical precision to a single digit after the decimal place. This decreases the required computation time and is inconsequential because a resolution of less than tens of kilometers in the horizontal with the raw Level-2 SSS data are not possible without a downscaling technique. This reduced numerical precision allows us to save memory and effectively bin the data. Next, the SMAP data were gridded at 50 km by 50 km resolution by averaging the values over 8 days (chosen because this is the satellite's repeat interval) and aggregating all of the data over 50 km by 50 km boxes. Then, for each grid point, the 8-day averages of SMAP SSS had their trends and seasonal cycles removed over their respective time periods, by using a Generalized Additive Model (GAM—Section 2.4) to fit SSS with a smooth function of time. The residual time series at each grid point had a variance and skewness computed. Lastly, we calculate two statistics from SSS data to examine their spatial distributions relative to marginal ice zones. We compute an anomalously large SSS value statistic at each point on the grid by counting the number of times where the median SSS is exceeded by more than three times the SSS standard deviation. We do not show the same for anomalously small SSS values because the distribution of SSS values tend to be negatively skewed (i.e., the distribution's tail is longer towards smaller SSS values). We also compute mixing length scales according to an established theory [28]. For these mixing length scales, we supplement SMAP data with a gridded product for horizontal spatial gradients of SSS, which comes from the Level-3 daily Earth & Space Research SMOS data.

We compare Level-2 SMAP data with in-situ data, described in the following subsection, from the top-5 m north of 55°N. We do this by searching for data points from SMAP that were within 50 km and 3.5 days of each in-situ observation's horizontal location and sampled time. These selected points, spatio-temporally local to the in-situ data, are used as training data for the statistical model described below.

2.2. In-Situ Data

In-situ data provide evidence that skin-salinity values from satellites should be converted into bulk-salinity values in multiple regions around the world. Observations from the SPURS-2 campaign demonstrate that precipitation (or other freshwater flux) into the sea surface have an influence on the skin salinity within the top half-meter of the water column, finer than the vertical resolution of most ocean models. Whether this skin-to-bulk salinity conversion is necessary at high latitudes is currently unknown. However, an additional issue that may be more important at high-latitudes is the error associated with retrieval algorithms for satellite-derived SSS due to low signal-to-noise ratios in cold brightness temperature environments [5].

In order to determine whether the skin-effect and/or biases in satellite-derived high-latitude SSS need to be corrected, we need to use in-situ data in the top-5 m of the water column in subpolar and Arctic Ocean locations. We make use of multiple in-situ data sets, including the salinity and pressure observations from Saildrone [29], Oceans Melting Greenland (OMG) [30], ship-based CTD hydrographic transects, and NOAA's National Centers for Environmental Information (NCEI) Surface Underway Marine Database (SUMD; "Underway" hereafter). The Saildrone sent to the Arctic in 2019 is a wind-powered, unmanned surface water vehicle. The data the Saildrone collect are transmitted via satellite and are available to both researchers and the public. The Underway data comprises uniformly, quality-controlled in-situ sea-surface measurements from thermosalinographs, involving more than 450 ships and unmanned surface vehicles. These data are so extensive that, even when we include all data sets available over the length of the SMOS Arctic time series, the number of data points in the Underway database are orders of magnitude larger than any other data sets used here. The OMG data comprise both CTD and Airborne eXpendable CTD (AXCTD) (CTD probes dropped from aircraft) data collected during the summer months, 2015 to the present, with about 250 probes being dropped each year. Ninety-two ship-based CTD hydrographic transect data sets are used here. The OMG and ship-based CTD hydrographic transect data are subsampled such that we only use data within 5 m of the sea surface.

2.3. OAFlux Air-Sea Forcing/Flux Data

We supplement the in-situ salinity data with air-sea forcing/flux fields from the OAFlux product [31,32]. To get the wind stress, data from six Special Sensor Microwave/Imager (SSM/I) sensors, two Special Sensor Microwave Imager/Sounder (SSMIS) sensors, Advanced Microwave Scanning Radiometer for EOS (AMSR-E), WindSat, QuikSCAT, and Advanced Scatterometer (ASCAT-A) were used [33]. The footprint resolution various across the SSM/I sensors is finer with higher frequencies (along × cross-track): 69 km × 43 km at 19 GHz, 50 km × 40 km at 22 GHz, 37 km × 28 km at 37 GHz, and 15 km × 13 km at 85 GHz. One-hundred twenty-six buoy time series were used to calibrate different SSM/I sensors due to known issues with drift. The footprint resolution of the conically scanning SSMIS varies from 14 km × 13 km at 183 GHz to 70 km × 42 km at 19 GHz; for ASMR-E varies from 75 km × 43 km at 6.9 GHz to 6 km × 4 km at 89 GHz; and for WindSat is 40 km × 60 km at 6.8 GHz, 25 km × 38 km at 10.7 GHz, 15 km × 13 km at 18.7 GHz, 12 km × 20 km at 23.8 GHz, and 8 km × 13 km at 37 GHz. The elliptical footprint size of the antenna for QuikSCAT is about 24 km × 31 km at inner beam. For ASCAT, an operational product at spatial resolutions of 25–34 or 50 km can be generated on a nodal grid of 12.5 or 25 km. Rain-contaminated retrievals of wind from microwave sensors were discarded because of known problems under rainy conditions. Surface winds from the European Centre for

Medium-Range Weather Forecasts (ECMWF) Re-Analysis (ERA) interim project [34] and the Climate Forecast System Reanalysis (CFSR) from the National Centers for Environmental Prediction (NCEP) [35] were used to as background data in the synthesis. Sensible and latent heat fluxes were similarly derived using satellite observations (the advanced microwave sounding unit A (AMSU-A) and the Special Sensor Microwave Imager) and reanalyses where and when satellite observations were not available [36], except surface fluxes were computed from the COARE bulk flux algorithm [37]. Evaporation is directly proportional to the latent heat flux and scaled by the inverse product of the density of sea water and the latent heat of vaporization. To get the surface humidity and temperature fields, brightness temperature observations from four vertically polarized channels at 19, 22, and 37 GHz from SSM/I and SSMIS and 52 GHz from AMSU-A were used and related to buoy observations of surface humidity and temperature at 2–3 m above the sea surface [38,39]. The surface humidity and temperature fields were height-adjusted to 2 m using the COARE algorithm [37]. Sea surface temperatures, derived from the global operational NOAA product at 25 km based on AMSR-E and the advanced very high resolution radiometer (AVHRR) [40], were used as constraints for the synthesis of surface humidity and temperature.

The theory of the least-variance linear statistical estimation [41,42] was the basis for the methodology of the OAFlux objective synthesis, using all of the above data constraints. This approach allows the formulation of a least squares estimator (i.e., the cost function) to include both data from different sources and a priori information. For the optimization of each of the turbulence flux fields, a conjugate-gradient method was used [31]. The 25 km resolution of the OAFlux product was chosen as a compromise between being able to satisfy the cost function and the data coverage.

2.4. Generalized Additive Model

We use a machine-learning-based approach to convert the satellite skin salinity observations to bulk near-surface salinity that match the salinities measured with in-situ instruments while accounting for high-latitude retrieval biases. The significance of particular terms in the regression equation used will yield evidence of whether the skin-effect and/or biases need to be corrected. Our algorithm of choice is a Generalized Additive Model (GAM) [43]. This machine-learning-based approach, in particular, has a history rooted in statistical regression techniques (e.g., [44]). Ultimately, predictions are made by using predictors (described below) as inputs, just as other statistical regression-based approaches would do. One primary difference between a general linear-regression technique and a GAM is that the latter aims to achieve a balance between the bias and variance of its predictions through a regularization term. This regularization term prevents the machine-learning method from over-fitting to a particular training data set, permitting the approach to be applied to other data sets for prediction purposes. To guarantee that the machine-learning model does not over-fit to the training data, a cross-validation is applied by excluding some of the observations from the training data set, predicting those data, verifying that those predictions are accurate, and then repeating this procedure for different subsets of the training data set.

Instead of estimating the bulk surface salinity, we use a GAM to estimate the bulk surface salinity bias plus skin effect in the satellite-derived SSS data,

$$\Delta SSS_{bulk} = f_0 + f_1(t) + f_2(\Delta SSS) + f_3(SSS_{skin}) \tag{1}$$
$$+h(SSS_{skin}, SST, \lambda, w_{inv}, Q_{sens}, Q_{lat}, E, q_{hum}, \Delta SSS),$$

where Table 1 describes what each term means.

Table 1. Descriptions of each term in Equation (1).

Term	Description
$f_i(\cdot)$	Smoother functions for $i = 0\ldots3$
$h(\cdot)$	Tensor product of pairwise variables
SSS_{skin}	Satellite-derived SSS from D_{sat}
t	Julian day relative to January 1 of 1970
z	Depth of the in-situ observations
λ	$= 6(1 + (16(Q_{sens} + Q_{lat}(1 + S\beta c_p/(\alpha L_e) + 0.99 \times 5.67 \times 10^{-8}(SST + 273.16)^4)g\alpha\rho c_p v^3 w_{inv}^4/k^2)^{3/4})^{1/3}$ an empirical coefficient as in [10]
Q_{sens}	Sensible heat flux from OAFlux [31]
Q_{lat}	Latent heat flux from OAFlux [31]
SST_{bulk}	Sea-surface temperature in Celsius from OAFlux [31]
L_e	Latent heat of evaporation calculated using TEOS-10 [45]
α	Thermal expansion coefficient calculated using TEOS-10 [45]
β	Haline contraction coefficient calculated using TEOS-10 [45]
c_p	Specific heat of seawater calculated using TEOS-10 [45]
v	$= 1.4 \times 10^{-6}$ is the kinematic viscosity of seawater
p	Pressure
k	$= 0.5715(1 + 0.003SST_{bulk} - 1.025 \times 10^{-5}SST_{bulk}^2 + 6.53 \times 10^{-4}p + 0.00029SSS_{bulk})$ thermal conductivity of seawater [46] (in W m^{-1} K^{-1})
g	$= 9.806$ m s^{-2} is the acceleration due to gravity
τ	wind stress from OAFlux [32]
ρ	in-situ density calculated using TEOS-10 [45]
w_{inv}	$= (\tau/\rho)^{-1/2}$ is a function of the inverse wind stress
E	Evaporation from OAFlux [31]
q_{hum}	Near-surface humidity from OAFlux [31]
ΔSSS	$= f_c SSS_{skin}\lambda E w_{inv}$ bias correction, with proportionality constant f_c [10]; f_c is determined with the GAM

The Julian day, t, is the most important term to include for the satellite-derived SSS data because it aligns satellite observations with when the in-situ observations were taken. ΔSSS is important to include in the GAM because it at least partially corrects for the skin effect seen in the satellite data; the remaining terms correct for biases. The correlation between ΔSSS from the co-located SMAP-derived SSS and the in-situ salinity observations in the top-5 m is 0.37, which is significant to the 95% level. However, the skin-effect correction associated with including ΔSSS in our algorithm reduces the RMSE by less than 10%. The majority of the decrease in RMSE between our algorithmically-calculated bulk salinities and the in-situ observations in the top-5 m can be explained by the other GAM terms, which are associated with bias-correction. The equivalent correlation for the co-located Barcelona Expert Center SMOS-derived SSS [47,48] is higher (0.46), suggesting that the GAM will be different for different data products.

We derive the corrected bulk surface salinities with the following order of operations. At each location and time, we predict the bulk surface salinity biases, $\Delta SSS_{bulk}(x,y,t)$. We then average these biases over the entire time satellite data period to get $\Delta SSS_{bulk}(x,y) = \overline{\Delta SSS_{bulk}(x,y,t)}$. We then add this temporally-averaged bias correction term to the satellite-derived SSS to get $SSS_{bulk}(x,y,t) = SSS_{sat}(x,y,t) + \Delta SSS_{bulk}(x,y)$. The order of these operations is important because t explains some variability that isn't simply related to the seasonal cycle and/or trend. The RMSE between the BEC SMOS SSS_{skin} and the in-situ data in the top-5 m is larger than that between the SMAP SSS_{skin} and the in-situ data in the top-5 m.

An important, but subtle, detail is that both w_{inv} and λ depend upon SSS_{bulk} and we will not know SSS_{bulk} everywhere when using the GAM for prediction. If we assume that we know SSS_{bulk} to calculate w_{inv} and λ, then our GAM can explain 100% of the deviance (with a RMSE of about 0.04%), but SSS_{bulk} is what we aim to predict. If we assume that we know SSS_{bulk}, then we would only be able to calculate SSS_{bulk} at the points where we have in-situ data, so we use SSS_{skin} to calculate w_{inv} and λ employing TEOS-10. We then estimate the values of SSS_{bulk} with the GAM. We could then iterate by recalculating w_{inv} and λ using the predicted values of SSS_{bulk} and subsequently estimate new values for SSS_{bulk}, reducing the RMSE with respect to in-situ data, but the time variability of the resulting SSS_{bulk} is not realistic. Thus, we use a single iteration. It is important to include a minimal number of tensor product terms in $g(\cdot)$ because the data close to the coast have large biases, due to land contamination, making the GAM over-fit to the data, resulting in large bias estimates in most locations outside of the training data.

While ocean state estimate outputs suggest that the difference between sea-surface height and ocean-bottom pressure anomalies could be a good proxy for SSS in many locations within the Arctic [49], operational use of coinciding Level-2 sea-surface height and ocean-bottom pressure data with Level-2 SSS data would be limited. Further, sea-surface heights and ocean-bottom pressures decrease the RMSE of the GAM by less than 0.1%; thus, we use the GAM specified in Equation (1).

3. Results

We first assess the biases in the satellite-derived SSS products relative to in-situ observations in the top-5 m. Next, we characterize the statistics (mean, standard deviation, seasonal cycle magnitude, skewness, horizontal gradient trends, large anomaly counts, and mixing lengths) of high-latitude satellite-derived SSS observations. Then we use our algorithm to convert the SMAP-derived (skin) salinities to near-surface (bulk) salinities that can be used for data assimilation and characterize the statistics of the skin-effect and bias-corrected surface salinities. We lastly co-locate in-situ observations and the skin-effect and bias-corrected surface salinities to examine whether our algorithm improves the fidelity of the satellite-derived SSS.

3.1. Satellite SSS and In-Situ Salinity Comparisons

We sample the satellite-derived SSS within 50 km and 3.5 days of all publicly available in-situ observations [50] of the top-5 m north of 55°N. The number of match-up observations for SMAP data is fewer than that for SMOS; so, for the SMAP observations, there are less data for training the GAM. For example, there are no Marine Mammals Exploring the Oceans Pole to Pole (MEOP) Conductivity, Temperature, and Depth (CTD) [51,52] observations in the top-5 m within 50 km and 3.5 days of SMAP data (Figure 1). There are very few ship-based CTD hydrography and OMG observations that can be compared with the SMAP data. The number of co-located SMAP-derived SSS data points with in-situ salinity observations in the top-5 m are: 2929 observations with ship-based CTD hydrography, 1,710,428 observations with Saildrone, 8,640,999 observations with Underway, and 3219 observations with OMG. For the available ship-based CTD hydrography and Saildrone match-ups with the SMAP data, their disagreement is smaller than in the comparison between OMG and SMAP data. As shown in Figure 1, the Underway data comparisons with SMAP data have at least two distinct clusters of salinities, one around 32 pss in the North Pacific Ocean and the other around 35 pss in the North Atlantic Ocean. There may be a third cluster of points in the North Sea at salinities between 26–28 pss in the Underway data but much saltier in the SMAP data.

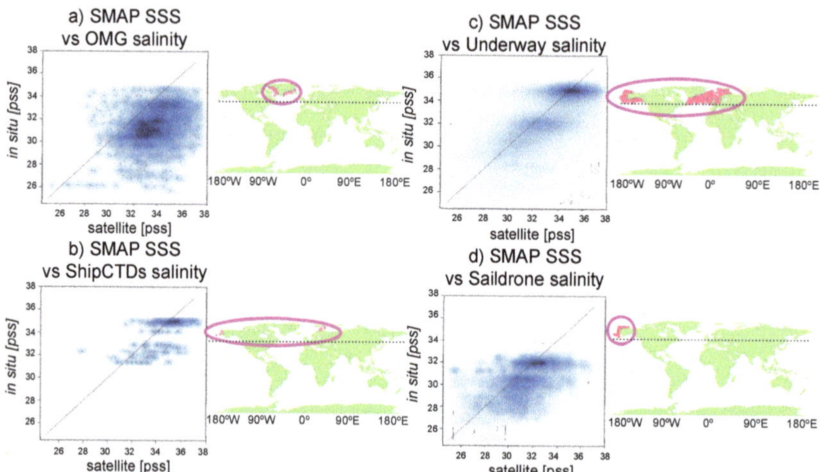

Figure 1. SMAP Level-2 skin SSS for April 2015 to December 2020 (abscissa) sampled within 50 km and 3.5 days of all in-situ observations in the top-5 m versus the bulk SSS from in-situ observations in the top-5 m (ordinate) from the Oceans Melting Greenland (OMG—panel **a**), ship-based CTD hydrography (panel **b**), Underway (panel **c**), and Saildrone (panel **d**) campaigns. The darker the shade of blue, the greater the number of points in the scatterplots; outliers are shown with single black dots. Additionally shown next to each scatterplot are the locations where the comparisons between the SMAP SSS product and the in-situ observations are made (pink dots, regions circled); the dashed black line indicates where 55°N is.

We next compare the SMAP data with the aggregated data from all in-situ data campaigns and inspect whether there is any depth-structure to the biases. The Underway data comparisons are very representative of the scatter between the satellite and in-situ data sets (Figure 2a) because they comprise most of the in-situ data. While the differences between the SMAP-derived SSS relative to the in-situ data have many more outliers in the top two meters, there is no statistically distinguishable depth-structure to the differences between the data sets (Figure 2b). The SMAP product has an overall 4.63% (1.54 pss) RMSE relative to the aggregate in-situ data, which is fairly consistent with a similar comparison with in-situ data north of 65°N tabulated in a previous study [53]. This value is relatively small compared to an overall 6.88% (2.29 pss) RMSE between SMOS, different from the product compared in the same previous study, and the aggregate in-situ data. Our values contrast with the ones the previous study reported because of differences in our domains, our versions of the SMAP and SMOS products, and our in-situ data.

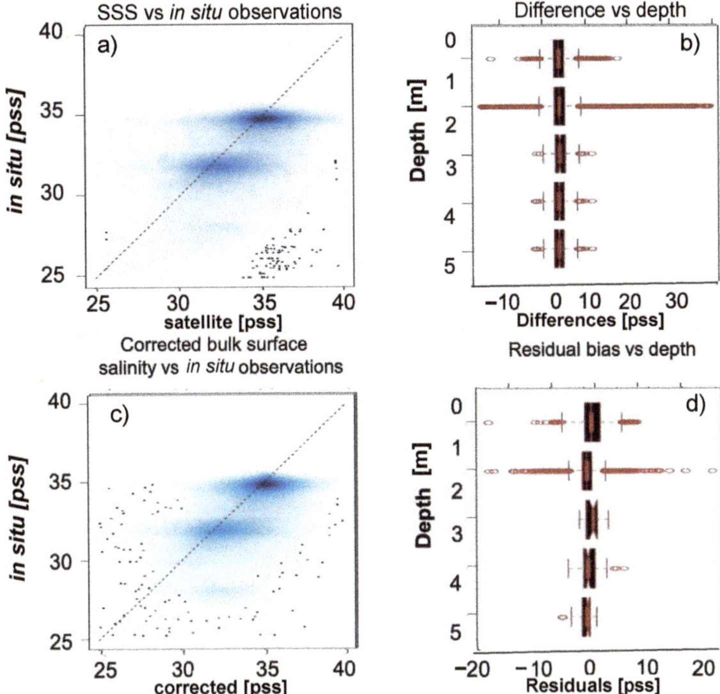

Figure 2. Comparison to in-situ salinity observations in the top-5 m (ordinate, panels **a** and **c**): (abscissa, panel **a**) SMAP-derived SSS and (abscissa, panel **c**) corrected bulk surface salinity field with a Generalized Additive Model—GAM, see Equation (1)) using Level-2 SMAP-derived SSS and OAFlux data. The darker the shade of blue, the greater the number of points in the scatterplots; outliers are shown with single black dots. Additionally shown are boxplots of the SSS minus in-situ observations (panel **b**) and corrected bulk surface salinity minus in-situ observations (panel **d**), each as a function of binned depths in the top-5 m. The first depth bin is for 0–1 m, the second depth bin is for 1–2 m, ..., and the deepest depth bin is for 4–5 m.

After applying our algorithm (Equation (1)) to the satellite and air-sea flux/forcing data sampled at the in-situ data locations and times, we can directly compare the in-situ data with the converted skin-to-bulk salinity data for each satellite data product separately. When trained on 75% of the in-situ data and predicted on the remaining 25%, the RMSE of the skin-to-bulk converted SMAP product relative to the in situ data are reduced to 2.43% (0.81 pss), explaining 71.6% of the deviance. We tested our GAM by training it on 50% of the in-situ data as well, with nearly identical results because the same portion of the phase space with salinity and air-sea forcing/flux factors gets spanned with this training data. We achieved this result by balancing the need to reduce the RMSE relative to random subsets of the in-situ salinity data with the need to not over-fit the GAM. It is possible to achieve a smaller RMSE using more combinations of predictors, but this increases the generalized cross-validation score, suggesting that the algorithm is less capable of estimating bulk surface salinities outside of the in-situ data set. In our final product, there is slightly more variability as a function of depth for our skin-effect and bias-corrected bulk surface salinities (Figure 2d) than for the SMAP-derived SSS (Figure 2b) in comparison with in-situ data over the top-5 m. However, there is no statistically significant depth-structure to the remaining bias in the skin-effect and bias-corrected bulk surface salinities (Figure 2d). As with the comparison of the satellite observations to in-situ observations (Figure 2a), the converted bulk surface salinity comparisons with in-situ observations (Figure 2c) display the clusters of salinities and generally lie along the one-to-one line. If we, instead, train a single GAM using an indicator function on the skin salinity term for SMOS, versus SMAP data, the RMSE is larger, but comparable (2.50% or 0.83 pss). Although not shown, applying the same Figure 2 analysis to the SMOS converted bulk surface salinities produces generally the same descriptions.

3.2. Temporal Statistics of Arctic SSS

Before comparing the in-situ near-surface (bulk) salinities with the satellite-derived (skin) salinities, we present the temporal statistics of the satellite-derived SSS from the SMAP product. When the BEC SMOS data are included, by eye the figures are identical. The SSS is, on average, typically between 33–35 pss, but can be lower to the east of Svalbard (Figure 3a). In regions with lower SSS, the SSS standard deviations (Figure 3b), after detrending and removing the seasonal cycle (Figure 3d), can be as high as 4–5 pss. The standard deviations of SSS tend to get smaller with distance from the perennial, sea-ice-covered regions. The same is true for the SSS skewness (Figure 3c), except the skewness values tend to be negative, indicating a long, relatively fresh SSS tail closer to sea ice and far northern coasts. These large, negative skewnesses could be due to ice melt and/or run-off, unless precipitation events affect SSS more at high northern latitudes than elsewhere. However, these skewnesses are impacted by SSS biases because the skewness is a function of the average SSS. Additionally, the SMAP SSS uncertainties in the Level-2 JPL product, which are estimated errors in the retrievals, are largest in high-latitude regions (Figure 4a). At high northern latitudes, these uncertainties reach 1.5 pss, with standard deviations and seasonal cycle magnitudes at about 0.5 pss (Figure 4b,d). The SMAP SSS uncertainty skewness is most negative in regions affected by ice melt (Figure 4c). While the SMAP SSS uncertainties are smaller than their biases in many locations in the high-latitude oceans, it is likely that the SMAP SSS uncertainties are too small to represent the true uncertainties in high-latitude regions. Our algorithm quantifies the functional uncertainty associated with our model specification, which are standard errors from the GAM, can then be added to the SMAP SSS uncertainties.

Before presenting the bulk surface salinities after conversion and some bias correction, we present an apparent relationship between SSS and sea-ice melting/refreezing to explain the spatial patterns in SSS statistics (Figure 3). Greater temporal fluctuations in SSS near sea ice (Figure 3b–d) can be explained by retreating sea ice leaving relatively fresh water behind as well as by more frequent absence of sea-ice cover resulting from greater salinity values, which have a colder freezing temperature. The seasonal cycle of SSS is largest near the perennial sea-ice edges (Figure 3d), but that has been removed to calculate the standard deviation and skewness. The trend in SSS is a mixture of increasing and decreasing salinity, with no large-scale pattern trend that is significant to the 95% level (not shown). However, anomalously large SSS, found by counting the number of 8-day averages where the average SSS is exceeded by more than three times the SSS standard deviation (see Section 2.1), align close with the marginal ice zones (Figure 5a). Further, although surface forcing dominates eddy stirring, theoretical estimates [28] suggest that the regions with statistically significant horizontal SSS gradients or anomalously high SSS values always occur where the mixing lengths are small (Figure 5b). The fact that mixing lengths are smaller in marginal ice zones is consistent with previously published theory [54]. These results suggest that the biases in satellite-derived SSS in marginal ice zones are not random and may even provide valuable constraints on ocean-sea ice data assimilation systems; this further motivates our skin effect and bias-correction procedure.

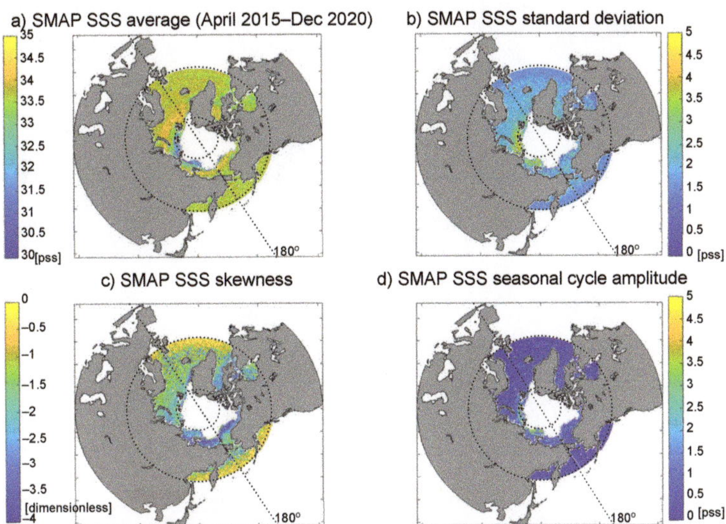

Figure 3. Statistics for SMAP Level-2 sea-surface salinity (SSS) product for the period April 2015 to December 2020: (**a**) SSS average, (**b**) SSS standard deviation, (**c**) SSS skewness, and (**d**) SSS seasonal cycle amplitude. The standard deviation and skewness are computed after the removal of the seasonal cycle and trend. The maps synthesize the SMAP data without interpolation, but average all data over each nearest 50 km by 50 km grid point and over each 8-day time period. Overlaid on top are geographical coordinates indicating where the 0° and 180° meridians, as well as the 55°N and 80°N latitudes are.

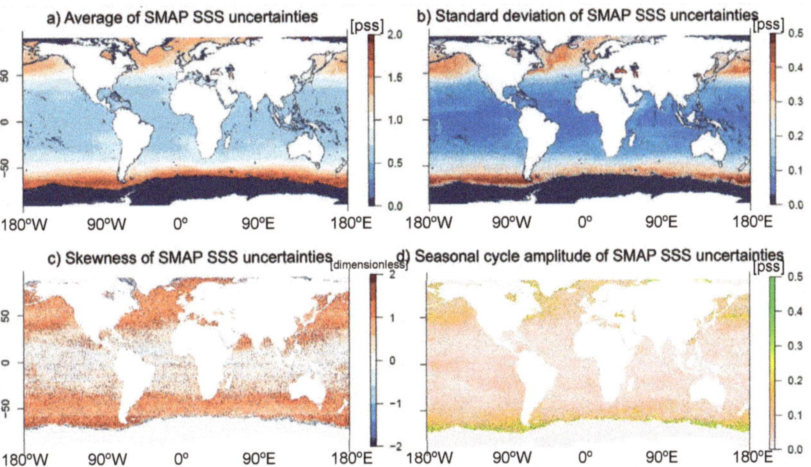

Figure 4. Statistics for SMAP Level-2 sea-surface salinity (SSS) product uncertainty for the period April 2015 to December 2020: (**a**) SSS uncertainty average, (**b**) SSS uncertainty standard deviation, (**c**) SSS uncertainty skewness, and (**d**) SSS uncertainty seasonal cycle amplitude. The standard deviation and skewness are computed after the removal of the seasonal cycle and trend. The maps synthesize the SMAP data without interpolation, but averages all data over each nearest 50 km by 50 km grid point and over each 8-day time period.

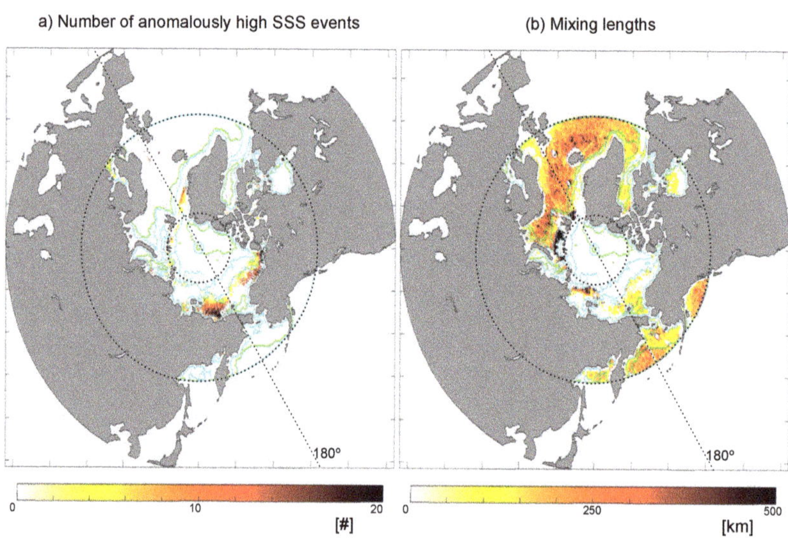

Figure 5. High-latitude satellite sea-surface salinity (SSS) anomalies and trends shown in heat-map colors: (**a**) anomalously high SMAP Level-2 SSS product sea-surface salinity (SSS) for the period April 2015 to December 2020, computed by counting the number of 8-day averages where the average SSS is exceeded by more than three times the SSS standard deviation; and (**b**) the mixing length scales calculated as the ratio of the temporal standard deviations from Level-2 SMAP SSS data (April 2015 to December 2020) to the horizontal spatial gradients of SSS from the Level-3 daily Earth & Space Research SMOS product (January 2011 to December 2020). The cyan (green) contours in each panel indicate the minimum (maximum) sea ice extent over all winters between 2015–2019. Overlaid on top are geographical coordinates indicating where the 0° and 180° meridians, as well as the 55°N and 80°N latitudes are.

In Figure 6, we repeat the temporal statistic calculations (Figure 3) using the SMAP data that has been corrected with our algorithmic approach for the skin effect and biases. Figure 6a–d are fairly similar maps to those in Figure 3a–d, but there are some important differences. The average corrected bulk surface salinity values are fresher in the North Pacific, Bering Sea, Chukchi Sea, Davis Strait, Hudson Bay, and coastal Greenland regions and saltier in the subpolar North Atlantic Ocean, Norwegian Sea, and Barents Sea regions (Figures 3a and 6a). The fresher corrected bulk surface salinity values near the Greenland coasts are in better agreement with the in-situ data from the OMG campaign than the satellite-derived SSS values. The corrected bulk surface salinity standard deviations and skewnesses have large magnitudes only for narrow bands near the perennial sea ice and the coasts (Figure 6b,c), as opposed to a larger area over the marginal ice zones (Figure 3b,c). Relative to the satellite-derived SSS seasonal cycle amplitudes (Figure 3d), there are large corrected surface salinity seasonal cycle amplitudes for a greater proportion of the marginal ice zones on the Siberian Shelf (Figure 6d). The algorithm we apply to calculate the corrected bulk surface salinities improves their agreement with in-situ data, but we need additional tests to determine whether the corrected bulk surface salinities are more realistic than the satellite-derived SSS.

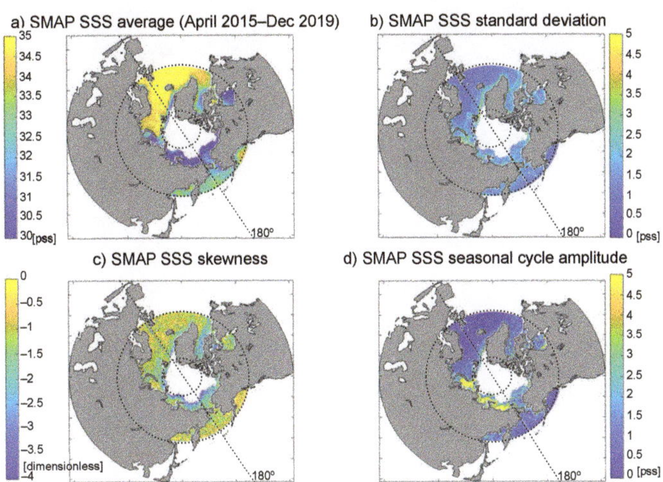

Figure 6. Statistics for SMAP Level-2 sea-surface salinity (SSS) product, corrected for the skin effect and bias (April 2015 to December 2019 due to the time range of the available OAFlux data): (**a**) SSS average, (**b**) SSS standard deviation, (**c**) SSS skewness, and (**d**) SSS seasonal cycle amplitude. The standard deviation and skewness are computed after the removal of the seasonal cycle and trend. The product synthesizes the products without interpolation, but averages all data over each nearest 50 km by 50 km grid point and over each 8-day time period, the same as for the statistics shown in Figure 3. Overlaid on top are geographical coordinates indicating where the 0° and 180° meridians, as well as the 55°N and 80°N latitudes are.

One test of how realistic the corrected bulk surface salinities are is to repeat the in-situ data comparisons shown in Figure 1. Because we only correct the time-mean bias at each horizontal location, it is possible that the instantaneous disagreements between the corrected bulk surface salinities and the near-surface in-situ data are about the same or worse; however, the skin-effect and bias corrections do not degrade the accuracy of the salinities relative to the in-situ data (Figure 7). The disagreements between the corrected bulk surface salinity product from SMAP and each of these in-situ data sets are typically less than 2 pss (<1 pss overall RMSE), with negligible bias overall, but disagreements near the coasts, where there is freshening from ice sheet melt, can be much greater. For example, there remains a 2 pss bias in the corrected bulk surface salinity product from SMAP SSS data relative to the OMG data (not shown). These biases are well within the uncertainties associated with the SMAP SSS product (~1.5 pss; Figure 4a) plus with the uncertainties associated with our algorithm (1–2 pss; not shown). After skin-effect and bias-correction, the corrected bulk surface salinities show three distinct clusters of salinities relative to the Underway data: (1) between 26–27 pss, (2) around 32 pss, and (3) around 35 pss. Overall, the corrected bulk surface salinities are improved relative to the Saildrone and Underway in-situ data sets, which comprise the shallowest data relative to any other data sets, but near-coastal satellite data points, where there is potential land contamination, should be removed.

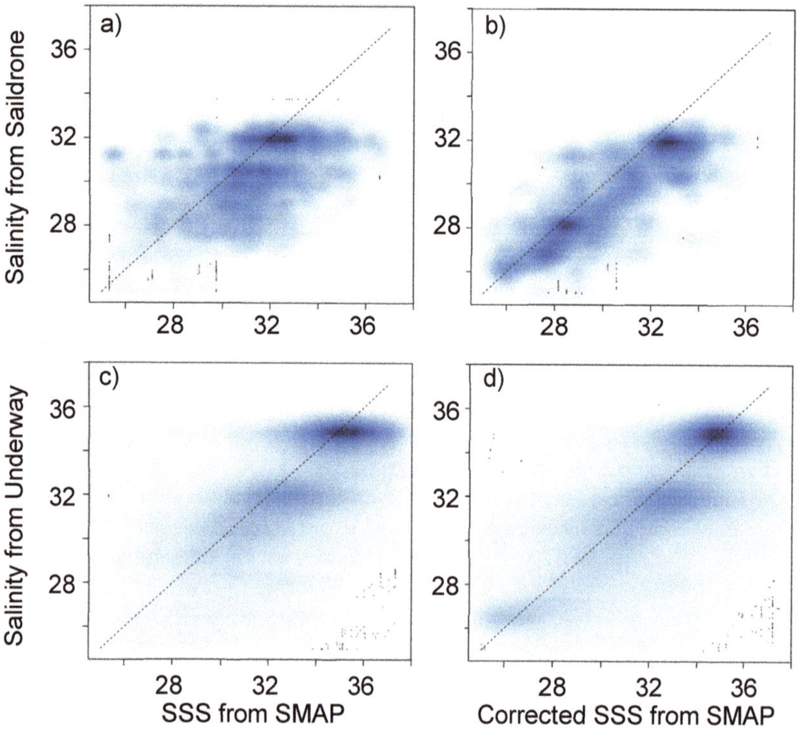

Figure 7. Comparison of in-situ and Level-2 SMAP SSS observations (sampled within 50 km and 3.5 days of in-situ bulk surface salinity observations) for the period of April 2015 to December 2019 (abscissa = SMAP, ordinate = in-situ): Level-2 SMAP (panels **a,c**) and skin-effect and bias-corrected Level-2 SMAP (panels **b,d**) data; Saildrone (panels **a,b**) and Underway (panels **c,d**). The darker the shade of blue, the greater the number of points in the scatterplots; outliers are shown with single black dots.

4. Discussion

One motivation for developing our algorithm is to convert skin salinities to bulk salinities to allow for data assimilation of satellite-derived salinity data in regions without Argo data, but the more important alteration of the satellite-derived SSS for assimilation purposes is its inaccuracy at high-latitudes. Our algorithm includes terms for multiple air-sea forcing/flux fields, including the wind stress, bulk SST, and the implicit exclusion of regions covered by sea-ice, by making use of the OAFlux product. One interpretation of our results is that these fields provide corrections to the equivalent ancillary fields used in SMAP-derived SSS retrievals. However, an alternative interpretation of our results is that the relevant terms in the Yu (2010) [10] theory and/or additional terms in our algorithm, together, are proxies for inaccuracies in the SMAP-derived SSS retrievals. Each of these interpretations are supported by the facts that: (1) relative to available in situ data, our algorithm reduces the RMSE of the corrected SMAP salinities, and (2) in order to minimize the RMSE, compared with other possible GAM term combinations, our algorithm requires the same terms for both SMAP and SMOS data.

Where they overlap, the spatial distributions of high-latitude surface salinity statistics from our algorithm are consistent with those presented in other observational product comparison studies [4,53,55–57]. Each of these products are within the ~2.0–3.5 pss uncertainties we find in high-latitude regions, after skin-effect and bias-correction. The spatial patterns are also consistent with previously presented products of surface salinity. Like our product, other Aquarius-based products, SMOS-based products, and the World Ocean Atlas suggest there is more fresh surface water near the Pacific and saltier surface water near the Atlantic [53,56]. Additionally, like our product, these other products suggests there are larger seasonal cycles in surface salinity closer to the Pacific than to the Atlantic. In subpolar regions, the seasonal cycle magnitudes of SSS are small (<1 pss), as shown in previous studies [4,57], especially in comparison with the seasonal cycles both other observational products and our algorithmically-derived product suggest near the Arctic Eurasian and Canadian coasts. The primary differences we find with our algorithmically-derived product in the spatial distribution of the surface salinity statistics are in the relative magnitudes of the standard deviations versus seasonal cycles in these Arctic coastal regions. This is conceivably a result of the temporal resolution of the OAFlux product. However, in these coastal regions and in other locations in the marginal ice zone, a seasonal cycle cannot be accurately estimated due to the seasonal aliasing of the satellite-derived SSS so estimates of the seasonal cycle magnitudes and higher-order statistics that rely on extraction of the seasonal cycle (e.g., standard deviations) are not reliable in these places.

5. Conclusions

This study presented a method to convert satellite skin salinity observations to bulk salinity for assimilation into modeling systems. The temporal statistics in a satellite-derived data set of SSS reveal likely influence from sea-ice melt in marginal ice zones. Trends were not detectable over the 5-year period of the data record. Data collected from the SPURS-1 campaign (not shown here) suggested that there could be non-constant structure to salinity profiles, even within the upper-50 centimeters of the water column. Point-by-point comparisons of the satellite SSS with several different sources of in-situ observations in the top-5 m for northern high-latitude regions demonstrated that different geographic regions have different clusters of salinity values and that the satellite-derived data do not agree well enough with the in-situ data for data assimilation purposes. The disagreements between the satellite and in-situ data exceed 1.5 pss, which can be greater than the temporal variability in the satellite data. We presented an algorithm, based on machine learning and trained on the in-situ salinity data and air-sea flux/forcing data, to convert skin-salinities to bulk-salinities. This algorithm for corrected bulk surface salinities cut the disagreement with the in-situ data down by at least half from the comparison between the satellite and in-situ data. The algorithm can reduce the disagreement to a level of less than 1 pss and can produce uncertainties that are simply propagated along with the Level-2 product uncertainties.

The algorithm to convert skin salinities to bulk salinities and correct for biases can be improved and used in multiple applications. First, we can repeat the application of our algorithm with an improved SMAP product after the removal of sea-ice contamination [58] as well as using a SMOS product. We also expect the upcoming SPURS-3 (Salinity and Stratification at the Sea Ice Edge or SASSIE) campaign to enhance the accuracy of our algorithm by making available a greater amount of variance data in the near-surface salinity field. Future air-sea forcing/flux data potentially will be provided in the future by using an observing system called FluxSat that could reduce air-sea flux observational errors by 50% [59]. Our algorithm can be applied to different depth ranges for the in-situ data, as well as in the Antarctic region, depending upon the resolution and domain of the modeling system assimilating the corrected bulk salinities. Both OAFlux and in-situ data would be required for these domains as well, with no additional requirements. The assimilation of bulk salinities can potentially constrain the salinity field at high latitudes, allowing models to evaluate the sensitivities of the surface salinity field to other model processes and parameters. A more realistic surface salinity field in the Arctic could enable better simulation/representation of sea-ice formation/melt, allowing coupled ocean-sea ice-atmosphere models to improve their representation of heat and moisture fluxes. Our algorithm and potential refinements allow for the possibility of these studies and more.

Future observing system evaluation (OSE) studies need to demonstrate improvements in representing upper-ocean hydrography and sea-ice properties to conclude that corrected bulk salinity data have value for assimilation purposes.

Author Contributions: Conceptualization, D.T. and E.B.; methodology, D.T.; software, D.T.; validation, D.T.; formal analysis, D.T.; investigation, D.T.; resources, E.B.; data curation, E.B.; writing—original draft preparation, D.T.; writing—review and editing, D.T. and E.B.; visualization, D.T.; supervision, E.B.; project administration, E.B.; funding acquisition, E.B. All authors have read and agreed to the published version of the manuscript.

Funding: This research was funded by the NOAA/NESDIS Center for Satellite Applications and Research (STAR).

Institutional Review Board Statement: Not applicable.

Informed Consent Statement: Not applicable.

Data Availability Statement: Sea ice extent: https://nsidc.org/data/G10017/versions/1, accessed on 19 March 2021 (U.S. National Ice Center. 2020. U.S. National Ice Center Daily Marginal Ice Zone Products, Version 1. [northern]. Boulder, Colorado USA. NSIDC: National Snow and Ice Data Center. https://doi.org/10.7265/ggcq-1m67.) Underway: https://www.ncei.noaa.gov/access/surface-underway-marine-database/, accessed on 6 January 2021 (Wang, Zhankun; NOAA National Centers for Environmental Information (2017). Quality-controlled sea surface marine physical, meteorological and other in-situ measurements from the NCEI Surface Underway Marine Database (NCEI-SUMD). [sea surface salinity]. NOAA National Centers for Environmental Information. Dataset at https://www.ncei.noaa.gov/archive/accession/NCEI-SUMD.). OMG: https://podaac.jpl.nasa.gov/dataset/OMG_L2_AXCTD, accessed on 26 February 2021 (OMG. 2019. OMG AXCTD Profiles. Ver. 1. PO.DAAC, CA, USA. Datase at https://doi.org/10.5067/OMGEV-AXCT1) and https://podaac.jpl.nasa.gov/dataset/OMG_L2_CTD, accessed on 26 February 2021 (OMG. 2020. OMG CTD Conductivity Temperature Depth. Ver. 1. PO.DAAC, CA, USA. Dataset at https://doi.org/10.5067/OMGEV-CTDS1). WOCE/CLIVAR/GO-SHIP: https://cchdo.ucsd.edu/search?bbox=-180,55,180,90, accessed on 1 March 2021 ([Cutter, G., Thierry, V., Jeansson, E., Gary, S. F., Lee, C., Ivanov, V., Ashik, I., Schauer, U., Kadko, D., Gobat, J., King, B. A., Holliday, N. P., Olsen, A., MacDonald, A., Mecking, S., Bullister, J. L., Baringer, M. O., Kieke, D., Yashayaev, I., Griffiths, C. R., Skagseth, ø., Fernández Rios, A., Beszczynska-Möller, A., McGrath, G., Read, J. F.]. [2021]. [CTD] data from cruise [33RR20180918, 35A320170715, 58GS20160802, 74EQ20160607, 316N20150906, RUB320150819, 06AQ20150817, 33HQ20150809, 33RO20150525, 58GS20150410, 74JC20140606, 316N20130914, 33RO20130803, 58GS20130717, 06M220130509, 18HU20130507, 740H20130506, 58HJ20120807, 74E320120731, 29AH20120622, 06AQ20120614, 18MF20120601, 45CE20120105, 316N20111002, 06MT20110624, 74E320110511, 18HU20110506, 45CE20110103], [NetCDF]. Accessed from CCHDO [https://cchdo.ucsd.edu/search?bbox=-180,55,180,90]. [n/a].). Saildrone Arctic: https://podaac.jpl.nasa.gov/dataset/SAILDRONE_ARCTIC, accessed on 20 November 2020 (Saildrone. 2020. Saildrone Arctic NOPP-MISST Field Campaign Products. Ver. 1.0. PO.DAAC, CA, USA. Dataset at https://doi.org/10.5067/SDRON-NOPP0). MEOP SEaOS: http://www.meop.net/database/meop-databases/density-of-data.html, accessed on 16 November 2020. OAFlux: https://oaflux.whoi.edu/data-access/, accessed in 16 November 2020. ESR SMOS SSS gradient: https://salinitydata.org/files/data/daily/Salinity/SMOS/locean_debiasedSSS_09days_v5/, accessed on 10 March 2020. JPL Level-2 SMAP SSS: https://podaac.jpl.nasa.gov/dataset/SMAP_JPL_L2B_SSS_CAP_V5, accessed on 15 April 2021 (JPL. 2020. JPL CAP SMAP Sea Surface Salinity Products. Ver. 5.0. PO.DAAC, CA, USA. Dataset at https://doi.org/10.5067/SMP50-2TOCS). The data used to generate our figures are available at https://doi.org/10.5281/zenodo.6353521.

Acknowledgments: The authors thank NOAA STAR IT for their support, and the reviewers of this manuscript for their suggestions. The scientific results and conclusions, as well as any views or opinions expressed herein, are those of the authors and do not necessarily reflect those of NOAA or the Department of Commerce.

Conflicts of Interest: The authors declare no conflict of interest.

References

1. Lee, T. Consistency of Aquarius sea surface salinity with Argo products on various spatial and temporal scales. *Geophys. Res. Lett.* **2016**, *43*, 3857–3864. [CrossRef]
2. Reul, N.; Grodsky, S.A.; Arias, M.; Boutin, J.; Catany, R.; Chapron, B.; D'Amico, F.; Dinnat, E.; Donlon, C.; Fore, A.; et al. Sea surface salinity estimates from spaceborne L-band radiometers: An overview of the first decade of observation (2010–2019). *Remote Sens. Environ.* **2020**, *242*, 111769. [CrossRef]
3. Tang, W.; Yueh, S.; Yang, D.; Fore, A.; Hayashi, A.; Lee, T.; Fournier, S.; Holt, B. The potential and challenges of using soil moisture active passive (SMAP) sea surface salinity to monitor Arctic ocean freshwater changes. *Remote Sens.* **2018**, *10*, 869. [CrossRef]
4. Yu, L. Variability and uncertainty of satellite sea surface salinity in the subpolar North Atlantic (2010–2019). *Remote Sens.* **2020**, *12*, 2092. [CrossRef]
5. Supply, A.; Boutina, J.; Vergely, J.-L.; Kolodziejczyk, N.; Reverdin, G.; Reul, N.; Tarasenko, A. New insights into SMOS sea surface salinity retrievals in the Arctic Ocean. *Remote Sens. Environ.* **2020**, *249*, 112027. [CrossRef]
6. Dong, S.; Volkov, D.; Goni, G.; Lumpkin, R.; Foltz, G.R. Near-surface salinity and temperature structure observed with dual-sensor drifters in the subtropical South Pacific. *J. Geophys. Res. Oceans* **2017**, *122*, 5952–5969. [CrossRef]
7. Anderson, J.E.; Riser, S.C. Near-surface variability of temperature and salinity in the near-tropical ocean: Observations from profiling floats. *J. Geophys. Res. Oceans* **2014**, *119*, 7433–7448. [CrossRef]
8. Bingham, F.M.; Tsontos, V.; de Charon, A.; Lauter, C.J.; Taylor, L. The SPURS-2 eastern tropical Pacific field campaign data collection. *Oceanography* **2019**, *32*, 142–149. [CrossRef]
9. Saunders, P. The temperature at the ocean-air interface. *J. Atmos. Sci.* **1967**, *24*, 267–273. [CrossRef]
10. Yu, L. On sea surface salinity skin effect induced by evaporation and implications for remote sensing of ocean salinity. *J. Phys. Oceanogr.* **2010**, *40*, 85–102. [CrossRef]
11. Polyakov, I.V.; Pnyushkov, A.V.; Alkire, M.B.; Ashik, I.M.; Baumann, T.M.; Carmack, E.C.; Coszczko, I.; Guthrie, J.; Ivanov, V.V.; Kanzow, T.; et al. Greater role for Atlantic inflows on sea-ice loss in the Eurasian Basin of the Arctic Ocean. *Science* **2017**, *356*, 285–291. [CrossRef] [PubMed]
12. Polyakov, I.V.; Rippeth, T.P.; Fer, I.; Alkire, M.B.; Baumann, T.M.; Carmack, E.C.; Ingvaldsen, R.; Ivanov, V.V.; Janout, M.; Lind, S.; et al. Weakening of cold halocline layer exposes sea ice to oceanic heat in the eastern Arctic Ocean. *J. Clim.* **2020**, *33*, 8107–8123. [CrossRef]
13. Davis, P.E.D.; Lique, C.; Johnson, H.L.; Guthrie, J.D. Competing effects of elevated vertical mixing and increased freshwater input on the stratification and sea ice cover in a changing Arctic Ocean. *J. Phys. Oceanogr.* **2016**, *46*, 1531–1553. [CrossRef]
14. Halloran, P.R.; Hall, I.R.; Menary, M.; Reynolds, D.J.; Scourse, J.D.; Screen, J.A.; Bozzo, A.; Dunstone, N.; Phipps, S.; Schurer, A.P.; et al. Natural drivers of multidecadal Arctic sea ice variability over the last millennium. *Nat. Sci. Rep.* **2020**, *10*, 688. [CrossRef] [PubMed]
15. Jahn, A.; Holland, M.M. Implications of Arctic sea ice changes for North Atlantic deep convection and the meridional overturning circulation in CCSM4-CMIP5 simulations. *Geophys. Res. Lett.* **2013**, *40*, 1206–1211. [CrossRef]
16. Liu, W.; Fedorov, A.; Sévellec, F. The mechanisms of the Atlantic meridional overturning circulation slowdown induced by arctic sea ice decline. *J. Clim.* **2019**, *32*, 977–996. [CrossRef]
17. Stouffer, R.J.; Yin, J.; Gregory, J.M.; Dixon, K.W.; Spelman, M.J.; Hurlin, W.; Weaver, A.J.; Eby, M.; Flato, G.M.; Hasumi, H.; et al. Investigating the causes of the response of the thermohaline circulation to past and future climate changes. *J. Clim.* **2006**, *19*, 1365–1387. [CrossRef]
18. Doddridge, E.W.; Meneghello, G.; Marshall, J.; Scott, J.; Lique, C. A three-way balance in the Beaufort Gyre: The ice-ocean governor, wind stress, and eddy diffusivity. *J. Geophys. Res. Ocean.* **2019**, *124*, 3107–3124. [CrossRef]
19. Haine, T.W.N. A conceptual model of polar overturning circulations. *J. Phys. Oceanogr.* **2021**, *51*, 727–744. [CrossRef]
20. Jensen, M.F.; Nilsson, J.; Nisancioglu, K.H. The interaction between sea ice and salinity-dominated ocean circulation: Implications for halocline stability and rapid changes of sea ice cover. *Clim. Dyn.* **2021**, *47*, 3301–3317. [CrossRef]
21. Meneghello, G.; Marshall, J.C.; Timmermans, M.-L.; Scott, J. Observations of seasonal upwelling and downwelling in the Beaufort Sea mediated by sea ice. *J. Phys. Oceanogr.* **2018**, *48*, 795–805. [CrossRef]
22. Proshutinsky, A.; Krishfield, R.; Timmermans, M.-L.; Toole, J.; Carmack, E.; McLaughlin, F.; Williams, W.J.; Zimmermann, S.; Itoh, M.; Shimada, K. Beaufort Gyre freshwater reservoir: State and variability from observations. *J. Geophys. Res.* **2009**, *114*, C00A10. [CrossRef]
23. Sedláček, J.; Knutti, R.; Martius, O.; Beyerle, U. Impact of a reduced Arctic sea ice cover on ocean and atmospheric properties. *J. Clim.* **2012**, *25*, 307–319. [CrossRef]
24. Kolodziejczyk, N.; Hamon, M.; Boutin, J.; Vergely, J.-L.; Reverdin, G.; Supply, A.; Reul, N. Objective Analysis of SMOS and SMAP Sea Surface Salinity to Reduce Large-Scale and Time-Dependent Biases from Low to High Latitudes. *J. Atmos. Ocean. Technol.* **2021**, *38*, 405–421. [CrossRef]
25. Mecklenburg, S.; Drusch, M.; Kaleschke, L.; Rodriguez-Fernandez, N.; Reul, N.; Kerr, Y.; Font, J.; Martin-Neira, M.; Oliva, R.; Daganzo-Eusebio, E.; et al. ESA's Soil Moisture and Ocean Salinity mission: From science to operational applications. *Remote Sens. Environ.* **2016**, *180*, 3–18. [CrossRef]
26. Vinogradova, N.; Lee, T.; Boutin, J.; Drushka, K.; Fournier, S.; Sabia, R.; Stammer, D.; Bayler, E.; Reul, N.; Arnold, G.; et al. Satellite salinity observing system: Recent discoveries and the way forward. *Front. Mar. Sci.* **2019**, *6*, 243. [CrossRef]

27. Fore, A.G.; Yueh, S.H.; Tang, W.; Stiles, B.W.; Hayashi, A.K. Combined Active/Passive Retrievals of Ocean Vector Wind and Sea Surface Salinity With SMAP. *IEEE Trans. Geosci. Remote Sens.* **2016**, *54*, 7396–7404. [CrossRef]

28. Cole, S.T.; Wortham, C.; Kunze, E.; Owens, W.B. Eddy stirring and horizontal diffusivity from Argo float observations: Geographic and depth variability. *Geophys. Res. Lett.* **2015**, *42*, 3989–3997. [CrossRef]

29. Mordy, C.W.; Cokelet, E.D.; De Robertis, A.; Jenkins, R.; Kuhn, C.E.; Lawrence-Slavas, N.; Berchok, C.L.; Crance, J.L.; Sterling, J.T.; Cross, J.N.; et al. Advances in ecosystem research: Saildrone surveys of oceanography, fish, and marine mammals in the Bering Sea. *Oceanography* **2017**, *30*, 113–115. [CrossRef]

30. Fenty, I.; Willis, J.K.; Khazendar, A.; Dinardo, S.; Forsberg, R.; Fukumori, I.; Holland, D.; Jakobsson, M.; Moller, D.; Morison, J.; et al. Oceans Melting Greenland: Early results from NASA's ocean-ice mission in Greenland. *Oceanography* **2016**, *29*, 72–83. [CrossRef]

31. Yu, L.; Jin, X.; Weller, A. *Multidecade Global Flux Datasets from the Objectively Analyzed Air-Sea Fluxes (OAFlux) Project: Latent and Sensible Heat Fluxes, Ocean Evaporation, and Related Surface Meteorological Variables*; WHOI OAFlux Project Technical Report (OA-2008-01); WHOI OAFlux: Falmouth, MA, USA, 2008.

32. Yu, L.; Jin, X. *Satellite-Based Global Ocean Vector Wind Analysis by the Objectively Analyzed Air-Sea Fluxes (OAFlux) Project: Establishing Consistent Vector Wind Time Series from July 1987 Onward through Synergizing Microwave Radiometers and Scatterometers*; WHOI OAFlux Project Technical Report; WHOI OAFlux: Falmouth, MA, USA, 2010.

33. Yu, L.; Jin, X. Insights on the OAFlux ocean surface vector wind analysis merged from scatterometers and passive microwave radiometers (1987 onward). *J. Geophys. Res. Ocean.* **2014**, *119*, 5244–5269. [CrossRef]

34. Dee, D.P.; Uppala, S.M.; Simmons, A.J.; Berrisford, P.; Poli, P.; Kobayashi, S.; Andrae, U.; Balmaseda, M.A.; Balsamo, G.; Bauer, P.; et al. The ERA-Interim reanalysis: Configuration and performance of the data assimilation system. *Q. J. R. Meteorol. Soc.* **2011**, *137*, 553–597. [CrossRef]

35. Saha, S.; Moorthi, S.; Pan, H.-L.; Wu, X.; Wang, J.; Nadiga, S.; Tripp, P.; Kistler, R.; Woollen, J.; Behringer, D.; et al. The NCEP climate forecast system reanalysis. *Bull. Am. Meteorol. Soc.* **2010**, *91*, 1015–1057. [CrossRef]

36. Jin, X.; Yu, L. Assessing high-resolution analysis of surface heat fluxes in the Gulf Stream region. *J. Geophys. Res. Ocean.* **2013**, *118*, 5353–5375. [CrossRef]

37. Fairall, C.W.; Bradley, E.F.; Hare, J.E.; Grachev, A.A.; Edson, J.B. Bulk parameterization of air-sea fluxes: Updates and verification for the COARE algorithm. *J. Clim.* **2003**, *16*, 571–591. [CrossRef]

38. McPhaden, M.J.; Busalacchi, A.J.; Cheney, R.; Donguy, J.-R.; Gage, K.S.; Halpern, D.; Ji, M.; Julian, P.; Meyers, G.; Mitchum, G.T.; et al. The Tropical Ocean-Global Atmosphere observing system: A decade of progress. *J. Geophys. Res.* **1998**, *103*, 14169–14240. [CrossRef]

39. Colbo, K.; Weller, R.A. The accuracy of the IMET sensor package in the subtropics. *J. Atmos. Ocean. Technol.* **2009**, *26*, 1867–1890. [CrossRef]

40. Reynolds, R.W.; Smith, T.M.; Liu, C.; Chelton, D.B.; Casey, K.S.; Schlax, M.G. Daily high-resolution blended analyses for sea surface temperature. *J. Clim.* **2007**, *20*, 5473–5496. [CrossRef]

41. Daley, R. *Atmospheric Data Analysis*; Cambridge Univ. Press: Cambridge, UK, 1991; 457p.

42. Talagrand, O. Assimilation of observations, an introduction. *J. Meteorol. Soc. Jpn.* **1997**, *75*, 191–209. [CrossRef]

43. Wood, S.N. *Generalized Additive Models: An Introduction with R*; Chapman and Hall/CRC Press: Boca Raton, FL, USA, 2006.

44. Hastie, T.; Tibshirani, R.; Friedman, J. *The Elements of Statistical Learning: Data Mining, Inference, and Prediction*; Springer: Berlin, Germany, 2001.

45. McDougall, T.J.; Barker, P.M. *Getting Started with TEOS-10 and the Gibbs Seawater (GSW) Oceanographic Toolbox*; WG127; SCOR/IAPSO: Paris, France, 2011.

46. Sharqawy, M.H.; Lienhard, V.J.H.; Zubair, S.M. Thermophysical properties of seawater: A review of existing correlations and data. *Desalin. Water Treat.* **2010**, *16*, 354–380. [CrossRef]

47. Martínez, J.; Gabarró, C.; Turiel, A. Arctic Sea Surface Salinity L2 Orbits and L3 Maps (v.3.1) [Dataset]. DIGITAL.CSIC 2019. Available online: http://hdl.handle.net/10261/229624 (accessed on 1 January 2021).

48. Olmedo, E.; Gabarró, C.; González-Gambau, V.; Martínez, J.; Ballabrera-Poy, J.; Turiel, A.; Portabella, M.; Fournier, S.; Lee, T. Seven years of SMOS sea surface salinity at high latitudes: Variability in Arctic and sub-Arctic regions. *Remote Sens.* **2018**, *10*, 1772. [CrossRef]

49. Fournier, S.; Lee, T.; Wang, X.; Armitage, T.W.K.; Wang, O.; Fukumori, I.; Kwok, R. Sea surface salinity as a proxy for Arctic ocean freshwater changes. *J. Geophys. Res. Ocean.* **2020**, *125*, e2020JC016110. [CrossRef]

50. Schanze, J.J.; Le Vine, D.M.; Dinnat, E.P.; Kao, H.-Y. Comparing satellite salinity retrievals with in-situ measurements: A recommendation for Aquarius and SMAP. *Zenodo* **2020**. [CrossRef]

51. Roquet, F.; Williams, G.; Hindell, M.A.; Harcourt, R.; McMahon, C.; Guinet, C.; Charrassin, J.-B.; Reverdin, G.; Boehme, L.; Lovell, P.; et al. A Southern Indian Ocean database of hydrographic profiles obtained with instrumented elephant seals. *Nat. Sci. Data* **2014**, *1*, 140028. [CrossRef] [PubMed]

52. Treasure, A.M.; Roquet, F.; Ansorge, I.J.; Bester, M.N.; Boehme, L.; Bornemann, H.; Charrassin, J.-B.; Chevallier, D.; Costa, D.P.; Fedak, M.A.; et al. Marine mammals exploring the oceans pole to pole: A review of the MEOP consortium. *Oceanography* **2017**, *30*, 132–138. [CrossRef]

53. Fournier, S.; Lee, T.; Tang, W.; Steele, M.; Olmedo, E. Evaluation and Intercomparison of SMOS, Aquarius, and SMAP Sea Surface Salinity Products in the Arctic Ocean. *Remote Sens.* **2019**, *11*, 3043. [CrossRef]

54. Manucharyan, G.E.; Thompson, A.F. Submesoscale sea ice-ocean interactions in marginal ice zones. *J. Geophys. Res. Ocean.* **2017**, *122*, 9455–9475. [CrossRef]

55. Brucker, L.; Dinnat, E.P.; Koenig, L.S. Weekly Gridded Aquarius L-band Radiometer/Scatterometer Observations and Salinity Retrievals Over the Polar Regions—Part 1: Product Description. *Cryosphere* **2014**, *8*, 905–913. [CrossRef]

56. Brucker, L.; Dinnat, E.P.; Koenig, L.S. Weekly Gridded Aquarius L-band Radiometer/Scatterometer Observations and Salinity Retrievals Over the Polar Regions—Part 2: Initial Product Analysis. *Cryosphere* **2014**, *8*, 915–930. [CrossRef]

57. Liu, C.; Liang, X.; Chambers, D.P.; Ponte, R.M. Global patterns of spatial and temporal variability in salinity from multiple gridded Argo products. *J. Clim.* **2020**, *33*, 8751–8766. [CrossRef]

58. Meissner, T.; Manaster, A. SMAP salinity retrievals near the sea-ice edge using multi-channel AMSR2 brightness temperatures. *Remote Sens.* **2021**, *13*, 5120. [CrossRef] [PubMed]

59. Gentemann, C.L.; Clayson, C.A.; Brown, S.; Lee, T.; Parfitt, R.; Farrar, J.T.; Bourassa, M.; Minnett, P.J.; Seo, H.; Gille, S.T.; et al. Fluxsat: Measuring the ocean–atmosphere turbulent exchange of heat and moisture from space. *Remote Sens.* **2020**, *12*, 1796. [CrossRef]

 *remote sensing*

Article

Super-Resolving Ocean Dynamics from Space with Computer Vision Algorithms

Bruno Buongiorno Nardelli [1,*], Davide Cavaliere [2], Elodie Charles [3] and Daniele Ciani [2]

1 Consiglio Nazionale delle Ricerche, Istituto di Scienze Marine (CNR-ISMAR), 80133 Naples, Italy
2 Consiglio Nazionale delle Ricerche, Istituto di Scienze Marine (CNR-ISMAR), 00133 Rome, Italy; davide.cavaliere@artov.ismar.cnr.it (D.C.); daniele.ciani@cnr.it (D.C.)
3 Collecte Localisation Satellites, 11 Rue Hermès, Parc Technologique du Canal, 31520 Ramonville Saint-Agne, France; echarles@groupcls.com
* Correspondence: bruno.buongiornonardelli@cnr.it

Abstract: Surface ocean dynamics play a key role in the Earth system, contributing to regulate its climate and affecting the marine ecosystem functioning. Dynamical processes occur and interact in the upper ocean at multiple scales, down to, or even less than, few kilometres. These scales are not adequately resolved by present observing systems, and, in the last decades, global monitoring of surface currents has been based on the application of geostrophic balance to absolute dynamic topography maps obtained through the statistical interpolation of along-track satellite altimeter data. Due to the cross-track distance and repetitiveness of satellite acquisitions, the effective resolution of interpolated data is limited to several tens of kilometres. At the kilometre scale, sea surface temperature pattern evolution is dominated by advection, providing indirect information on upper ocean currents. Computer vision techniques are perfect candidates to infer this dynamical information from the combination of altimeter data, surface temperature images and observing-system geometry. Here, we exploit one class of image processing techniques, super-resolution, to develop an original neural-network architecture specifically designed to improve absolute dynamic topography reconstruction. Our model is first trained on synthetic observations built from a numerical general-circulation model and then tested on real satellite products. Provided concurrent clear-sky thermal observations are available, it proves able to compensate for altimeter sampling/interpolation limitations by learning from primitive equation data. The algorithm can be adapted to learn directly from future surface topography, and eventual surface currents, high-resolution satellite observations.

Keywords: earth observations; ocean dynamics; satellite altimetry; sea surface temperature; artificial intelligence; machine learning; deep learning; neural networks

Citation: Buongiorno Nardelli, B.; Cavaliere, D.; Charles, E.; Ciani, D. Super-Resolving Ocean Dynamics from Space with Computer Vision Algorithms. *Remote Sens.* **2022**, *14*, 1159. https://doi.org/10.3390/rs14051159

Academic Editors: Veronica Nieves, Ana B. Ruescas and Raphaëlle Sauzède

Received: 17 January 2022
Accepted: 24 February 2022
Published: 26 February 2022

Publisher's Note: MDPI stays neutral with regard to jurisdictional claims in published maps and institutional affiliations.

1. Introduction

In the last decade, technological progress has opened new prospects for the application of deep-learning techniques in a wide range of fields. This revolutionary change originated from the concurrent increase of computational power at widely affordable costs and impressive growth of openly available data. Computer vision is one specific branch of artificial intelligence (AI) that is driving significant improvements thanks to the possibility to design and implement complex model architectures based on deep convolutional neural networks (CNN). Computer vision originally aimed to emulate the human capability to immediately discriminate objects and features in a picture or video, as well as to extrapolate/predict relevant information from partial or degraded input, either for recreational, medical, security or other commercial uses, e.g., for automated focusing on specific subjects in consumer and professional cameras, for semantic/instance segmentation and anomaly detection in medical imagery or in support of self-driving automated vehicles.

The Earth system research community is increasingly exploring and developing AI technologies to solve complex data processing and analysis problems and go beyond the present limitations of numerical models (see also [1]). Indeed, to discover the laws governing Earth system processes and better predict their evolution over several spatial and temporal scales, a combination of precise observations and theoretical/numerical models is needed [2]. In fact, even considering the significant increase in the number of acquisitions by remote sensing platforms and autonomous instruments, it will never be possible to describe and predict the state of the Earth system at all scales (or even of just one of its subsystems, such as the ocean) only through observed data. Satellite observations over the ocean, for example, only measure surface properties, with distinct temporal sampling and coverage depending on the sensor and mission. Conversely, information on the vertical distribution of properties along the water column can only be provided through in situ sensors that clearly lack the ability to simultaneously provide large spatial coverage and high space-time resolution. Empirical/statistical methodologies and historical data are thus often used to interpolate or reconstruct approximated 2D or 3D descriptions of ocean dynamics from a limited number of observed state variables, with poor to no physically driven constraints (e.g., [3]). On the other hand, full descriptions of the ocean state evolution (over a predefined set of scales) can be obtained through numerical models, still requiring a prior knowledge/guess of the initial state and of the forcings over time, as well as the parameterization of sub-grid physics. Due to the uncertainties in the initial conditions and parameterizations and the non-linearity of the dynamics, model predictions easily drift away from what is seen in the observations, unless observations are ingested within the simulation itself through data assimilation (DA), e.g., [4]. At present, DA is mostly based on probabilistic approaches, and it is also not rigorously tractable due to the huge number of variables and nonlinear processes involved, as well as the difficulty in simultaneously and properly characterizing model and observation errors.

Despite some scepticism due to the generally limited interpretability and explainability of complex neural networks, deep learning methods are perfect candidates to cope with the high-dimensional spaces, multiple processes, non-linear relations and noisy data involved in Earth system observations and models [5]. While it is beyond our scope to provide a comprehensive list of the ever-growing applications of AI algorithms to Earth system science, it is worth citing some of the most relevant objectives, which span from hybrid modelling approaches, such as the development of new sub-grid-scale parameterizations [6,7] and the improvement of DA techniques to be used in classical numerical circulation models [8,9], to the downscaling of low resolution models [10], to supervised learning approaches for data augmentation, filtering, interpolation or prediction [11–18], to the detection of dynamical features [19], to the set-up of neural networks for partial differential equation solution/identification and modelling of latent dynamics, e.g., [20–24]. Indeed, whenever sufficient information is available, physically informed neural network models can be designed to explicitly include physical constraints, for example, by building custom loss-functions that enforce the structure of the network to obey a known governing equation (through automatic differentiation), and/or by exploiting the similarities between residual networks and the numerical schemes used to integrate the ordinary differential equations that govern dynamical systems. However, sometimes sequential observations can merely resolve the large-scale dynamics, as high-resolution spatial snapshots are available only episodically and provide only limited information to describe (and directly learn) the evolution of small-scale processes. In those cases, alternative approaches can be tested to improve the dynamical reconstruction and eventually recover high-resolution information from existing data.

Actually, many advanced computer vision algorithms can be adapted to geoscientific analysis, making the most of past efforts dedicated to addressing similar problems. One such example is given by a specific class of techniques, known as super-resolution (SR), that aims to recover high-resolution (HR) details from low-resolution images [25]. In the single-image super-resolution algorithms, deep convolutional neural networks are optimized to identify the features of an image by looking at its different channels and learning how to recover the original features from their degraded versions. This is done by combining the translation invariance and locality properties of convolutional layers with the impressive learning properties of deep architectures coupled with non-linear activations. In computer vision applications, this problem is inherently ill-posed since multiple HR images could correspond to a single LR image, and the performance of the models is highly dependent on the extensiveness of the catalogue of images it was trained with. State-of-the-art super-resolution algorithms do achieve impressive results in the processing of blurred/low-resolution photographs, though, and some attempts to adapt them to Earth observation problems have already been carried out: DeepSD (Deep Statistical Downscaling), a stacked SR-CNN algorithm based on the early three-layer model by [26], has been applied to downscale Earth system model simulations, and CNN models have been tested for sea surface temperature (SST) and wind field downscaling, with positive results [27,28]).

Here, we aim to recover high-resolution sea surface dynamical features by combining low-resolution ocean absolute dynamic topography (ADT) fields based on satellite altimetry (resolving $O(100$ km) wavelengths) and high-resolution SST acquired by thermal imaging spaceborne radiometers (between $O(1$ km) and $O(10$ km) depending on the sensor). This represents a different problem with respect to simple model output downscaling or single variable super-resolution, as we want to combine the information provided by channels at both original and degraded resolution in a multi-channel image, taking advantage of the physical relations among the variables we include in the different channels. In fact, even if it only implies a weak constraint, our strategy includes physical considerations in the choice of the predictor variables, building upon the role of surface water mass advection in the local evolution of the SST [29], but we also aim to exploit the repetitiveness of satellite observing system geometries. We thus consider also the temporal SST variation and the ADT mapping error as additional predictors.

High-resolution observations of ocean dynamic topography through imaging sensors (thus 'natively' 2D), however, will not be available before the launch of the joint NASA/CNES/CSA ASC/UK Space Agency surface water and ocean topography (SWOT) mission, expected by the end of 2022, while more direct observations of surface currents could be provided by ESA Earth Explorer 10 Harmony mission only after 2027–2028, if present phase A, namely the design consolidation and feasibility studies, proves success-ful [30], or even later by SEASTAR ESA Earth Explorer 11 candidate mission [31]. As such, we rely here on an observing system simulation experiment (OSSE). In practice, we use the output of an ocean general circulation numerical model to simulate both predictor and target variables, considering that, in the future, our network can be trained directly with remotely sensed data. After training with OSSE data, our model can be applied to real altimeter-derived ADT and SST data in the test/prediction phase. In practice, learning first from primitive equation simulations and known observing-system geometry and succes-sively testing over true observation-based products can also be interpreted as a means to assimilate model physics in our data-driven reconstruction. Presently, core estimates of ocean surface currents are obtained by measuring absolute dynamic topography (i.e., the surface height referenced to an empirical geoid) through radar altimeters installed on a constellation of polar-orbiting satellite platforms. Sea level observations are acquired by altimeters along a discrete number of tracks, and surface geostrophic currents are obtained by first interpolating the ADT onto a regular 2D grid [32] and then computing ADT gradi-ents (geostrophy implies velocities are perpendicular to the pressure gradients associated with sea surface level differences, with an inverse dependence on the Coriolis parameter). ADT-interpolated products reach an effective resolution of $O(100$ km) at mid-latitudes [33], but recent studies [34,35] also revealed that many unresolved structures are aliased into larger structures and that the gridded altimetry products contain an unrealistic number of large mesoscale eddies. Hence, even for large scale eddies, having a typical wavelength larger than 100 km, the standard altimetry may be biased, with such large-scale bias mainly occurring in cyclonic eddies.

Altimeter-derived ADT maps can thus be thought of as a deformed view of true surface elevation, obtained through a transformation that combines the satellite observation geometry and the space-time surface elevation evolution effectively captured by the interpolation algorithm. Our objective is to set up a neural network that is able to learn the inverse mapping from our limited input ADT to the true sea surface elevation. To do that, we explicitly include as tentative predictors the low-resolution ADT field, the SST field and its temporal derivative, $\partial SST/\partial t$, as well as the formal interpolation error (ΔADT) that is associated with the input ADT product (retrieved as part of the optimal interpolation algorithm), and we set the high-resolution ADT as our target. In fact, at the large scale, SST responds to air-sea fluxes and related upper layer mixing with the deeper oceanic layers, but when getting close to the mesoscale and sub-mesoscale dynamical range (namely at scales between a few kms and a few tens of km, over day-to-weeks timescales), a significant contribution to the local SST variations is given by horizontal advection (though vertical advection is also expected to play a significant role, especially at the sub-mesoscale, e.g., [36]). SST products obtained from thermal images provide synoptic high-resolution data up to (nominal) 1 km spatial resolution over wide portions of the ocean surface [37,38]. Even if their effective spatial resolution rarely exceeds a few kilometres to tens of kilometres, they allow an almost continuous monitoring of SST changes at daily intervals and longer timescales. As such, several past attempts to improve surface current retrieval have been based on the use of the sequential information provided by SST products, either through maximum cross-correlation techniques [39] or by directly considering tracers' advection equation [40–43].

Our work exploits the data prepared for an OSSE that was originally designed for different objectives in the framework of the European Space Agency ocean CIRculation from ocean COLour observations (CIRCOL) project [44]. They consist of one year of synthetic, daily ADT and surface geostrophic currents data over the Mediterranean Basin. Full details explaining how we simulate the observing system geometry are reported in Section 2, where all pre-processing steps to prepare our training and test datasets are described. It must be stressed that for this OSSE, SST data have been assumed to be void-free and error-free, which is clearly not true, especially when looking at kilometre-scale features; so our work must be considered as a first exploratory step that will need to be significantly expanded for eventual operational applications.

2. Materials and Methods

2.1. Primitive Equation Model Data

The Mediterranean Forecasting System (MFS) is a hydrodynamic model for the Mediterranean Basin and the Atlantic Ocean off the Strait of Gibraltar [45]. Monthly to 15-minute instantaneous outputs of 3D horizontal currents and sea surface height (SSH), as well as monthly to hourly estimates of 3D temperature and salinity fields are available via the Copernicus Marine Service web portal (Product ID: MEDSEA-ANALYSIS-FORECAST-PHY-006-013). For the present study, we relied on daily outputs of SSH and SST, extracting information within the boundaries of the Mediterranean Basin (30 to 46° N and −6 to 37° E). These fields are provided on a 1/24° regular grid and 125 unequally spaced vertical levels. The simulations are based on the NEMO model (Nucleus for European Modelling of the Ocean) used in combination with Wave Watch-III for the wave component. The MFS simulations also account for data-assimilation of 2D satellite-derived SST, salinity vertical profiles, as well as along-track sea-level anomaly observations.

2.2. Satellite Absolute Dynamic Topography

The sea surface geostrophic currents were obtained from the Copernicus Marine Service and are derived from optimally interpolated absolute dynamic topography data merging observations from a constellation of radar altimeters. Such a constellation is composed of four to six altimeters in the 2008–2019 temporal range [32]. The geostrophic currents are provided as daily fields with nominal 1/8° horizontal resolution. The 2008–2019 time series was extracted. The corresponding Copernicus Marine Service product and dataset ID are SEALEVEL_MED_PHY_L4_REP_OBSERVATIONS_008_051/dataset-duacs-rep-medsea-merged-allsat-phy-l4, respectively (accessed on 1 March 2021 and now included as part of the SEALEVEL_EUR_PHY_L4_MY_008_068/cmems_obs-sl_eur_phy-ssh_my_allsat-l4-duacs-0.125deg_P1D dataset).

2.3. Satellite Sea Surface Temperature Data

Remotely sensed SST data are taken from the European Copernicus Marine Service (https://marine.copernicus.eu/access-data, last accessed 14 January 2022). They are Level-4 (L4) products, which means they provide gap-free estimates on a regular grid and are operationally produced and freely distributed in near-real time. We have used here 11 years (2008–2019) of the ultra-high-spatial-resolution (UHR) Mediterranean dataset, reaching a nominal $0.01° \times 0.01°$ resolution (Product ID: SST_MED_SST_L4_NRT_OBSERVATIONS_010_004_c_V2). This SST dataset is retrieved by first combining the night-time images collected by multi-platform infrared sensors, after specific quality control and cloudy pixel removal, and by successively running a two-step optimal interpolation algorithm [37]. Before using satellite SST L4 data to build our predictor tensor, we had to map them on the same grid used for the model training. To obtain a consistent prediction, we preliminary assessed the effective spatial scales resolved by the model SST, and eventually filtered the UHR to remove scales that have never been seen by the network. This was achieved by applying a low-pass Lanczos 2D filter (with window size = 9 and cutoff = 1/8) directly to the UHR data before remapping on the final $1/24°$ grid (through a basic bilinear interpolation). As discussed in the Results section, this pre-processing has the drawback of further smoothing the SST field in areas not covered by concurrent infrared satellite measurements due to cloud contamination or other coverage issues.

2.4. Sea Surface Drifter Data

In situ measurements of sea surface currents were obtained from autonomous Lagrangian drifting buoys that are passively transported by the ocean surface currents [46,47]. During the drifting buoy evolution, the data on the position are interpolated at uniform intervals (~30 minutes) relying on the kriging interpolation method developed by [47]. The velocities are finally computed via a finite-difference method of the interpolated positions and provided with six-hourly temporal resolution. The data covering the period of our study have originally been provided by the Italian Institute of Oceanography and Experimental Geophysics (OGS) for the purposes of the ESA-CIRCOL project. The timeseries are accessible via http://doi.org/10.6092/7a8499bc-c5ee-472c-b8b5-03523d1e73e9, last accessed 14 January 2022); buoy-derived surface current values are only retained if the buoy is equipped with a drogue: a device that guarantees the buoy evolution to be driven by the ocean currents rather than by surface winds [48].

2.5. Simulating Altimeter-like ADT Maps

One year (2017) of synthetic altimeter-derived ADT maps was obtained from the outputs of the Copernicus Marine Service MFS hydrodynamic simulation, using the data unification and altimeter combination system (DUACS) mapping method. The different steps are detailed below. Firstly, sea level anomaly (SLA) was computed from model outputs by means of Equation (1):

$$SLA = SSH - (MDT - 0.344) \tag{1}$$

where the mean dynamic topography (MDT) is provided as a static field together with the model outputs. A 0.344 constant (expressed in m) allows us to adjust the SLA values in the Mediterranean Sea to guarantee that the spatio–temporal average of SLA is zero during 2017. The large-scale, high-frequency variability, usually removed by applying a dynamic atmospheric correction (DAC) [49] is filtered out of these synthetic data by applying a Loess filter. The SLA is then sampled along the actual tracks of a synthetic constellation composed of four radar altimeters: Jason-3, Sentinel-3A, SARAL/Altika and Cryosat-2 missions. This step is achieved by running the SWOT simulator software [50], which allows us to account for the actual orbits, errors and noise that characterize each mission. The chosen four-satellite constellation is representative of the constellation ingested in Copernicus Marine Service processing during 2017. Such along-track synthetic measurements are then ingested by the DUACS processing chain to produce L4 SLA maps. The optimal interpolation (OI) scheme follows the DUACS DT2018 (Delayed Time) configuration for the Mediterranean area, described in [32]. The reconstructed small-scale maps are then recombined with the filtered large-scale maps. Such data are provided on a daily basis and over a regular $1/8°$ grid (more details available in [44]).

2.6. Preparation of Training and Test Datasets for Deep Convolutional Learning

The OSSE data were simulated starting from year one of the primitive equation model daily output described above. The original input images cover the entire Mediterranean domain at $1/24°$ spatial resolution, leading to an individual image size of 380×1000 pixels. We randomly chose 40 dates (~11% of the total) to be kept aside as independent test data, and we successively resampled the original images, extracting much smaller tiles (76×100), which were later used as input to the network training. Despite the random holdout strategy being a standard, different choices of the test dataset could also be done. However, the test on OSSE prediction only serves here to assess the relative performance of the different network architectures, not its absolute performance, which is not relevant per se, when looking at simulated input data. The tiles are then extracted by going through a double loop on latitude and longitude, imposing a spatial overlap of 50%, so that a total of 42,250 samples is finally available for the training. The dimension of the tiles has been chosen to simplify the pre-processing and reduce the memory required by the training steps. In fact, all tiles are normalized before entering the network. In the case of the SST, ADT and $\partial SST/\partial t$, they are first transformed into anomalies estimated with respect to the tile spatial mean and are successively scaled by dividing the anomalies by the maximum value (in absolute value) recorded throughout the series. The ΔADT only goes through the normalization step. As the tiles cover an area of approximately 300 km $\times$ 400 km, the anomaly computation is indeed serving as a high-pass filter, removing the background variability associated with basin scale processes and seasonal variations (e.g., steric and thermal variations driven by large-scale, air–sea interactions), which are not relevant to reveal the impact of mesoscale processes on SST evolution related to horizontal advection. This filtering is consistent with the tests described in [44]. After the test/prediction, the tiles are merged together to compute a weighted average on overlapping areas.

2.7. Deep Convolutional Models Learning Strategy and Configuration

All deep convolutional models considered in this work (Section 3) have been written in Python using the open-source library *Keras*. They are trained adopting an early stopping rule to avoid overfitting and minimize the generalization error. In practice, the original training dataset is randomly split into a proper training set (85% of data) and a validation set (15%) (not to be confused with the fully independent test dataset described above, which is never seen during the training) based on which both training (hindcast) and validation losses are updated during the network optimization. The validation loss, in particular, is used as an estimate of the generalization error, and early stopping consists in terminating the iterative learning as soon as its values start to increase. As the estimations can be rather noisy, early stopping admits a "patience" parameter, which defines the number of epochs to be completed before the loss function minimum can be considered such. Here we have set the patience equal to 20 for the SRCNN model (whose computations are very fast but require many more epochs to converge) and reduced it to five for all deeper models. The adaptive moment estimator, Adam, is applied for the stochastic optimization of models' parameters [51], with the learning rates (l_r) and numerical stability constants (ε) kept as in the original implementations of the baseline networks (i.e., $l_r = 3 \times 10^{-4}$ and $\varepsilon = 10^{-7}$ for

SRCNN; $l_r = 10^{-4}$ and $\varepsilon = 10^{-8}$ for the other networks). The latter values are also adopted for dADR-SR. Within the dADR-SR model, we have also tested the implementation of a DropBlock strategy [52] to improve the network regularization with minimal performance differences.

2.8. Automatic Eddy Detection

The angular momentum for eddy detection and tracking algorithm (AMEDA) is freely available software for the detection and tracking of oceanic eddies from 2D gridded fields of surface currents and/or sea surface height [53]. It is based on the computation of eddy local normalized angular momentum (LNAM) and on the observations of closed streamlines around the LNAM extrema. The algorithm was successfully applied to remotely sensed 2D fields and model outputs, e.g., [35,53,54], and enables the determination of eddies' contour and trajectories as well as eddy merger/splitting events. In this study, AMEDA was used to identify eddy shapes seen by standard altimetry products (described in Section 2.2) and the fields obtained from the combination of satellite altimetry ADT and high-resolution satellite SSTs. We relied on the AMEDA default configuration, accounting for the expected perturbation lengths in the Mediterranean area (i.e., considering the typical Mediterranean Rossby deformation radii).

3. Results

3.1. Testing Single-Image, Super-Resolution Configurations and Designing a Multi-Scale Adaptive Model

A large variety of neural network architectures have been proposed to achieve single-image super-resolution, even considering only those dealing with input–output images of the same size [25]. Comparing all of them is clearly beyond the scope of our work, which is rather to demonstrate whether the super-resolution class of techniques can be efficiently used for quantitatively accurate dynamical retrievals based on multiple variables and observation types. As such, we have implemented here four different models, three of them basically reproducing already published (baseline) networks, and a third one that includes elements from different models but represents an original network architecture. All models are trained considering the mean-squared error as the reference loss function.

The first model is the Super-Resolution Convolutional Neural Network (SRCNN) proposed by [26]. It consists of three 2D convolutional layers: the first one includes 128 filters with a 9×9 kernel size, the second one with 64 filters and a 3×3 kernel size and the third one with a single 5×5 filter. The first two layers include a nonlinear activation (rectified linear unit, ReLU), and zero-padding is applied in every layer to keep the original image size end-to-end (Figure 1A). Each layer represents a specific operation in the conceptual explanation of the SRCNN given by [26]: overlapping patches extraction from the low-resolution image (where the patches have the same size as the kernel) and representation into a high-dimensional vector (feature mapping, with vector dimensions equal to the number of filters); non-linear mapping of each high-dimensional vector onto another high-dimensional vector comprising a second set of high-resolution feature maps; these are directly linked to the final image in the third step (reconstruction). We applied SRCNN in four different configurations, namely considering SST, $\partial SST/\partial t$, ADT and ΔADT in input (all together) and alternately removing either $\partial SST/\partial t$, ΔADT or both $\partial SST/\partial t$ and ΔADT from the predictor variables. After training the networks (see details in Section 2), we used the fully independent test data to assess the accuracy of the prediction over the entire Mediterranean Basin, comparing the root-mean-squared differences (RMSD) between the altimeter-like ADT and the "true" ADT, as well as those between the super-resolved ADT field and our simulated "ground truth". SCRNN gave us some first indications (Figure 1B–E): overall, we could not find an improvement in the ADT reconstruction by incorporating all predictors, but what appeared to be actually detrimental was the inclusion of the ADT interpolation error. In fact, excluding ΔADT from the input already improves the accuracy of the simple SRCNN's reconstruction. Conversely, including $\partial SST/\partial t$ always appeared beneficial. Even if our initial hypothesis on the relevance of the information on the observing geometry (as provided by the ADT interpolation error) seemed wrong, we must stress that the features that dominate the patterns of ΔADT are much larger than the scales of the dynamical features we want to super-resolve, and SRCNN is plausibly a too shallow/simple network to deal with such different scales due to a very limited ability to learn complex interdependencies between channels (it actually contains only around 110k trainable parameters). Thus, we are still confident that a properly defined network architecture would be able to exploit the information on where the altimeter-like ADT field is expected to be more accurate and where it deserves stronger corrections. As such, we decided to test all successive (and gradually more complex) network architectures with both the configuration with the four predictors and the one without the ADT error.

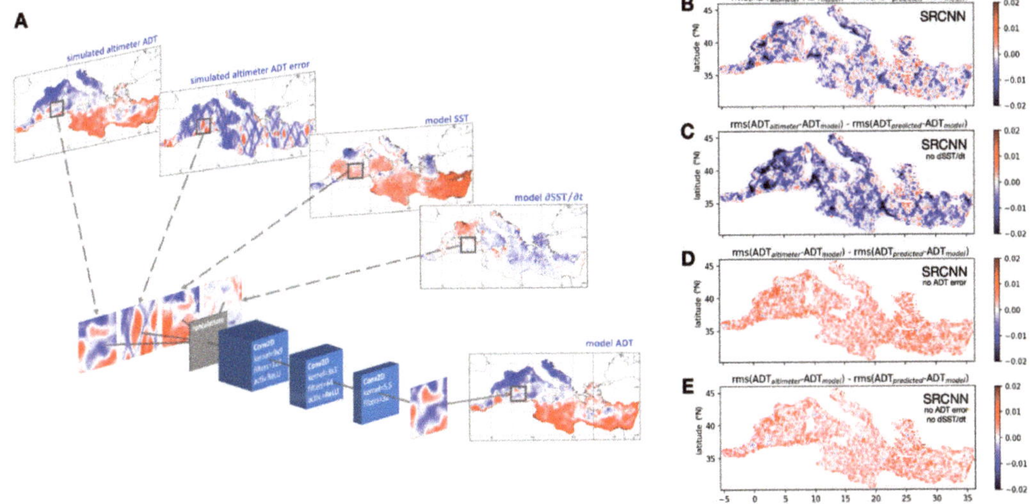

Figure 1. Super-Resolution Convolutional Neural Network (SRCNN) adapted to the reconstruction of absolute dynamic topography from multiple channel inputs. (**A**) SRCNN network architecture. (**B–E**) Relative performance of the SRCNN reconstructions, assessed on the independent test data as the difference between the RMSD between the altimeter-like and original model output and the RMSD between the super-resolved absolute dynamic topography (ADT) and the original model ADT (red indicates smaller RMSD from model predictions). The panels show the performance of the models trained/tested considering: (**B**) the full set of predictor variables; (**C**) removing $\partial SST/\partial t$ from the predictors; (**D**) removing ΔADT from the predictors; (**E**) removing both $\partial SST/\partial t$ and ΔADT. ADT and related RMSD values are expressed in m.

The second network we have implemented is the baseline Enhanced Deep Residual network for Super-Resolution (EDSR) proposed by [55]. EDSR was designed to exploit the possibility to significantly deepen the networks (i.e., to increase the number of layers) opened by residual learning frameworks. Instead of learning fully unreferenced functions, residual networks (ResNet) define the layers as residual functions (actually they are based on residual blocks, including different convolutional and batch normalization layers, and activation functions, where the residual is referenced to the block input) and have been proven much easier to optimize than conventional networks, allowing users to train considerably deeper networks and obtain significantly better accuracies [56]. EDSR simplified the network architecture with respect to models based on original ResNet by reducing the number of parameters employed in each residual block (Figure 2A). In its baseline formulation, it includes a first 2D convolutional layer made up of 64 filters with 3 × 3 kernel size, followed by 16 residual blocks, increasing the number of trainable parameters to approximately 1.2 M, one order of magnitude higher than SRCNN. The first layer output and the output of the last residual block in the sequence are also summed up (skip connection) before entering the output convolutional layer (including a single 3×3 filter), which connects to the target image. Residual blocks include two convolutional layers with the number of filters equal to the input channels (64) and a 3×3 kernel size. These two layers are connected through a non-linear activation (ReLU). The outputs of the second convolutional layer within the residual block are summed to the input channels to obtain the residual, after applying them a fixed scaling factor of 0.1. Notably, the scaling strategy applied within the EDSR residual block was formerly proven to stabilize the training of complex networks, allowing users to safely increase the number of filters [57].

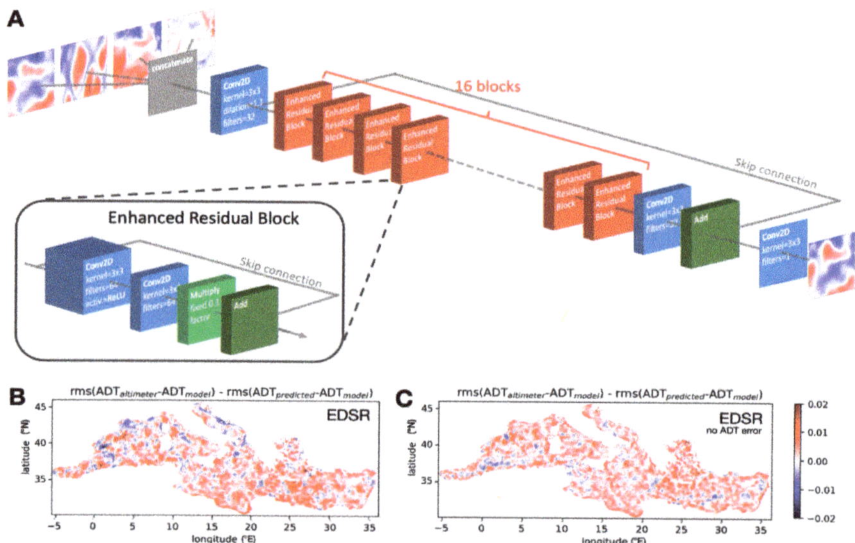

Figure 2. Enhanced Deep Super-Resolution (EDSR) baseline model tested for the reconstruction of high-resolution absolute dynamic topography from multiple channel inputs. (**A**) EDSR network architecture. EDSR is based on a specific residual block design. (**B,C**) Relative performance of the EDSR reconstructions, assessed on the independent test data as the difference between the RMSD between the altimeter-like and original model output and the RMSD between the super-resolved ADT and the original model ADT (red indicates smaller RMSD from model predictions). The panels show the performance of the models trained/tested considering: (**B**) EDSR and the full set of predictor variables; (**C**) EDSR removing ΔADT from the predictors. ADT and related RMSD values are expressed in m.

The minimum of the loss function reached during the model training was around ~4.5×10^{-3} for SRCNN, whatever the configuration, with small differences between hindcast and validation. The same numbers would indicate a much better performance of ESDR, with validation loss values of around 1.8×10^{-3} for the configuration excluding the ADT error from the predictors and values close to 1.6×10^{-3} for the full predictors set. The minimum hindcast loss got close to 1.1×10^{-3} in both EDSR configurations. However, the test run on the independent data clearly indicated that the improvement only occurred in some parts of the domain, while worse reconstructions can be obtained in dynamically relevant areas, both including the ADT error or not in the predictor list (Figure 2B,C). Not too surprisingly, in the first case, higher RMSD are found along some of the repeated altimeter tracks, well visible as diamond/rhomboid shapes (Figure 2B), which again indicates that the network is not able to efficiently exploit the information on the

ADT interpolation error. In the latter, though, higher errors in the reconstruction are found also in some areas that are well known for being dynamically very active (e.g., offshore the Algerian coast, Figure 2C).

The third network considered is the Adaptive Deep Residual Network for Super-Resolution (ADR-SR) proposed by [58]. ADR-SR represents an interesting evolution of the EDSR. Its main improvement consists in substituting the learned feature fixed scaling applied within the EDSR residual block with an adaptive scaling, obtained by introducing a squeeze-and-excitation (SE) module [59]. Within the adaptive residual block (ARB), channel-wise feature responses are adaptively recalibrated through an SE module before summing them to the block input, allowing the network to more efficiently model complex interdependencies between the learned feature channels (Figure 3A). Conceptually, we might thus expect it to drive a substantial advance with respect to simpler networks.

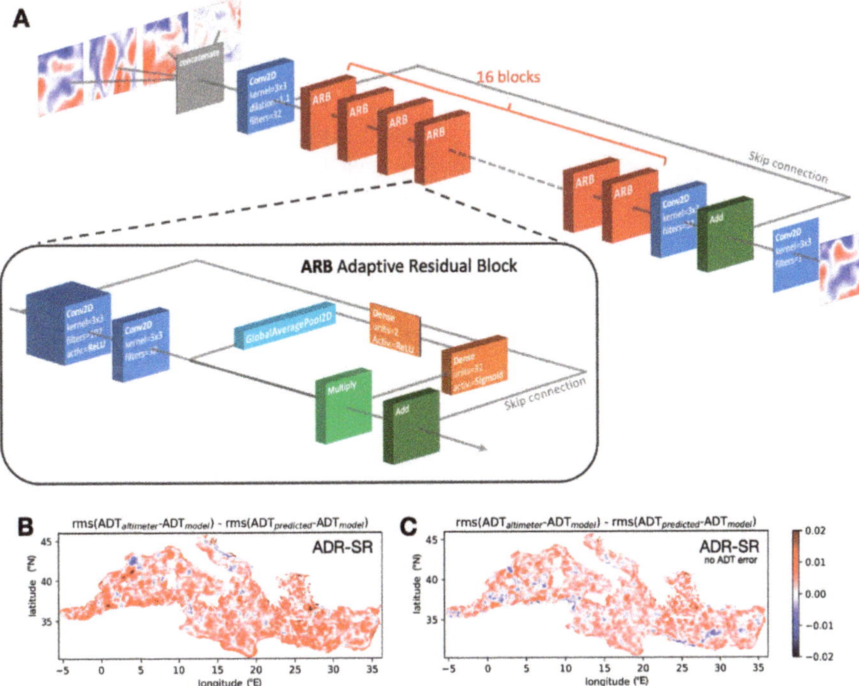

Figure 3. Adaptive Super-Resolution (ADR-SR) baseline model tested for the reconstruction of high-resolution absolute dynamic topography from multiple channel inputs. (**A**) ADR-SR network. ADR-SR is based on the inclusion of a squeeze-and-excitation module within its residual block design. (**B,C**) Relative performance of the ADR-SR reconstructions, assessed on the independent test data as the difference between the RMSD between the altimeter-like and original model output and the RMSD between the super-resolved ADT and the original model ADT (red indicates smaller RMSD from model predictions). The panels show the performance of the models trained/tested considering: (**B**) ADR-SR and the full set of predictor variables; (**C**) ADR-SR removing ΔADT from the predictors. ADT and related RMSD values are expressed in m.

The SE module first reduces all 2D feature channels into 1D values through global average pooling (squeeze). The excitation operation then consists in learning a weight vector, built through a self-gating mechanism that takes the output of the global average pooling as input and provides per-channel modulation weights. The self-gating consists in a bottleneck with two, fully connected layers (including non-linear activations), the first one reducing the dimensionality by a predefined factor (and including a ReLU activation), and the second one increasing it back to the number of channels in the input to the module (followed by a sigmoid activation). These weights are successively used to suppress or enhance individual channel features (feature recalibration) before summation to get the residuals. With respect to baseline EDSR, ADR-SR increases the number of filters within each block from 64 to 192 but limits the number of feature channels in input to the residual block from 64 to 32. The ADR-SR network implemented here employs 16 residual blocks and finally includes slightly less than 1.8M trainable parameters.

ADR-SR performance assessed on the test dataset significantly improved with respect to that of the other networks, and, for the first time, including the information on the low-resolution ADT interpolation error leads to a tangible reduction of the RMSD over most of the basin (Figure 3B,C). Still, a lower accuracy is found close to some of the altimeter repeated tracks (visible as diamond shapes), which might be due to the limited ability either of ADR-SR and of the previously tested networks to correctly handle the information provided at different spatial scales by the input predictor variables. This specifically reflects the larger scale of ADT mapping error patterns with respect to the geophysical variables.

To overcome this issue, we have developed here a novel deep convolutional architecture, which combines the successful developments of previously tested super-resolution models with the dilated-convolution-based multi-scale information learning inception module proposed by [60]. Dilated convolution allows users to extract information at different scales and significantly expands the network's receptive field even without enlarging the kernel size [61]. Choosing a dilation rate, r, r-adjacent pixels are skipped by the convolution kernel, so that related weights refer to samples taken at tuneable distances. In the inception module designed by [60], the channels input to each module first pass through three parallel dilated convolution layers with kernel size 3×3 and dilation factor of 1, 2 and 3, respectively. Then, all convolution outputs are concatenated and passed to the successive layers.

We have named the new model "dilated Adaptive Deep Residual Network for Super-Resolution (dADR-SR)". Its architecture is depicted in Figure 4. In the first step, dADR-SR input channels are passed to three parallel convolutional layers, each one with ten filters and a 3×3 kernel size, but with an increasing dilation factor of 1, 3 and 5, respectively. The output of the three convolutional layers is then concatenated into a single multiscale feature tensor, which represents the input to a sequence of multiscale adaptive residual blocks. Indeed, within each residual block, the same multiscale parallel feature extraction is carried out, thus defining a multiscale adaptive residual block (M-ARB). To avoid excessively increasing the number of parameters to train, the number of residual blocks is here kept to 12 (four fewer than in the EDSR/ADR-SR baseline), and the number of filters included in the two sets of convolutional layers inside each M-ARB is chosen as 120 and 10, respectively. Within the M-ARB, after concatenating the learned multiscale features, a SE module is included (with a predefined dimensionality reduction factor of 10 instead of 16, so that the bottleneck in dADR-SR is shaped 30-3-30, instead of 32-2-32). The final number of trainable parameters in the dADR-SR model is slightly below 1.6 M.

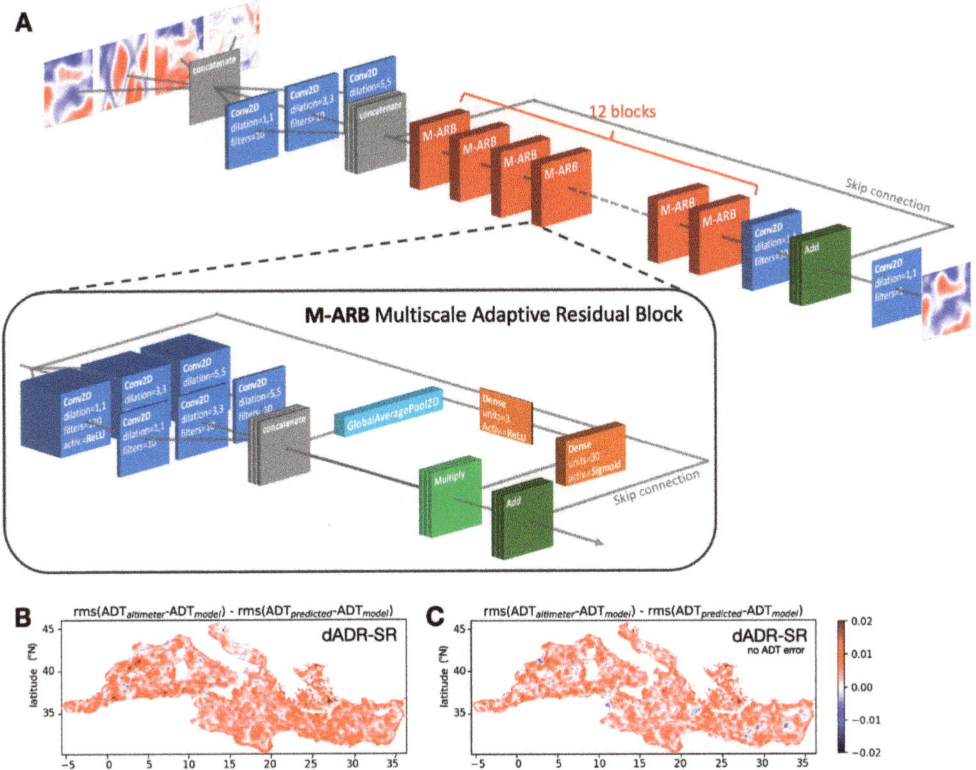

Figure 4. The dilated Adaptive Super-Resolution (dADR-SR) model developed to reconstruct high-resolution absolute dynamic topography from multiple channel inputs. (**A**) The dADR-SR network architecture; dADR-SR is based on the inclusion of dilated, convolution-based learning inception modules in the core layers of ADR-SR. (**B,C**) Relative performance of the dADR-SR reconstructions, assessed on the independent test data as the difference between the RMSD between the altimeter-like and original model output and the RMSD between the super-resolved ADT and the original model ADT (red indicates smaller RMSD from model predictions). The panels show the performance of the models trained/tested considering: (**B**) dADR-DR and the full set of predictor variables; (**C**) dADR-SR removing ΔADT from the predictors. ADT and related RMSD values are expressed in m.

The dADR-SR model outperforms any of the previous networks when tested on the independent dataset, displaying a marked reduction of the RMSD over the entire basin (Figure 5), with only extremely few and very small spots showing a minimal degradation (Figure 4B,C). The information captured at the different scales by including all predictors thus further enhances the accuracy of the reconstruction with respect to the model that does not consider the low-resolution ADT interpolation error.

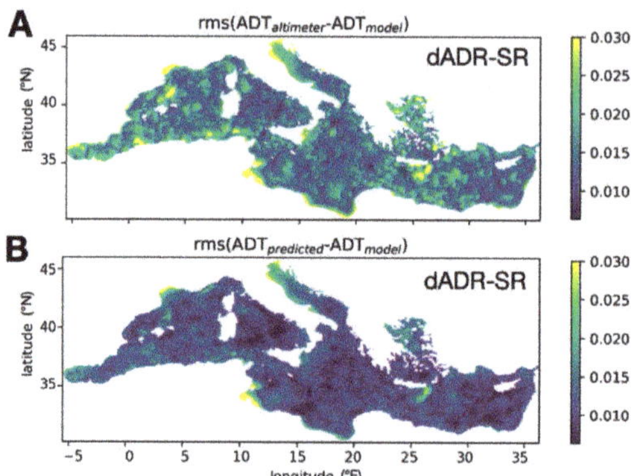

Figure 5. The dADR-SR model performance compared to simulated standard altimetry. (**A**) RMSD between the altimeter-like and original model output and (**B**) RMSD between the super-resolved absolute dynamic topography (ADT) obtained with dADR-SR (using full predictors set) and the original model ADT. ADT and related RMSD values are expressed in m.

3.2. Applying Dilated Adaptive Residual Super-Resolution Trained on Simulated Data to Real Satellite Observations

Our successive analysis is aimed to verify to what extent we can use the features learned from observing system simulations based on primitive equation modelling to improve present-day, data-driven reconstructions, but also to identify eventual limitations of the present OSSE set up and issues related to real observation-based products. We have thus applied the model trained on OSSE data to predict high-resolution ADT starting from properly pre-processed, satellite-altimeter-based, low-resolution ADT and high-resolution SST products (e.g., Figure 6). For both variables, we have actually taken optimally interpolated data (also known as Level 4 (L4), see Section 2), covering the 2008–2019 period. Assessing the accuracy of the observation-based, super-resolved maps is not trivial, as large-coverage, high-resolution observations of the surface topography/currents are not presently available. As such we followed a double approach: on the one hand, we performed a qualitative analysis of the reconstructed patterns, looking at areas of intense mesoscale activity visible in the satellite SST L4 data and comparing the (sub)mesoscale eddies identified by the AMEDA automatic detection algorithm in original and super-resolved data; on the other hand, we built a match-up database with the surface current estimates provided by surface drifters (see Section 2) and used it to compute the statistics of the differences with respect to the geostrophic currents estimated from standard altimeter ADT and super-resolved ADT field.

The dADR-SR model reveals impressive potential to resolve mesoscale turbulent features that are generally smeared out, often misplaced or even totally missed by standard altimetry, when clear-sky thermal data are present. Figure 7 presents a wonderful example of such a turbulent field, with many mesoscale eddies, dipoles and current meanderings well visible, especially in the western Mediterranean Basin, detaching from the Algerian current towards the centre of the basin along the North Balearic front and Liguro-Provençal current. Surface geostrophic currents estimated from the dADR-SR ADT (Figure 6B) not only appear much sharper than those depicted by low-resolution altimetric data (Figure 6A) but prove also able to reconstruct dynamical features that were completely absent in the standard product (Figure 7).

Specifically, dADR-SR recovers the strong cyclonic eddies associated with two mushroom-like dipoles along the Algerian coast, marked with the letters "A" and "B" in the zoomed panels of Figure 7, displaying much more consistent shapes and intensities. The cyclonic circulation in (B) is actually described as two eddies by the AMEDA eddy contours estimated from the super-resolved field, which is much more consistent with the

SST patterns, while a single and almost rectangular shape is found in standard altimeter estimates. Consistency is not meant here as a perfect alignment between the SST gradients and the geostrophic currents—which is not at all to be expected in non-stationary current fields—but to the impossibility of having so many isotherms being crossed by very large-scale currents, considering the well-developed structures found in the SST field, even assuming a strong chaotic stirring. The dADR-SR prediction also recovers much more reliable patterns associated with the weaker meanders and smaller-scale recirculations in the centre of the sub-basin (C and D) and also the highly asymmetric, strong dipole visible east of Menorca Island (E). Remarkably, it is also capable of identifying the winding north-eastward current close to Corsica (F) that is seen as a rather straight and uniform flow in low-resolution altimetry and completely missed by corresponding AMEDA, while being detected as a small dipole in super-resolved AMEDA contours.

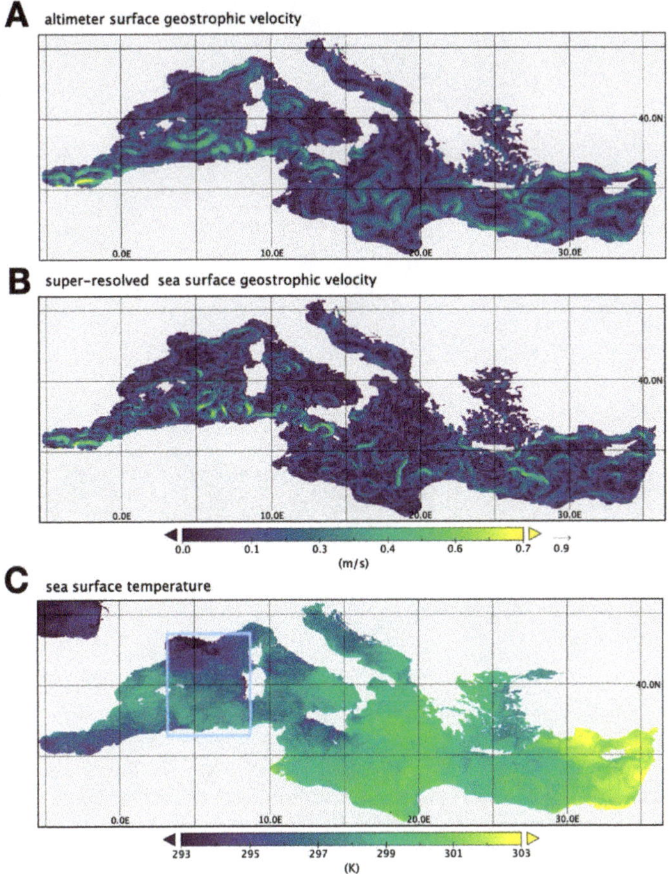

Figure 6. The dADR-SR prediction from real satellite-derived absolute dynamic topography (ADT) and sea surface temperature data (SST) for one example date (17-07-2016). (**A**) Original altimeter-based surface geostrophic currents (obtained from the ADT gradients); (**B**) super-resolved surface geostrophic currents; (**C**) satellite SST field. The cyan box in (**C**) identifies the area plotted in Figure 7.

The situation looks very different when the original thermal-infrared data are masked by clouds, because in these cases SST data interpolation leads to much smoother SST structures and gradients than what is observed in clear-sky conditions. This filtering unfortunately reflects on the structures retrieved by the dADR-SR model as well, evidently dumping surface current intensities and also eventually clearing out several of the mesoscale features.

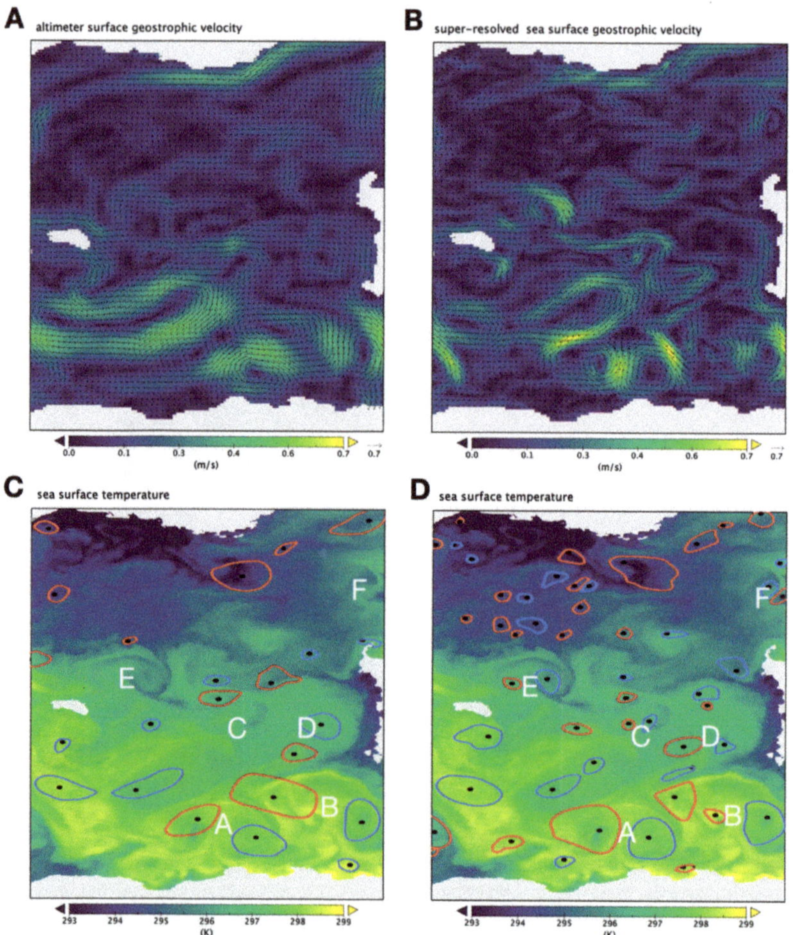

Figure 7. Dynamical structures reconstructed by dADR-SR prediction from real satellite-derived absolute dynamic topography (ADT) and sea surface temperature data (SST) (zoomed from Figure 6). (**A**) Original altimeter-based surface geostrophic currents; (**B**) super-resolved surface geostrophic currents; (**C**,**D**) satellite SST field with overplot of the eddy contours identified through AMEDA detection algorithm (red = cyclonic, blue = anticyclonic, black dots stand for automatically detected eddy centres) applied to original altimeter currents (**C**) and to super-resolved field (**D**). A–F letters serve to more easily locate the dynamical features that are recovered by dADR-SR and missed/misplaced by standard altimetry products (discussed in the text).

One such example is given in Figure 8, which shows the interpolated SST and the associated nominal interpolation error on the 1st of July 2014 and the same fields taken three days apart. During these three days, clouds gradually moved into the southwestern Mediterranean from Morocco, completely hiding the sea surface to the satellite infrared radiometers on the second date (Figure 8B,E). The corresponding SST field, presenting very clear and distinct structures on the first day, is dramatically blurred by the interpolation in data-void areas (Figure 9). Similar to the example presented in Figure 7, the dADR-SR reconstruction is able to recover much more consistent surface current patterns in the presence of clear-sky thermal observations, e.g., aligning the currents to the true shape/position of the jet found close to 6°E (Figure 9G) along the Algerian coast. This jet is actually feeding a mushroom-like mesoscale structure whose cyclonic eddy is much better retrieved in the super-resolved image and significantly misplaced by the standard product (Figure 9E). Likewise, SR fields also reconstruct the small-scale features observed in the Almeria–Oran front region (i.e., to the north of the SST front found around the Greenwich meridian), which is seen as a unique and quite large cyclone in altimetry maps (consistent with what was noticed by [35]). Notably, however, the geostrophic current field based on altimetry only suffers minor evolutions after three days (Figure 9F), while dADR-SR actually appears to have smeared out most of the features observed previously. Current intensities are also significantly and unrealistically reduced on the 3rd of July, especially in the intense, anticyclonic meander observed along the Algerian coast at 6°E (Figure 9H).

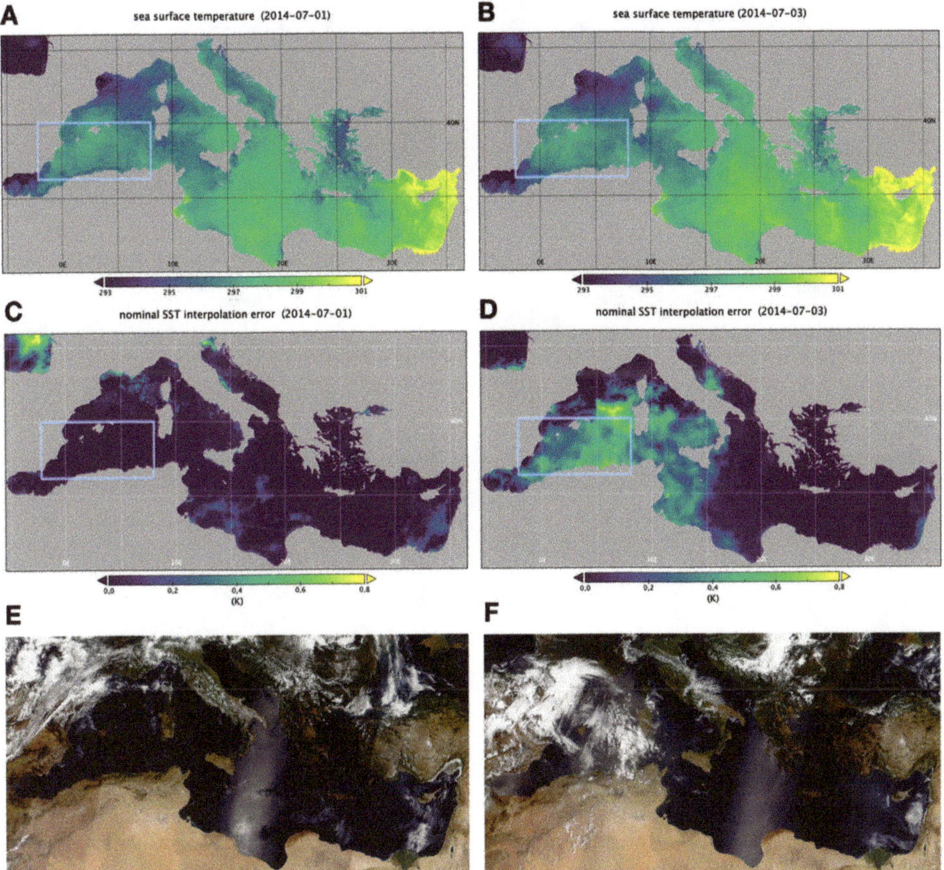

Figure 8. Impact of cloud cover on satellite SST interpolated data. Interpolated field; (**A,B**), related nominal interpolation error; (**C,D**), MODIS Terra pseudo-true colour images; (**E,F**) from NASA Worldview (https://worldview.earthdata.nasa.gov, last accessed on 14 January 2022). On the first date, clear-sky conditions (**E**) lead to extremely clear and distinct SST patterns and low interpolation errors in all the Mediterranean (**A,C**). Three days later, clouds arriving from Morocco (**F**) prevent the reconstruction of small-scale dynamical features in the SST field and lead to increased interpolation errors (**B,D**). The thin cyan box identifies the area zoomed in Figure 9.

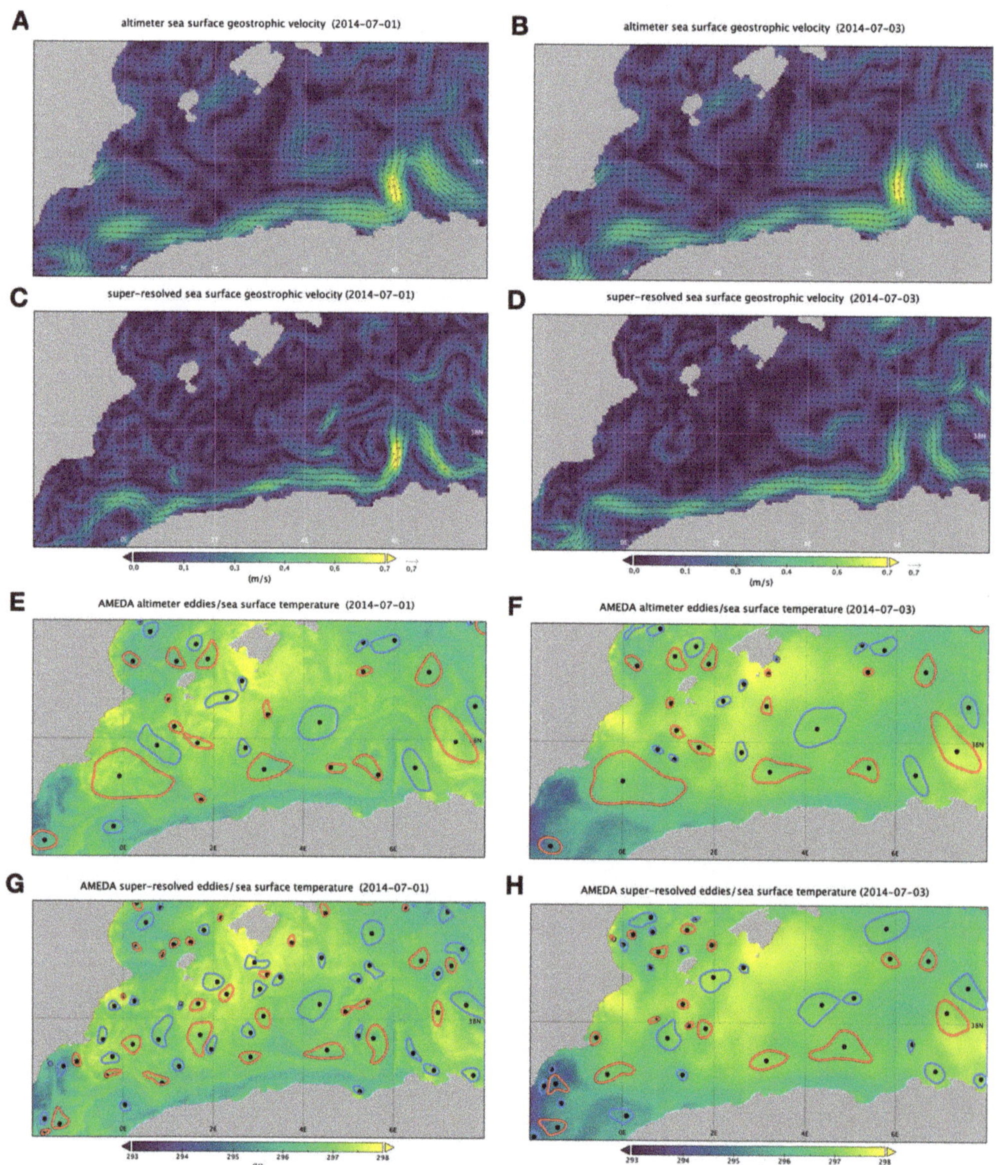

Figure 9. Impact of the smoothing introduced by SST data interpolation on the dynamical reconstruction (zoomed from Figure 8). (**A,B**) original altimeter-based surface geostrophic currents; (**C,D**) super-resolved surface geostrophic currents; (**E–H**) SST L4 field with overplot of the eddy contours identified through AMEDA detection algorithm (red = cyclonic, blue = anticyclonic, black dots stand for automatically detected eddy centres) applied to original altimeter currents (**E,F**) and to super-resolved field (**G,H**). Current vectors are overplot in (**A–D**) plots.

Several similar cases can be picked up by looking at the entire time series, but our qualitative understanding of the power and limitations of the model and observations analysed in this first work are fully confirmed also by the successive quantitative assessment. This latter exercise was carried out by matching the independent estimates of surface currents obtained from in situ drifting floats with co-located/concurrent data-driven modelling and altimeter data. Actually, the error associated with the super-resolved and standard altimetric estimates (assessed here as the absolute value of the difference with respect to drifter data) does not display a unique behaviour along the individual drifters' trajectories, with alternating improvements and degradations of the current velocity components, which do not present marked geographical characterization (Figure 10). Overall, though, SR fields seem to provide slightly more accurate values in some (mostly offshore) areas of the western basin, while, on average, they seem to perform worse in the easternmost part of the Levantine basin (close to Israel/Lebanon coasts).

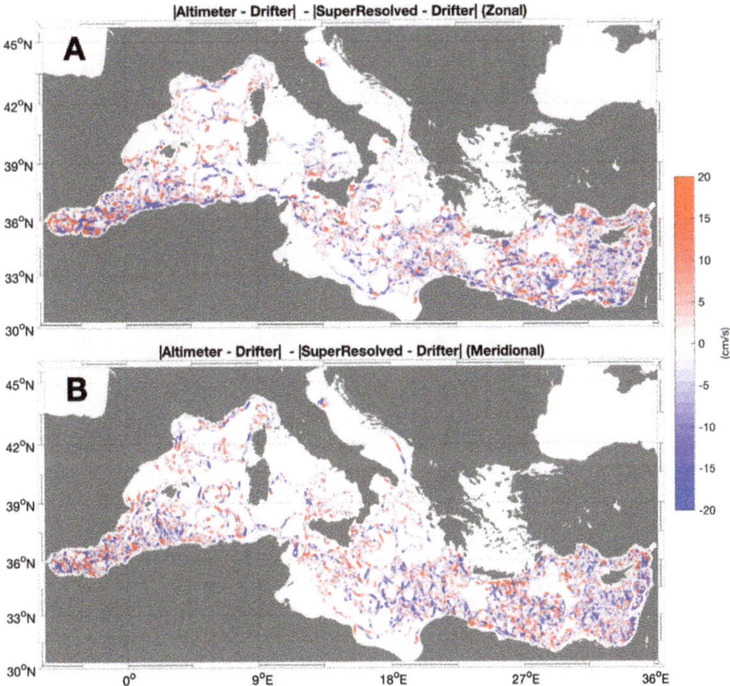

Figure 10. Performance of dADR-SR reconstruction of the geostrophic currents with respect to standard altimeter L4 data along drifter trajectories. The plots show the difference between the absolute error of altimetry and that of super-resolved currents, estimated vs the drifter velocities: (**A**) zonal component; (**B**) meridional component. Positive values indicate an improvement with respect to altimetry.

4. Discussion

While our findings could be of general interest to a broad range of scientists, we expect them to stimulate further investigations and suggest new ways to efficiently combine Earth system data from multiple sensors and models. In fact, the adaptation and development of AI and deep-learning tools for Earth observation encompasses an ever-growing number of potential applications. We have tested here a novel approach that not only allows users to take advantage of neural network techniques for the combination of multi-sensor, remotely sensed data, but proposes an innovative way to merge satellite observations and numerical models, building an observation-based product that implicitly includes knowledge of the physics of the system in a way different from classical statistical interpolation or data assimilation.

We performed an observing system simulation experiment to test the applicability of computer vision algorithms (originally designed for single image super resolution) for the improved estimation of ocean absolute dynamic topography. Usually, OSSEs provide a controlled testing environment, targeted to better tune observing system design or to set up the retrieval algorithms for future satellite missions. Indeed, one interesting aspect of the experiment carried out here is that, while it will hopefully be possible in the future to train our super-resolution model directly from observations (as soon as SWOT, Harmony or SEASTAR measurements become available), it is already possible to apply it to super-resolve present altimeter products, provided the area and period under study are effectively covered by cloud-free/noise-free infrared images.

In fact, we have identified a number of issues and aspects that will need further investigation and new developments. First of all, a dedicated effort would be needed to take into account the limitations of present, satellite-based SST L4 data, which are obtained through spatio-temporal statistical interpolation of eventually cloud-contaminated infrared images and thus provide uneven spatial spectral information (depending on the persistence of cloud cover and interpolation strategy adopted) [41]. One possible strategy to expand the applicability of algorithms based on super-resolution would thus be to also fully simulate SST interpolation and related error in an improved OSSE. This could be achieved by adding realistic data voids to model data and implementing the same algorithm used operationally for SST interpolation. The interpolated SST and related formal interpolation error could then be included as predictors rather than the original model SST. As an alternative, one could test the applicability of the model to non-interpolated SST data (also known as Level 3), starting already from the training phase. Moreover, this approach requires a specific OSSE and dedicated tests.

One additional aspect that deserves to be mentioned is that we carried out our tests in the Mediterranean Sea, which is likely one of the most complex areas in terms of upper ocean dynamics (though cloud coverage is less critical than elsewhere) [62]. This is due to the small Rossby radius of deformation (i.e., the scale at which buoyancy and rotation effects are comparable, which set the characteristic size of dominant flow instabilities) and the occurrence of intense, small-scale air–sea interactions (modulated by highly variable orography) that drive complex coastal dynamics that can significantly affect the source/sink terms in the upper-ocean-temperature-evolution equation. This may undermine our working hypothesis that small scale changes are substantially dominated by advection. It would thus be interesting to train (or even only to test) our model in other areas where intense mesoscale activity is observed (e.g., western boundary currents, Antarctic Circumpolar Current, etc.), eventually starting from different numerical simulations. Additionally, we plan to further improve the network architecture by implementing and testing the latest modules and ideas emerging from computer vision research (e.g., convolutional block attention module (CBAM)), [63]. Major advances might then come from the design of network architectures that admit the explicit use of sequences of ADT and SST data as predictors, which could then allow users to define physically informed loss functions (e.g., by enforcing some approximate potential vorticity/tracer conservation) to jointly improve the reconstruction of both current and tracer evolution over time.

Author Contributions: Conceptualization, B.B.N.; methodology, B.B.N., D.C. (Daniele Ciani), D.C. (Davide Cavaliere) and E.C.; investigation, B.B.N. and D.C. (Daniele Ciani); visualization, B.B.N. and D.C. (Daniele Ciani); supervision, B.B.N.; writing—original draft, B.B.N. and D.C. (Daniele Ciani); writing—review and editing, B.B.N. and D.C. (Daniele Ciani), D.C. (Davide Cavaliere) and E.C. All authors have read and agreed to the published version of the manuscript.

Funding: This work has been partly supported by the Copernicus Marine Environment Monitoring Service Multi Observations Thematic Assembly Center (CMEMS MOB TAC), funded through subcontracting agreement no. CLS-SCO-18-0004 between Consiglio Nazionale delle Ricerche and Collecte Localisation Satellites (CLS), the latter of which is presently leading the CMEMS MOB TAC. The 83-CMEMS-TAC-MOB contract was funded by Mercator Ocean as part of its delegation agreement with the European Union, represented by the European Commission, to set up and manage CMEMS.

The data used for the observation system simulation experiment have been provided by the ocean CIRculation from ocean COLour observations (CIRCOL) project, funded by the European Space Agency via the contract grant no. 4000128147/19/I-DT.

Institutional Review Board Statement: Not applicable.

Informed Consent Statement: Not applicable.

Data Availability Statement: The dADR-SR code is available at: https://github.com/bbuong/dADR-SR (last accessed 14 January 2022) and training/test datasets are available at: https://doi.org/10.5281/zenodo.5815330 (last accessed 14 January 2022). The AMEDA code and description are freely accessible through https://github.com/briaclevu/AMEDA (last accessed 14 January 2022).

Acknowledgments: We thank Emanuela Clementi for providing support on the numerical simulation data used for the setup of the observing system simulation experiment and Milena Menna for providing support on the drifter database. We also wish to thank Claudia Cesarini for critical reading of the paper and useful suggestions on the presentation of the results. We acknowledge the use of imagery from the NASA Worldview application (https://worldview.earthdata.nasa.gov, last accessed 14 January 2022), part of the NASA Earth Observing System Data and Information System (EOSDIS).

Conflicts of Interest: The authors declare no conflict of interest.

References

1. Schultz, M.G.; Betancourt, C.; Gong, B.; Kleinert, F.; Langguth, M.; Leufen, L.H.; Mozaffari, A.; Stadtler, S. Can deep learning beat numerical weather prediction? *Philos. Trans. R. Soc. Lond. Ser. A Math. Phys. Eng. Sci.* **2021**, *379*, 20200097. [CrossRef] [PubMed]
2. Schneider, T.; Lan, S.; Stuart, A.; Teixeira, J. Earth System Modeling 2.0: A Blueprint for Models That Learn From Observations and Targeted High-Resolution Simulations. *Geophys. Res. Lett.* **2017**, *44*, 12396–12417. [CrossRef]
3. Mulet, S.; Rio, M.-H.; Mignot, A.; Guinehut, S.; Morrow, R. A new estimate of the global 3D geostrophic ocean circulation based on satellite data and in-situ measurements. *Deep Sea Res. Part II: Top. Stud. Oceanogr.* **2012**, *77–80*, 70–81. [CrossRef]
4. Moore, A.M.; Martin, M.J.; Akella, S.; Arango, H.G.; Balmaseda, M.A.; Bertino, L.; Ciavatta, S.; Cornuelle, B.; Cummings, J.; Frolov, S.; et al. Synthesis of Ocean Observations Using Data Assimilation for Operational, Real-Time and Reanalysis Systems: A More Complete Picture of the State of the Ocean. *Front. Mar. Sci.* **2019**, *6*. [CrossRef]
5. Reichstein, M.; Camps-Valls, G.; Stevens, B.; Jung, M.; Denzler, J.; Carvalhais, N.; Prabhat. Deep learning and process understanding for data-driven Earth system science. *Nature* **2019**, *566*, 195–204. [CrossRef] [PubMed]
6. Bolton, T.; Zanna, L. Applications of Deep Learning to Ocean Data Inference and Subgrid Parameterization. *J. Adv. Model. Earth Syst.* **2019**, *11*, 376–399. [CrossRef]
7. Brajard, J.; Carrassi, A.; Bocquet, M.; Bertino, L. Combining data assimilation and machine learning to emulate a dynamical model from sparse and noisy observations: A case study with the Lorenz 96 model. *J. Comput. Sci.* **2020**, *44*, 101171. [CrossRef]
8. Ruckstuhl, Y.; Janjić, T.; Rasp, S. Training a convolutional neural network to conserve mass in data assimilation. *Nonlinear Process. Geophys.* **2021**, *28*, 111–119. [CrossRef]
9. Storto, A.; De Magistris, G.; Falchetti, S.; Oddo, P. A Neural Network-Based Observation Operator for Coupled Ocean-Acoustic Variational Data Assimilation. *Mon. Weather Rev.* **2021**, *149*, 1967–1985. [CrossRef]
10. Vandal, T.; Kodra, E.; Ganguly, S.; Michaelis, A.; Nemani, R.; Ganguly, A.R. DeepSD: Generating high resolution climate change projections through single image super-resolution. *arXiv* **2017**, arXiv:1703.03126.
11. Barth, A.; Alvera-Azcárate, A.; Licer, M.; Beckers, J.-M. DINCAE 1.0: A convolutional neural network with error estimates to reconstruct sea surface temperature satellite observations. *Geosci. Model Dev.* **2020**, *13*, 1609–1622. [CrossRef]
12. Sammartino, M.; Nardelli, B.B.; Marullo, S.; Santoleri, R. An Artificial Neural Network to Infer the Mediterranean 3D Chlorophyll-a and Temperature Fields from Remote Sensing Observations. *Remote Sens.* **2020**, *12*, 4123. [CrossRef]
13. Nardelli, B.B. A Deep Learning Network to Retrieve Ocean Hydrographic Profiles from Combined Satellite and In Situ Measurements. *Remote Sens.* **2020**, *12*, 3151. [CrossRef]
14. Han, Z.; He, Y.; Liu, G.; Perrie, W. Application of DINCAE to Reconstruct the Gaps in Chlorophyll-a Satellite Observations in the South China Sea and West Philippine Sea. *Remote Sens.* **2020**, *12*, 480. [CrossRef]
15. Sauzède, R.; Claustre, H.; Uitz, J.; Jamet, C.; Dall'Olmo, G.; D'Ortenzio, F.; Gentili, B.; Poteau, A.; Schmechtig, C. A neural network-based method for merging ocean color and Argo data to extend surface bio-optical properties to depth: Retrieval of the particulate backscattering coefficient. *J. Geophys. Res. Oceans* **2016**, *121*, 2552–2571. [CrossRef]
16. Sauzède, R.; Bittig, H.; Claustre, H.; De Fommervault, O.P.; Gattuso, J.-P.; Legendre, L.; Johnson, K.S. Estimates of Water-Column Nutrient Concentrations and Carbonate System Parameters in the Global Ocean: A Novel Approach Based on Neural Networks. *Front. Mar. Sci.* **2017**, *4*. [CrossRef]
17. Sinha, A.; Abernathey, R. Estimating Ocean Surface Currents With Machine Learning. *Front. Mar. Sci.* **2021**, *8*. [CrossRef]
18. Zheng, G.; Li, X.; Zhang, R.-H.; Liu, B. Purely satellite data–driven deep learning forecast of complicated tropical instability waves. *Sci. Adv.* **2020**, *6*, eaba1482. [CrossRef]

19. Liu, Y.; Zheng, Q.; Li, X. Characteristics of Global Ocean Abnormal Mesoscale Eddies Derived From the Fusion of Sea Surface Height and Temperature Data by Deep Learning. *Geophys. Res. Lett.* **2021**, *48*, e2021GL094772. [CrossRef]

20. Fablet, R.; Amar, M.; Febvre, Q.; Beauchamp, M.; Chapron, B. End-to- end physics-informed representation learning from and for satellite ocean remote sensing data. In Proceedings of the XXIV ISPRS 2021: Intenational Society for Photogrammetry and Remote Sensing Congress, Nice, France, 4–10 July 2021; p. hal-03189218.

21. Pannekoucke, O.; Fablet, R. PDE-NetGen 1.0: From symbolic partial differential equation (PDE) representations of physical processes to trainable neural network representations. *Geosci. Model Dev.* **2020**, *13*, 3373–3382. [CrossRef]

22. Raissi, M.; Perdikaris, P.; Karniadakis, G. Physics-informed neural networks: A deep learning framework for solving forward and inverse problems involving nonlinear partial differential equations. *J. Comput. Phys.* **2018**, *378*, 686–707. [CrossRef]

23. Ouala, S.; Nguyen, D.; Drumetz, L.; Chapron, B.; Pascual, A.; Collard, F.; Gaultier, L.; Fablet, R. Learning latent dynamics for partially observed chaotic systems. *Chaos Interdiscip. J. Nonlinear Sci.* **2020**, *30*, 103121. [CrossRef] [PubMed]

24. Fablet, R.; Beauchamp, M.; Drumetz, L.; Rousseau, F. Joint Interpolation and Representation Learning for Irregularly Sampled Satellite-Derived Geophysical Fields. *Front. Appl. Math. Stat.* **2021**, *7*, 655224. [CrossRef]

25. Wang, Z.; Chen, J.; Hoi, S.C.H. Deep Learning for Image Super-Resolution: A Survey. *IEEE Trans. Pattern Anal. Mach. Intell.* **2020**, *43*, 3365–3387. [CrossRef]

26. Dong, C.; Loy, C.C.; He, K.; Tang, X. Image Super-Resolution Using Deep Convolutional Networks. *IEEE Trans. Pattern Anal. Mach. Intell.* **2016**, *38*, 295–307. [CrossRef]

27. Ducournau, A.; Fablet, R. Deep learning for ocean remote sensing: An application of convolutional neural networks for super-resolution on satellite-derived SST data. In Proceedings of the 9th IAPR Workshop on Pattern Recogniton in Remote Sensing (PRRS), Cancun, Mexico, 4 December 2016; pp. 1–6. [CrossRef]

28. Höhlein, K.; Kern, M.; Hewson, T.; Westermann, R. A comparative study of convolutional neural network models for wind field downscaling. *Meteorol. Appl.* **2020**, *27*, e1961. [CrossRef]

29. Rio, M.-H.; Santoleri, R.; Bourdalle-Badie, R.; Griffa, A.; Piterbarg, L.; Taburet, G. Improving the Altimeter-Derived Surface Currents Using High-Resolution Sea Surface Temperature Data: A Feasability Study Based on Model Outputs. *J. Atmospheric Ocean. Technol.* **2016**, *33*, 2769–2784. [CrossRef]

30. Lopez-Dekker, P.; Biggs, J.; Chapron, B.; Hooper, A.; Kääb, A.; Masina, S.; Mouginot, J.; Buongiorno Nardelli, B.; Pasquero, C. The Harmony Mission: End of Phase-0 Science Overview. In Proceedings of the 2021 IEEE International Geoscience and Remote Sensing Symposium IGARSS, Brussels, Belgium, 11–16 July 2021; pp. 7752–7755.

31. Gommenginger, C.; Chapron, B.; Hogg, A.; Buckingham, C.; Fox-Kemper, B.; Eriksson, L.; Soulat, F.; Ubelmann, C.; Ocampo-Torres, F.; Burbidge, G. SEASTAR: A Mission to Study Ocean Submesoscale Dynamics and Small-Scale Atmosphere-Ocean Processes in Coastal, Shelf and Polar Seas. *Front. Mar. Sci.* **2019**, *6*, 457. [CrossRef]

32. Taburet, G.; Sanchez-Roman, A.; Ballarotta, M.; Pujol, M.-I.; Legeais, J.-F.; Fournier, F.; Faugere, Y.; Dibarboure, G. DUACS DT2018: 25 years of reprocessed sea level altimetry products. *Ocean Sci.* **2019**, *15*, 1207–1224. [CrossRef]

33. Ballarotta, M.; Ubelmann, C.; Pujol, M.-I.; Taburet, G.; Fournier, F.; Legeais, J.-F.; Faugère, Y.; Delepoulle, A.; Chelton, D.; Dibarboure, G.; et al. On the resolutions of ocean altimetry maps. *Ocean Sci.* **2019**, *15*, 1091–1109. [CrossRef]

34. Amores, A.; Jordà, G.; Arsouze, T.; Le Sommer, J. Up to What Extent Can We Characterize Ocean Eddies Using Present-Day Gridded Altimetric Products? *J. Geophys. Res. Ocean.* **2018**, *123*, 7220–7236. [CrossRef]

35. Stegner, A.; Le Vu, B.; Dumas, F.; Ghannami, M.A.; Nicolle, A.; Durand, C.; Faugere, Y. Cyclone-Anticyclone Asymmetry of Eddy Detection on Gridded Altimetry Product in the Mediterranean Sea. *J. Geophys. Res. Oceans* **2021**, *126*, e2021JC017475. [CrossRef]

36. Mahadevan, A.; Tandon, A. An analysis of mechanisms for submesoscale vertical motion at ocean fronts. *Ocean Model.* **2006**, *14*, 241–256. [CrossRef]

37. Nardelli, B.B.; Tronconi, C.; Pisano, A.; Santoleri, R. High and Ultra-High resolution processing of satellite Sea Surface Temperature data over Southern European Seas in the framework of MyOcean project. *Remote Sens. Environ.* **2013**, *129*, 1–16. [CrossRef]

38. Chin, T.M.; Vazquez-Cuervo, J.; Armstrong, E.M. A multi-scale high-resolution analysis of global sea surface temperature. *Remote Sens. Environ.* **2017**, *200*, 154–169. [CrossRef]

39. Bowen, M.M.; Emery, W.; Wilkin, J.; Tildesley, P.C.; Barton, I.J.; Knewtson, R. Extracting Multiyear Surface Currents from Sequential Thermal Imagery Using the Maximum Cross-Correlation Technique. *J. Atmospheric Ocean. Technol.* **2002**, *19*, 1665–1676. [CrossRef]

40. Ciani, D.; Rio, M.-H.; Menna, M.; Santoleri, R. A Synergetic Approach for the Space-Based Sea Surface Currents Retrieval in the Mediterranean Sea. *Remote Sens.* **2019**, *11*, 1285. [CrossRef]

41. Ciani, D.; Rio, M.-H.; Nardelli, B.B.; Etienne, H.; Santoleri, R. Improving the Altimeter-Derived Surface Currents Using Sea Surface Temperature (SST) Data: A Sensitivity Study to SST Products. *Remote Sens.* **2020**, *12*, 1601. [CrossRef]

42. Isern-Fontanet, J.; García-Ladona, E.; González-Haro, C.; Turiel, A.; Rosell-Fieschi, M.; Company, J.B.; Padial, A. High-Resolution Ocean Currents from Sea Surface Temperature Observations: The Catalan Sea (Western Mediterranean). *Remote Sens.* **2021**, *13*, 3635. [CrossRef]

43. Rio, M.-H.; Santoleri, R. Improved global surface currents from the merging of altimetry and Sea Surface Temperature data. *Remote Sens. Environ.* **2018**, *216*, 770–785. [CrossRef]

44. Ciani, D.; Charles, E.; Nardelli, B.B.; Rio, M.-H.; Santoleri, R. Ocean Currents Reconstruction from a Combination of Altimeter and Ocean Colour Data: A Feasibility Study. *Remote Sens.* **2021**, *13*, 2389. [CrossRef]

45. Clementi, E.; Pistoia, J.; Escudier, R.; Delrosso, D.; Drudi, M.; Grandi, A.; Lecci, R.; Cretí, S.; Ciliberti, S.; Coppini, G.; et al. *Mediterranean Sea Analysis and Forecast (CMEMS MED-Currents 2016–2019) (Version 1) [Data Set]*; Copernicus Monitoring Environment Marine Service (CMEMS): Ramonville-Saint-Agne, France, 2021. [CrossRef]
46. Poulain, P.-M.; Menna, M.; Mauri, E. Surface Geostrophic Circulation of the Mediterranean Sea Derived from Drifter and Satellite Altimeter Data. *J. Phys. Oceanogr.* **2012**, *42*, 973–990. [CrossRef]
47. Hansen, D.V.; Poulain, P.-M. Quality Control and Interpolations of WOCE-TOGA Drifter Data. *J. Atmospheric Ocean. Technol.* **1996**, *13*, 900–909. [CrossRef]
48. Menna, M.; Poulain, P.-M.; Bussani, A.; Gerin, R. Detecting the drogue presence of SVP drifters from wind slippage in the Mediterranean Sea. *Measurement* **2018**, *125*, 447–453. [CrossRef]
49. Carrère, L.; Lyard, F. Modeling the barotropic response of the global ocean to atmospheric wind and pressure forcing—comparisons with observations. *Geophys. Res. Lett.* **2003**, *30*. [CrossRef]
50. Gaultier, L.; Ubelmann, C.; Fu, L.-L. The Challenge of Using Future SWOT Data for Oceanic Field Reconstruction. *J. Atmospheric Ocean. Technol.* **2016**, *33*, 119–126. [CrossRef]
51. Kingma, D.P.; Ba, J.L. Adam: A method for stochastic optimization. In Proceedings of the International Conference on Learning Representations (ICLR), San Diego, CA, USA, 7–9 May 2015; pp. 1–15. Available online: https://hdl.handle.net/11245/1.505367 (accessed on 1 March 2021).
52. Ghiasi, G.; Lin, T.Y.; Le, Q.V. Dropblock: A regularization method for convolutional networks. *Adv. Neural Inf. Process. Syst.* **2018**, arXiv:1810.12890v1 [cs.CV], 10727–10737.
53. Le Vu, B.; Stegner, A.; Arsouze, T. Angular Momentum Eddy Detection and Tracking Algorithm (AMEDA) and Its Application to Coastal Eddy Formation. *J. Atmospheric Ocean. Technol.* **2018**, *35*, 739–762. [CrossRef]
54. Bagaglini, L.; Falco, P.; Zambianchi, E. Eddy Detection in HF Radar-Derived Surface Currents in the Gulf of Naples. *Remote Sens.* **2019**, *12*, 97. [CrossRef]
55. Lim, B.; Son, S.; Kim, H.; Nah, S.; Lee, K.M. Enhanced deep residual networks for single image super-resolution. In Proceedings of the IEEE Conference on Computer Vision and Pattern Recognition Workshops (CVPRW), Honolulu, HI, USA, 21–26 July 2017; pp. 136–144. [CrossRef]
56. He, K.; Zhang, X.; Ren, S.; Sun, J. Deep residual learning for image recognition. In Proceedings of the IEEE Conference on Computer Vision and Pattern Recognition (CVPR), Las Vegas, NV, USA, 27–30 June 2016.
57. Szegedy, C.; Ioffe, S.; Vanhoucke, V.; Alemi, A.A. Inception-v4, inception-ResNet and the impact of residual connections on learning. In Proceedings of the 31st AAI Conference on Artificial Intelligence, San Francisco, CA, USA, 4–9 February 2017; pp. 4278–4284.
58. Liu, S.; Gang, R.; Li, C.; Song, R. Adaptive deep residual network for single image super-resolution. Comput. *Vis. Media* **2019**, *5*, 391–401. [CrossRef]
59. Hu, J.; Shen, L.; Albanie, S.; Sun, G.; Wu, E. Squeeze-and-Excitation Networks. *IEEE Trans. Pattern Anal. Mach. Intell.* **2020**, *42*, 2011–2023. [CrossRef] [PubMed]
60. Shi, W.; Jiang, F.; Zhao, D. Single image super-resolution with dilated convolution based multi-scale information learning inception module. In Proceedings of the IEEE International Conference on Image Processing (ICIP), Beijing, China, 17–20 September 2017; pp. 977–981. [CrossRef]
61. Luo, W.; Li, Y.; Urtasun, R.; Zemel, R. Understanding the effective receptive field in deep convolutional neural networks. In Proceedings of the 30th Conference on Neural Information Processing Systems (NIPS 2016), Barcelona, Spain, 5–10 December 2016; pp. 4905–4913.
62. Malanotte-Rizzoli, P.; Artale, V.; Borzelli-Eusebi, G.L.; Brenner, S.; Crise, A.; Gacic, M.; Kress, N.; Marullo, S.; D'Alcalà, M.R.; Sofianos, S.; et al. Physical forcing and physical/biochemical variability of the Mediterranean Sea: A review of unresolved issues and directions for future research. *Ocean Sci.* **2014**, *10*, 281–322. [CrossRef]
63. Woo, S.; Park, J.; Lee, J.; Kweon, I.S. CBAM: Convolutional Block Attention Module. In Proceedings of the European Conference on Computer Vision, Munich, Germany, 8–14 September 2018. [CrossRef]

MDPI
St. Alban-Anlage 66
4052 Basel
Switzerland
www.mdpi.com

Remote Sensing Editorial Office
E-mail: remotesensing@mdpi.com
www.mdpi.com/journal/remotesensing

Disclaimer/Publisher's Note: The statements, opinions and data contained in all publications are solely those of the individual author(s) and contributor(s) and not of MDPI and/or the editor(s). MDPI and/or the editor(s) disclaim responsibility for any injury to people or property resulting from any ideas, methods, instructions or products referred to in the content.